Lecture Notes in Mathematics

Edited by A. Dold and B. Eckmann

997

Algebraic Geometry – Open Problems

Proceedings of the Conference
Held in Ravello, May 31 – June 5, 1982

Edited by C. Ciliberto, F. Ghione, and F. Orecchia

Springer-Verlag
Berlin Heidelberg New York Tokyo 1983

Editors

Ciro Ciliberto
Istituto di Matematica "R. Caccioppoli", Università di Napoli
Via Mezzocannone 8, 80100 Napoli, Italy

Franco Ghione
Dipartimento di Matematica, Università di Roma II
Tor Vergata, 00100 Roma, Italy

Ferruccio Orecchia
Istituto di Matematica "R. Caccioppoli", Università di Napoli
Via Mezzocannone 8, 80100 Napoli, Italy

AMS Subject Classifications (1980): 14-06

ISBN 3-540-12320-2 Springer-Verlag Berlin Heidelberg New York Tokyo
ISBN 0-387-12320-2 Springer-Verlag New York Heidelberg Berlin Tokyo

This work is subject to copyright. All rights are reserved, whether the whole or part of the material is concerned, specifically those of translation, reprinting, re-use of illustrations, broadcasting, reproduction by photocopying machine or similar means, and storage in data banks. Under § 54 of the German Copyright Law where copies are made for other than private use, a fee is payable to "Verwertungsgesellschaft Wort", Munich.

© by Springer-Verlag Berlin Heidelberg 1983
Printed in Germany

Printing and binding: Beltz Offsetdruck, Hemsbach/Bergstr.
2146/3140-543210

Introduction

The Conference on " Open problems in Algebraic Geometry " was held in Ravello (Salerno) during the week : May 31 st - June 5 th , 1982 . This volume contains most of the lectures and talks given during the Conference as well as papers grown up from discussions among participants . The only exception is paper n° 15 , which however fits very well with the theme of the Conference . We are extremely grateful to all participants and in particular to all contributors of this volume . We also thank :

- the " Consiglio Nazionale delle Ricerche " , the University of Naples , the " Banco di Napoli " , for their financial support ;
- the " Ente Provinciale per il Turismo di Salerno " for having allowed the use of the wonderful " Villa Rufolo " , where the Conference took place .

We finally thank all those persons who helped either in the organization or in the developement of the Conference .

Ciro Ciliberto

Franco Ghione

Ferruccio Orecchia

List of Lectures

A. Conte : " Enriques threefolds "

S. Greco : " Remarks on the singularities of algebraic surfaces "

C. Peskine : " Classification of curves in $\mathbb{P}^3$ "

W. Fulton : " Nodal curves "

E. Arbarello : " A few things about the variety of irreducible plane curves of given degree and genus "

D. Mumford : " The geometry of the moduli space of curves "

F. Catanese : " On the moduli space of surfaces of general type "

S. L. Kleiman : " The enumeration of varieties touching given ones "

A. Beauville : " Problems on rational and unirational varieties "

C. De Concini : " On complete symmetric varieties "

E. Sernesi : " On a problem of uniqueness for certain linear series "

J. Harris : " Problems in the projective geometry of curves "

List of Talks

V. Mehta : " Vector bundles on projective varieties and their restriction to curves "

C. Ceresa : " Remarks on algebraic equivalence "

P. Maroscia : " The Hilbert function of a finite set of points in $\mathbb{P}^n$ "

G. Van der Geer : " The geometry of a Siegel modular threefold "

R. Gupta : " Schubert calculus and geometry of representation theory "

R. Smith : " The branch locus of the Prym map "

M. Beltrametti : " Conic bundles on non rational surfaces "

H. Hulek : " The normal bundle of space curves "

D. Eisenbud : " Rational curves with cusps "

E. Stagnaro : " Constructing Enriques surfaces from quintics in $\mathbb{P}^3$ "

P. Craighiero : " Cubic surfaces in $A_{\mathbb{C}}^3$ whose curves are set-theoretic complete intersection "

List of Participants

A. Albano (University of Torino) , E. Ambrogio (Torino) , E. Arbarello (Roma I) , D. Arezzo (Genova) , M.G. Ascensi (Brandeis) , F. Baldassarri (Padova) , E. Ballico (Pisa) , U. Bartocci (Perugia) , A. Beauville (Ec. Polytechnique,Paris) , G. Beccari (Torino) , B. Bellaccini (Siena) , M. Beltrametti (Genova) , J.F. Boutot (Strasbourg) , M. Brundu (Genova) , M. Candilera (Padova), F. Catanese (Pisa) , M. Cavaliere (Genova) , G. Ceresa (Torino), L. Chiantini (Torino) , S. Chiaruttini (Padova) , Ciro Ciliberto (Napoli) , A. Collino (Torino) , A. Conte (Torino) , M. Contessa (Roma I) , M. Cornalba (Pavia) , P.C. Craighiero (Padova) , V. Cristante (Padova) , C. Cumino (Torino) , C. De Concini (Roma II) , A. Del Centina (Firenze) , P. De Vito (Napoli) ,A. Di Sante (Napoli) , D. Eisenbud (Brandeis) , P. Ellia (Nice) , D. Epema (Leiden) , G. Faltings (Wuppertal) , M. Fiorentini (Ferrara) , M. Formisano (Napoli) , P. Francia (Genova) , W. Fulton (Brown), S. Gabelli (Roma I) , R. Gattazzo (Padova) , F. Ghione (Roma II), A. Gimigliano (Firenze) , S. Greco (Torino) , Guerra (Perugia), R. Gupta (M.I.T.) , J. Harris (Brown) , D. Husemoller , K. Hulek (Erlangen) , M. Idà (Bologna) , S. Kleiman (M.I.T.) , W. Kleinert (East Berlin) , Kooler (Brandeis) , A. Lanteri (Milano) , E. Li Marzi (Messina) , R. Maggioni (Catania) , M. Manaresi (Bologna) , M.G. Marinari (Genova) , P. Maroscia (Roma I) , G. Martens (Erlangen) , C. Martinengo (Genova) , C. Massaza (Siena) , L. Mazzi (Torino) , V. Mehta (Bombay) , I. Morrison , D. Mumford (Harvard) , , G. Niesi (Genova) , F. Odetti (Genova) , P. Oliverio (Pisa) , A. Oneto (Genova) , F. Orecchia (Napoli) , M. Palleschi (Parma) , G. Paxia (Catania) , C. Pedrini (Genova) , U. Persson (Stockolm) , C. Peskine (Oslo) , L. Picco Botta (Torino) , A. Ragusa (Catania) , L. Ramella (Genova) , S. Recillas (Firenze) , L. Robbiano (Genova) , N. Rodinò (Firenze) , M. Roggero (Genova) , D. Romagnoli (Torino) , M.E. Rossi (Genova) , G. Sacchiero (Ferrara) , P. Salmon (Genova) , F.O. Schreyer (Brandeis) , E. Sernesi (Roma I) , M.E. Serpico (Genova) , J. Shah (Northeastern) , R. Smith (Georgia) , E. Stagnaro (Padova) , R. Strano (Catania) , E. Strickland (Roma I) , F. Sullivan (Padova) , G. Tamone (Genova) , G. Tedeschi (Torino) , C. Traverso (Pisa) , C. Turrini (Milano) , Ughi (Perugia) , G. Valla (Genova), G. Van de Geer (Amsterdam) , , B. Van Geemen (Utrecht) , G. Vecchio (Catania) , L. Verdi (Firenze) , A. Verra (Torino) , G.E. Welters (Barcelona) .

TABLE OF CONTENTS

E. BALLICO and Ph. ELLIA:
On degeneration of projective curves 1

A. BEAUVILLE:
Variétés rationelles et unirationelles 16

M. BELTRAMETTI and P. FRANCIA:
Conic bundles on non-rational surfaces 34

F. CATANESE:
Moduli of surfaces of general type 90

C. CILIBERTO:
On a proof of Torelli's theorem 113

A. CONTE:
Two examples of algebraic threefolds whose hyperplane
sections are Enriques surfaces 124

D. EISENBUD and J. HARRIS:
On the Brill - Noether theorem 131

G. FALTINGS:
Properties of Arakelov's intersection product 138

W. FULTON:
On nodal curves 146

W. FULTON, S. KLEIMAN and R. MACPHERSON:
About the enumeration of contacts 156

F. GHIONE:
Un probleme du type Brill-Noether pour les
fibrés vectoriels 197

S. GRECO and A. VISTOLI:
On the construction of rational surfaces with
assigned singularities 210

L. GRUSON and C. PESKINE:
Postulation des courbes gauches 218

K. HULEK:
    Projective geometry of elliptic curves 228

R. LAZARSFELD and P. RAO:
    Linkage of general curves of large degree 267

P. MAROSCIA:
    Some problems and results on finite sets of
    points in P^n 290

V.B. MEHTA and A. RAMANATHAN:
    Homogeneous bundles in characteristic p 315

I. MORRISON and U. PERSSON:
    The group of sections on a rational elliptic surface 321

D. MUMFORD:
    On the Kodaira dimension of the Siegel
    modular variety 348

F. ORECCHIA:
    Generalized Hilbert functions of Cohen-Macaulay varieties 376

L. ROBBIANO and G. VALLA:
    Some curves in P^3 are set-theoretic complete intersections....391

E. STAGNARO:
    Constructing Enriques surfaces from quintics in P^3_K 400

G. VAN DER GEER:
    Prym surfaces and a Siegel modular threefold 404

ON DEGENERATION OF PROJECTIVE CURVES

by E. BALLICO and Ph. ELLIA

In this paper we study the problem of embedded deformations of projective curves. A typical example is as follows : given in $\mathbb{P}^n$ a smooth curve C of degree d and a line L intersecting C at only one point (L not tangent to C) does there exists a scheme T and a subscheme Z of $\mathbb{P}^n_T$, flat over T , such that the generic fiber Z_t is smooth and the special fiber is C $\cup$ L ? In the affirmative case we will say that C $\cup$ L is smoothable.

If C $\cup$ L is smoothable then it is in Z(g, d+1; n) which denotes the closure in Hilb $\mathbb{P}^n$ of the set of nonsingular curves of genus g and degree d . By the way not every curve is smoothable, for example (see IV.2) there exists a smooth curve C of genus 9 and degree 8 , a line L intersecting C at only one point such that C $\cup$ L has no smooth embedded deformation. On the other hand it follows from our method that if $O_C(1)$ is non special then C $\cup$ L is in Z(g, d+1; n) . In fact we prove a stronger result by constructing the degeneration in small subschemes of Z(g, d+1; n) (see Thm III.5) . In the same way let C be a given curve of degree d in $\mathbb{P}^n$, $\mathbb{P}^k$ a linear subspace of $\mathbb{P}^n$ and denote by $Pr_d(C, \mathbb{P}^k)$ the closure in Hilb $\mathbb{P}^k$ of the set of general projections of C on $\mathbb{P}^k$. Then we are able to describe some reducible curves of $Pr_d(C, \mathbb{P}^k)$ (see II)

Indeed the main idea of this paper is as follows : take a

curve Y in $\mathbb{P}^{n+1}$, a line D intersecting Y at one point and an hyperplane H of $\mathbb{P}^{n+1}$. Consider the flat family of curves in H obtained by projecting Y from the points of D : the generic fiber will be a smooth curve in Pr(Y, H) and the special fiber (over the point Y $\cap$ D) will be a reducible curve C $\cup$ L (see I.1) . This is our starting point and all our results are just variations on this theme.

Proposition III.4 and Theorem III.5 are largely used in [B - E]. Finally we plane in a future paper to use the results of § II to solve Hartshorne's conjecture about generic projections of elliptic curves (see [Ha,2] 4.3.4)

NOTE : As this paper was finished we learned that Theorem III.5. had been proved by Tannenbaum with a different method ([T2]) .

NOTATIONS . We work over an algebraically closed field K with ch(K) = 0 . Two subschemes X and Y of $\mathbb{P}^n$ are said to be (quasi)-transversal at $x \in X \cap Y$ if they are nonsingular at x and if the natural map : $T_x X \oplus T_x Y \to T_x \mathbb{P}^n$ is surjective (resp of maximal rank) .

If C is a smooth curve of degree d in $\mathbb{P}^n$ and H is a linear subspace of $\mathbb{P}^n$ we denote by $Pr_d(C; H)$ the closure in Hilb H of the set of general projections of C in H . Note that $Pr_d(C; H)$ is irreducible. If X is a variety and F an O_X-module we put : $\check{F} := Hom_{O_X}(F, O_X)$, $h^0(F) := \dim H^0(X, F)$. If X is a smooth curve, D a divisor on X and L a line bundle on X we write : $L(D) := L \otimes_{O_X} \mathcal{O}(D)$. We denote by ω_X the canonical sheaf on X.

I . THE METHOD

In this section we prove Prop. I.1 which is our main tool to construct embedded deformations of curves of type C $\cup$ L . In the next sections we will get our results from little variations of Prop. I.1. Let us describe the typical situation we will consider.

In $\mathbb{P}^{n+1}$ ($n \geq 2$) take a non degenerate smooth curve C of degree d and genus g, a point x of C and an hyperplane H such that : $x \notin H$. We denote by $\pi_n : \mathbb{P}^{n+1} \dashrightarrow H$ the projection from x. Put $Y := \pi_x(C)$. We assume Y is non-singular (of degree d-1) and we denote by y the point of Y which is the image of x under π_x (i.e. : $\{y\} = T_x C \cap H$). In this situation we have :

I.1. **PROPOSITION**. Let L be a line in H intersecting Y only at y quasi-transversally, then $Y \cup L$ is in $Pr_d(L; H)$.

PROOF. Let D be a line intersecting C only at x, not tangent to C, and intersecting L in a point t distinct from y (see Fig 1). Now consider the projections of C from the points of $D \setminus \{x\}$. This yields a family of curves, $X \subseteq H_{D \setminus \{x\}}$ parametrized by $D \setminus \{x\}$. Since C is non degenerate we can choose D such that, disregarding if necessary a finite number of points, the fibers of X are smooth irreducible curves of degree d and genus g. This family is flat ([Ha], II.9.9).

Furthermore there exists a unique subscheme $\bar{X}$ of H_D, flat over D, whose restriction to $H_{D \setminus \{x\}}$ is X ([Ha] II 9.8). We claim that the fiber of $\bar{X}$ over x is $Y \cup L$.

Obviously the fiber contains Y and, since the degree is preserved by flatness, it also contains a line. This line is precisely L because every fiber has the double point ξ at x (see Fig 1), indeed ξ arises from the projection of the tangent $T_x C$. Now from the exact sequence : $o \to O_{Y \cup L} \to O_Y \oplus O_L \to O_{Y \cap L} \to o$ and by the assumption : $Y \cap L = \{y\}$ it follows that

$p_a(Y \cup L) = g$, therefore the set theoretic fiber $Y \cup L$ agrees with the schematic fiber $\bar{X}_x$

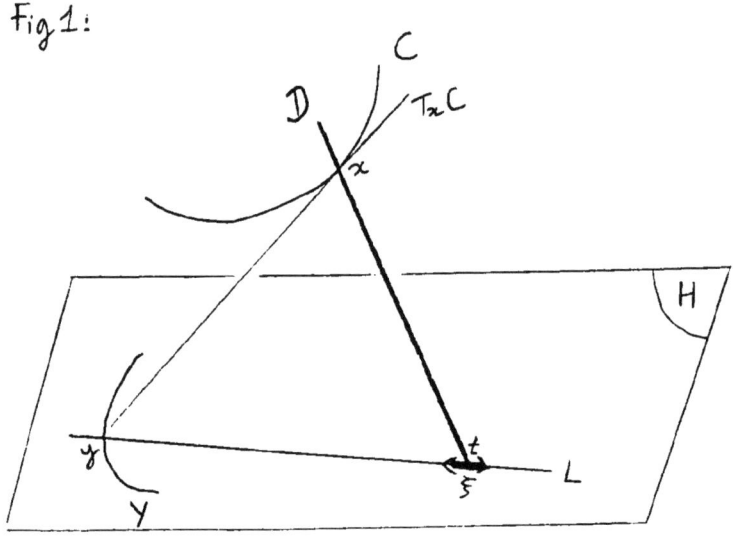

Fig 1:

For further applications of Prop. I.1. we will need also the following lemma :

I.2. **LEMMA** . Let X be a complete nonsingular curve of genus g and L a very ample line bundle on X . If L is non special then for every effective divisor D on X the line bundle $L(D)$ is very ample too.

PROOF . We have to show that for every $S, T \in X$: $h^0(L(D-S-T)) = h^0(L(D))-2$ which is equivalent, by Riemann-Roch, to : $h^0(\omega_X \otimes L^\vee(S+T-D)) = h^0(\omega_X \otimes L^\vee(-D))$ (*) . But L being very ample, in a similar way, we have $h^0(\omega_X \otimes L^\vee) = h^0(\omega_X \otimes L^\vee(S+T))$. Since L is non special $h^0(\omega_X \otimes L^\vee) = 0$ and (*) follows immediately ∎

II . DEGENERATIONS OF PROJECTIONS

We continue our study of reducible elements in $Pr_d(C; H)$. In fact Prop. II.1 is obtained from a repeated use of Prop. I.1 and Prop. II.2 is just a combination of I.1 and II.1.

II.0. PRELIMINARIES

Let X be a complete nonsingular curve of genus g and let d,n be integers satisfying $n \geq 3$ and $d > g+n$. Consider a non special very ample line bundle $\mathcal{L}$ of degree $g+n$ on X ([Ha] IV, 6.1), and take $d-g-n$ distinct points in X : $P_1,\ldots,P_{d-g-n}$. Define $L := \mathcal{L}(P_1 + \ldots + P_{d-g-n})$. The line bundle $\mathcal{L}$ gives an embedding of X in $\mathbb{P}^n$: $\varphi_\mathcal{L} : X \hookrightarrow \mathbb{P}^n$.

As $\mathcal{L}$ is non special, L is very ample (I.2) and non special. Thus L defines an embedding $\varphi_L : X \hookrightarrow \mathbb{P}^{d-g}$. Since the divisor $P_1 + \ldots + P_{d-n-g}$ induces an inclusion : $H^0(\mathcal{L}) \hookrightarrow H^0(L)$, we may consider : $\mathbb{P}^n \subset \mathbb{P}^{d-g}$.

II.1. PROPOSITION.

Consider $Y := \varphi_\mathcal{L}(X) \cup D_1 \cup \ldots \cup D_{d-g-n}$ where the D_i's are $d-g-n$ distinct lines not tangent to $\varphi_\mathcal{L}(X)$ and satisfying : $D_i \cap \varphi_\mathcal{L}(X) = \{\varphi_\mathcal{L}(P_i)\}$, $1 \leq i \leq d-g-n$. Then there exists a curve $\tilde{Y}$ in $Pr_d(\varphi_L(X), \mathbb{P}(H^0(\mathcal{L})^\vee))$ with same support and singular locus as Y.

PROOF. Suppose we have constructed

$$Y' = \varphi_{\mathcal{L}(P_1+\ldots+P_i)}(X) \cup \Delta'_{i+1} \cup \ldots \cup \Delta'_{d-g-n}$$

in $Pr_d(\varphi_L(X), \mathbb{P}^{n+i})$ where the (Δ'_j) are general lines through the points $\varphi_{\mathcal{L}(P_1+\ldots+P_i)}(P_j)$, $i+1 \leq j \leq d-g-n$ and where $\mathbb{P}^{n+i} \subset \mathbb{P}^{d-g}$ corresponds to the inclusion of $H^0(\mathcal{L}(P_1+\ldots+P_i))$ in $H^0(L)$ given by the divisor $P_{i+1} + \ldots + P_{d-g-n}$. Denote by x the point $\varphi_{\mathcal{L}(P_1+\ldots+P_i)}(P_i)$. Now consider a line D through x such that the plane spanned by D and the tangent to $\varphi_{\mathcal{L}(P_1+\ldots+P_i)}(X)$ at x does not contain any of the Δ'_j, $i+1 \leq j \leq d-g-n$. Arguing as in Prop. I.1 we get that

$$Y'' = \varphi_{\mathcal{L}(P_1+\ldots+P_{i-1})}(X) \cup \Delta''_i \cup \ldots \cup \Delta''_{d-g-n}$$ is in

$Pr_d(\varphi_L(X), \mathbb{P}(H^o(\mathcal{L}(P_1+\ldots+P_{i-1})^v)))$. After $d-g-n$ steps we have that $\varphi_{\mathcal{L}}(X) \cup \tilde{D}_1 \cup \ldots \cup \tilde{D}_{d-g-n}$ is in $Pr_d(\varphi_L(X), \mathbb{P}(H^o(\mathcal{L})^v))$ where the $\tilde{D}_i$'s, $1 \leq i \leq d-g-n$, are general lines through the points $\varphi_{\mathcal{L}}(P_i)$.

Next consider a variable line D parametrized by a smooth curve U such that :

$$\forall u \in U : D(u) \cap \varphi_{\mathcal{L}}(X) = \{\varphi_{\mathcal{L}}(P_1)\}$$

$D(u_o) = D_1$ and for general u in U : $D(u) \cap \tilde{D}_i = \emptyset$, $2 \leq i \leq d-g-n$.

The morphism $(U \times \varphi_{\mathcal{L}}(X)) \cup D \longrightarrow U$ is flat because it has no torsion. Its generic fiber is in $Pr_d(\varphi_L(X), \mathbb{P}^n)$ by the first part of the proof so the same occurs for the fiber over u_o. A repeated use of this process yields the proposition. ∎

II.2. **REMARK**. It may happen that $\tilde{Y} \neq Y$: this will be the case if the lines D_i intersect, then $\tilde{Y}$ will have embedded points at $\{D_i \cap D_j\}$. As an example of this situation we have :

II.3. **PROPOSITION**. Under the hypothesis of II.1 assume moreover that D_1 and D_2 intersect in a point y not in $Y \setminus (D_1 \cup D_2)$ then $Y \cup \chi_E(y)$ is a curve in $Pr_d(\varphi_L(X), \mathbb{P}(H^o(\mathcal{L})^v))$ ($\chi_E(y)$ is the first infinitesimal neighborhood of y in a three dimensional linear subspace of $\mathbb{P}(H^o(\mathcal{L})^v)$ containing D_1 and D_2).

The proof follows from II.1 because $D_1 \cup D_c \cup \chi_E(y)$ can be deformed in E into $D_1 \cup D_2'$ where D_2' is a line through $\varphi_{\mathcal{L}}(P_2)$ not meeting D_1.

Now we give other examples of reducible elements in $Pr_d(\varphi_L(X), \mathbb{P}^n)$ but this time the configuration of the lines is quite different : connected chains of lines linked to a smooth curve.

Let L_0 to be a nonsingular irreducible curve in $\mathbb{P}^n$ of degree $d-s$. Consider a reduced connected curve Y which is the union of L_0 with s distinct lines. Since Y is connected there exists an ordering of the lines in Y, $\{L_i\}$, $1 \leq i \leq s$, such that $L_0 \cup L_1 \cup \ldots \cup L_k$ is connected for $1 \leq k \leq s$. Such an ordering is called a compatible order of the lines in Y.

Let $\tau : \{1, \ldots, s\} \rightarrow \{0, \ldots, s-1\}$ be a function with $\tau(i) < i$ for every i.

II.4. <u>DEFINITION</u>. A bamboo of type τ for L_0 is a reduced connected union Y of L_0 with s lines endowed with a compatible order τ such that :

(1) for every $i \geq 1$, L_i intersects L_0 in at most one joint and quasi transversally

(2) $L_j \cap L_i = \emptyset$ if $j > i$ and $i \neq \tau(j)$.

A branch of the bamboo is a maximal connected union of lines in Y. The length of a branch is the number of its lines.

In the situation of II.0 we take $D := \sum_{i=1}^{r} n_i P_i$, $\deg(D) = d-g-n$.

II.5. <u>PROPOSITION</u>. Let Y be a bamboo of degree d for $\varphi_{\mathscr{L}}(X)$ with a branch of lenght n_i linked to $\varphi_{\mathscr{L}}(X)$ at $\varphi_{\mathscr{L}}(P_i)$, $1 \leq i \leq r$. Then Y is in $\text{Pr}_d(\varphi_L(X), \mathbb{P}(H^0(\mathscr{L})^{\vee}))$.

<u>PROOF</u>. For simplicity we assume $r = 1$ and $n_1 = 2$. From II.1 we get $\varphi_{\mathscr{L}(P_1)}(X) \cup \Delta_1'$ in $\text{Pr}_d(\varphi_L(X), \mathbb{P}(H^0(\mathscr{L}(P_1))^{\vee})$. Then projecting from the point $x = \varphi_{\mathscr{L}(P_1)}(P_1)$ and arguing as in II.1 one get the desired bamboo (see Fig. 2).

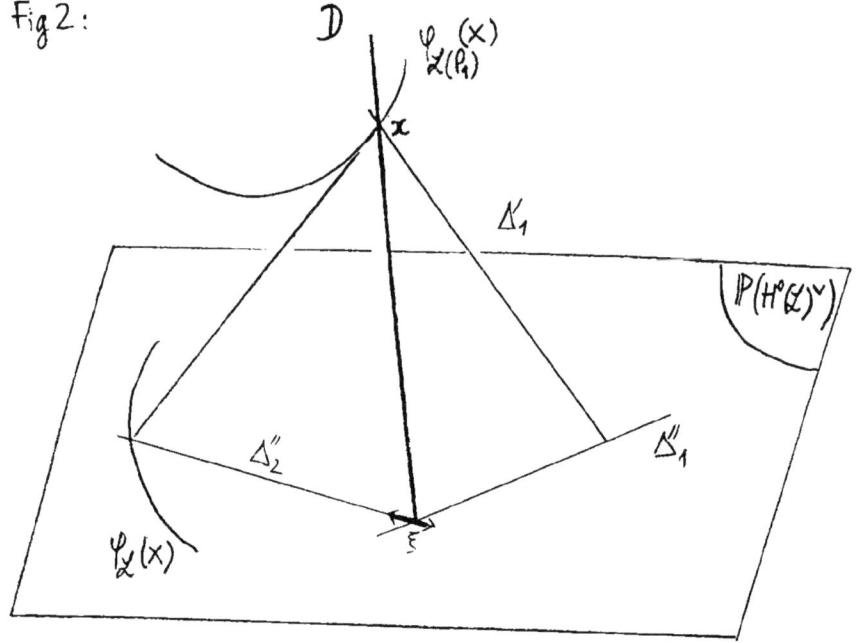

Fig 2:

III. FURTHER DEGENERATIONS

In this section we consider our problem from a different point of view : is a given curve Y in $\mathbb{P}^n$, of type $C \cup L$, smoothable ? We show (III.2) that a sufficient condition is that C is non degenerate and $O_C(1)$ is non special. Indeed in this case we prove that C can be obtained by projecting a curve C' in $\mathbb{P}^{n+1}$ from one of its points (III.1) . This enables us to describe reducible curves in some (small) closed subscheme of $H(d, g; \mathbb{P}^n)$, the Hilbert scheme of curves in $\mathbb{P}^n$ of degree d and genus g (III.5) . This result together with (III.4) is widely used in [B-E, 1] (see also [B-E; 2]) .

Finally we don't think that the condition $O_C(1)$ being non special is also necessary. However in IV we give an example of a curve $C \cup L$ in $\mathbb{P}^3$ which is not smoothable with C a non plane curve of genus 9 and degree 8 .

III.1. **LEMMA**. Let C be a smooth nondegenerate curve in $\mathbb{P}^n$ of genus g and degree d with $\mathcal{O}_C(1)$ non special. Take x a point of C. Then there exists a nonsingular curve Y in $\mathbb{P}^{n+1}$ of genus g and degree (d+1), a point y of Y such that C is the projection of Y from the point y, this latter having x as image.

<u>PROOF</u>. The proof is divided into two parts. First we prove that there exist an integer $N \geq n$, a curve C" in $\mathbb{P}^{N+1}$ of genus g and degree (d+1), a point ξ of C", a nonsingular curve C' in $\mathbb{P}^N$ of genus g and degree d such that :

1/ C' is the projection of C" from the point ξ of $\mathbb{P}^{N+1}$

2/ C is the image of C' under a projection of $\mathbb{P}^N$ into $\mathbb{P}^n$.

Then " reversing " the centers of the projections we will obtain the thesis of the lemma.

i) Let X be an abstract model of the curve and $\varphi : X \hookrightarrow \mathbb{P}^n$ an embedding with $\varphi(X) = C$. We put $L := \varphi^* \mathcal{O}_{\mathbb{P}^n}(1)$. Let S be the subvector space of $H^0(X, L)$ generated by the $\varphi^*(Z)$'s for Z in $H^0(\mathbb{P}^n, \mathcal{O}_{\mathbb{P}^n}(1))$. Since C is not contained in a hyperplane S has dimension n+1. Then $\varphi : X \longmapsto \mathbb{P}(S^*) \simeq \mathbb{P}^n$ sends $x \in X$ in the set of linear forms on S with kernel $\{s \in S / s(x) = 0\}$. Now L being very ample it defines an embedding : $\varphi_L : X \hookrightarrow \mathbb{P}(H^0(X, L)^*)$.

We put $N := h^0(X, L) - 1$. We have $N \geq n$. Now define $L':= L(P)$ where $P \in X$ satisfies $\varphi(P) = x$. By Lemma I.2. L' is a very ample line bundle and therefore it gives an embedding $\varphi_{L'} : X \hookrightarrow \mathbb{P}(H^0(X, L')^*)$. Denote by W^*, U^* the spaces of sections $H^0(X, L')$, $H^0(X, L)$.

Let π' be the projection of $\mathbb{P}(U)$ in $\mathbb{P}(S^*)$ from the center $\mathbb{P}(E^*)$ where E^* is defined by the exact sequence :

$$0 \longrightarrow E^* \longrightarrow U \longrightarrow S^* \longrightarrow 0 \qquad (1).$$

Let π'' be the projection of $\mathbb{P}(W)$ in $\mathbb{P}(U)$ from the point $\mathbb{P}(K^*)$ where K^* is given by the exact sequence :

$$0 \longrightarrow K^* \longrightarrow W \xrightarrow{h} U \longrightarrow 0 \qquad (2).$$

We put $\xi := \varphi_{L'}(P)$, $C' := \varphi_L(X)$, $C'' := \varphi_{L'}(X)$.

The first part of the lemma follows if we show :

    (a) $\pi'(C') = C$

    (b) $\pi''(C'') = C'$

    (c) π'' is the projection from the point ξ.

(a) et (b) follow because by definition we have the equalities : $\varphi = \pi' \circ \varphi_L$ and $\varphi_L = \pi'' \circ \varphi_{L'}$.

(c) is true because the natural map from L to $L(P)$ is the kernel of the evaluation at p.

ii) Look at the exact sequences (1) and (2) $\mathbb{P}(h^{-1}(E^*))$ intersects C'' only in a finite number of points because C'' generates $\mathbb{P}(W)$. Thus we may choose a subvector space D of W such that h maps D isomorphically onto E^* and such that we have $\mathbb{P}(D) \cap C'' = \emptyset$. Let K' be the image of K^* into W/D. Denote by p'' the projection of $\mathbb{P}(W)$ into $\mathbb{P}(W/D)$ from the center D and by p' the projection of $\mathbb{P}(W/D) \simeq \mathbb{P}^{n+1}$ into $\mathbb{P}^n$ from the center $\mathbb{P}(K')$. The rational maps $\pi' \circ \pi''$ and $p' \circ p''$ are defined and equal. We put $Y := p''(C'')$ and $y := p''(\xi)$. Then Y is isomorphic to C'' and has degree $(d+1)$ while p' is the projection from the center y ∎

III.2. <u>COROLLARY</u> . Let C be a non degenerate curve in $\mathbb{P}^n$ with $O_C(1)$ non special and let L be a line intersecting C only at one point x and quasi transversally then $C \cup L$ is smoothable.

III.3. **DEFINITION** . Let X be a complete nonsingular curve of genus g . We define $Z(X, d; n)$ as the closure in Hilb $\mathbb{P}^n$ of the set of nonsingular curves of degree d isomorphic to X . We also put $\delta := \text{Max}(g+n, 2g-1)$.

III.4. **PROPOSITION** . For $d \geq \delta$ $Z(X, d; n)$ is irreducible.

PROOF . From the assumption it follows that the very ample line bundles of degree d on X form a dense open subset U of $\text{Pic}^d(X)$. Now let P be the Poincaré bundle on $X \times U$ and π the projection on the second factor. Since $d \geq 2g - 1$ it follows that $\pi_* P$ is a vector bundle and we put $H := (n+1).\pi_* P$. Consider the closed subset Γ of $\mathbb{P}(H) \times X$ defined by $\Gamma := \{ [h], x)/h(x) = 0 \}$. Let Γ_1 be its projection on the first factor. Then $B := \mathbb{P}(H) \setminus \Gamma_1$ is open and dense in $\mathbb{P}(H)$. Consider the natural map : $h : B \times X \times X \longrightarrow \mathbb{P}^n \times \mathbb{P}^n$ given by evaluation. Let Z be the closure of $h^{-1}(\Delta) \setminus (B \times D)$ in $B \times X \times X$ (here Δ denote the diagonal of $\mathbb{P}^n \times \mathbb{P}^n$ and D the diagonal of $X \times X$) and T its projection in B . Consider the natural map $\alpha : B \setminus T \longrightarrow \text{Hilb } \mathbb{P}^n$ (evaluation at X). Let $U(X, d; n)$ be the set of points in Hilb $\mathbb{P}^n$ parametrizing the smooth curves of degree d isomorphic to X . Then $Z(X,d;n)$ is irreducible for α induces a dominant morphism : $f : \alpha^{-1}(R) \longrightarrow Z(X, d; n)$ where R is the open subset of smooth curves in $\mathbb{P}^n$. In fact the image of f is exactly $U(X, d; n)$ ∎

REMARK . From the proof it follows that if $d \geq \delta$ then $Z(X, d; n)$ is the closure of the set of non degenerate nonsingular curves of degree d isomorphic to X .

III.5. **THEOREM** . Let C be a curve in $Z(X, d; n)$ with $d \geq \delta$. Let D_i , $1 \leq i \leq s$, be distinct lines not contained in C . Suppose :

1/ $C \cup D_1 \cup \ldots \cup D_s$ is connected
2/ C is nonsingular along $C \cap (D_1 \cup \ldots \cup D_s)$.

Then there exists a curve Y in $Z(X, d+s; n)$ with same support and singular locus as $C \cup D_1 \cup \ldots \cup D_s$.

PROOF . We use induction on s .

For $s = 1$: we may assume C is nonsingular, non degenerate (of course with $0_C(1)$ non special). If D_1 inetrsects C only in one point and quasi transversally then it is III.2. Suppose D_1 intersects C in many points. Let x_0 be in $D_1 \cap C$. By assumption the regular part U of C contains x_0 . Consider a line $L(x)$ parametrized by a point x in U , intersecting C in x with $L(x_0) = D_1$ and $L(x) \cap C = \{x\}$ for generic $\{x\}$. The morphism $(U \times C) \cup L \longrightarrow U$ is flat because it has no torsion. Its generic fiber is a curve in $Z(X, d+1; n)$ so the same occurs for the fiber over x_0 .

Now note the following fact : there exists j , $1 \leq j \leq s$, such that $C \cup D_1 \cup \ldots \cup D_s \setminus D_j$ is connected. Indeed, define $m(i)$ to be the number of lines you need to go from D_i to C , pick j such that $m(j)$ is maximum then $C \cup D_1 \cup \ldots \cup D_s \setminus D_j$ is connected. Now suppose $s > 1$. We may assume $C \cup D_1 \cup \ldots \cup D_{s-1}$ is connected. The inductive hypothesis shows the existence of a curve Y' in $Z(X,d+s-1;n)$ with same support and singular locus as $C \cup D_1 \cup \ldots \cup D_{s-1}$.

If D_s intersects D_i (or C) consider a variable line parametrized by D_i (or U) and conclude as above. ■

III.6. REMARKS

(i) It may happen that Y is different from $C \cup D_1 \cup \ldots \cup D_s$ (see II.2, II.3).

(ii) From III.5 it follows that in $Z(X, d; n)$, $d \geq \delta+3$, there is any curve $C \cup D_1 \cup D_2 \cup D_3$ where C is isomorphic to X and $(D_i)_{1 \leq i \leq 3}$ are lines such that $D_1 \cap D_k \neq \emptyset$ and $D_k \cap C = \emptyset$, $2 \leq k \leq 3$. We don't know if such configuration of lines may be found in $Pr_d(\varphi_L(X), \mathbb{P}^{d-3})$ (compare with II.1, II.5) .

IV. (CONTRE)-EXAMPLES

We give examples of curves of type $C \cup L$ in $\mathbb{P}^3$ which are not smoothable. First recall the following result :

IV.1. PROPOSITION. Let C be a smooth plane curve of degree $d \geq 4$ and let L be a line intersecting C at only one point. Then $C \cup L$ is not smoothable.

PROOF. From Castelnuovo's bound on the genus of space curves it follows that there are no smooth curve of degree $d+1$, genus $(d-1)(d-2)/2$ for $d \geq 4$ (see [T]).

Now we give a similar example with a non plane curve C.

IV.2. PROPOSITION. There exists a smooth curve C in $\mathbb{P}^3$ of degree 8 and genus 9, a line L intersecting C at only one point such that $C \cup L$ is not smoothable.

PROOF. Let $H(d, g)$ denote the Hilbert secheme of one dimensional subschemes of $\mathbb{P}^3$ of degree d and arithemtic genus g. As usual let $H(d, g)_s$ be the open subscheme of $H(d; g)$ consisting of smooth connected curves.

A necessary condition for every curve Y of the type $Y = C \cup L$, C in $H(d, g)_s$, L a line intersecting C at one point, to be smoothable is : $\dim(H(d, g)_s + 3 < \dim H(d+1, g)_s$, since the choice of L is a three parameters choice.

So to conclude we have just to prove the following claim :

$$(*) \quad \begin{cases} (\alpha) & \dim H(8, 9)_s = 33 \\ (\beta) & \dim H(9, 9)_s = 36 \end{cases}$$

PROOF OF (*) :

(α) this is well known : one easily proves that every smooth curve

C of genus 9, degree 8 is a complete intersection $F_2.F_4$ and (α) follows.

(β) Let Y be any smooth curve of genus 9, degree 9.

First we observe that : $h^o(\mathbb{P}^3, I_Y(2)) = 0$ (I_Y denotes the ideal sheaf of Y in $\mathbb{P}^3$), this is just because $9 = a+b = (a-1)(b-1)$ has no integral solution. Then from the exact sequence :

$$0 \longrightarrow H^o(\mathbb{P}^3, I_Y(3)) \longrightarrow H^o(\mathbb{P}^3, O_{\mathbb{P}^3}(3)) \longrightarrow H^o(Y, O_Y(3))$$

and by Riemann-Roch, we get $h^o(\mathbb{P}^3, I_Y(3)) \geq 1$.

On the other hand : $h^o(\mathbb{P}^3, I_Y(3)) < 2$ because otherwise Y would be a complete intersection $F_3.F_3'$ which is impossible because of its genus. Now from the exact sequence :

$$0 \longrightarrow H^o(\mathbb{P}^3, I_Y(4)) \longrightarrow H^o(\mathbb{P}^3, O_{\mathbb{P}^3}(4)) \longrightarrow H^o(Y, O_Y(4))$$

we obtain : $h^o(\mathbb{P}^3, I_Y(4)) \geq 7$. We conclude that Y can be linked to a Cohen-Macaulay curve Y' by a complete intersection $F_3.F_4$. We have : deg (Y') = 3 $P_a(Y') = 0$. Since every such curve Y' is projectively normal ([ELLING] Ex 1 pg 430) the same occurs for Y. Furthermore from the resolution of the cone

$$C(Y') : 0 \longrightarrow R(-3)^{\oplus 2} \longrightarrow R(-2)^{\oplus 3} \longrightarrow R \longrightarrow C(Y') \longrightarrow 0$$

([ELLING] loc.cit)
and from the resolution of F_3,F_4 we get, as mapping cone the resolution of C(Y) :

$$0 \longrightarrow R(-5)^{\oplus 3} \longrightarrow R(-4)^{\oplus 3} \oplus R(-3) \longrightarrow R \longrightarrow C(Y) \longrightarrow 0 .$$

By [ELLING] thm. 2 we conclude that $H(9, 9)_s$ is smooth of dimension 36

BIBLIOGRAPHY

[B-E, 1] BALLICO, E. - ELLIA, Ph. : General curves of small genus in $\mathbb{P}^3$ are of maximal rank " Preprint

[Ha, 2] HARTSHORNE, R : " Stable vector bundles of rank 2 on $\mathbb{P}^3$ " Math. Ann. $\underline{238}$, 229-280 (1978

[ELLING] ELLINGSRUD, G. : " Sur le schéma de Hilbert des variétés de codimension 2 dans $\mathbb{P}^e$ à cône de Cohen-Macaulay " Ann. E.N.S. 4e série t. 8 fasc. 4 (1975) 423-431

[Ha] HARTSHORNE, R. : " Algebraic Geometry " Graduate texts in Mathematics, $\underline{52}$, Springer-Verlag (1977)

[Hi] HIRSCHOWITZ, A. : " Sur la postulation générique des courbes rationnelles " Acta Mat. $\underline{146}$ (1981) 209-230

[T] TANNENBAUM, A. : " On the geometric genera of projective curves " Math. Ann. $\underline{240}$, 213-221 (1979)

[T2] TANNENBAUM, A. : " Deformations of space curves " Arch. Mat. vol. $\underline{34}$, 37-42, (1980)

E. BALLICO
Scuola Normale Superiore
56100 PISA
Italy

Ph. ELLIA
C.N.R.S. LA 168
Université de Nice
Département de MATHEMATIQUES
Parc Valrose
06034 NICE CEDEX

VARIETES RATIONNELLES ET UNIRATIONNELLES

A. BEAUVILLE

Centre de Mathématiques de l'Ecole Polytechnique
F 91128 Palaiseau Cedex - France

"Laboratoire Associé au C. N. R. S. No 169"

1. ÉNONCÉ DU PROBLÈME.

Le problème dont je veux parler est souvent appelé le problème de Lüroth. Il peut s'exprimer en termes algébriques (rappelons qu'une extension de $\mathbb{C}$ est dite pure si elle est $\mathbb{C}$-isomorphe au corps des fractions rationnelles sur $\mathbb{C}$ en un nombre fini d'indéterminées) :
   Toute sous-extension d'une extension pure de $\mathbb{C}$ est-elle pure ?

Ce problème est en fait de nature géométrique. Introduisons deux définitions :

Définition : Soit X une variété algébrique complexe irréductible.
  a) On dit que X est unirationnelle s'il existe une application rationnelle dominante (c'est-à-dire génériquement surjective)
$f : \mathbb{P}^n \dashrightarrow X$.
  b) On dit que X est rationnelle s'il existe une application birationnelle $f : \mathbb{P}^n \dashrightarrow X$.

Puisque toute extension de type fini de $\mathbb{C}$ est le corps des fonctions d'une variété algébrique, le problème de Lüroth admet la formulation géométrique suivante :
   Toute variété unirationnelle est-elle rationnelle ?

Remarques : 1) Il est facile de voir que la restriction d'une application dominante $f : \mathbb{P}^n \dashrightarrow X$ à une sous-variété linéaire générale de $\mathbb{P}^n$, de dimension au moins égale à celle de X, est encore dominante. Dans la définition a), on peut donc supposer $n = \dim(X)$, ce que nous ferons désormais.

2) Il est clair que le problème ne dépend que de la classe d'équivalence birationnelle de la variété considérée : nous supposerons donc désormais que celle-ci est lisse.

2. LE PROBLÈME DE LÜROTH EN DIMENSION UN ET DEUX.

Lüroth a résolu affirmativement le problème en dimension un : toute courbe unirationnelle est rationnelle [L]. Sa démonstration est algébrique (alors qu'il donne l'énoncé sous forme géométrique). La démonstration géométrique est très facile : soit C une courbe unirationnelle, de sorte qu'il existe une application $f : \mathbb{P}^1 \to C$. L'espace $H^o(C, \Omega_C^1)$ est nul : en effet, si ω est une forme holomorphe sur C, la forme holomorphe $f^* \omega$ sur $\mathbb{P}^1$ est nulle, donc $\omega = 0$. Or toute courbe de genre nul est isomorphe à $\mathbb{P}^1$, d'où le résultat.

Le problème devient beaucoup plus difficile en dimension deux. Il a été résolu par Castelnuovo en 1894 [C] ; ce résultat constitue l'un des premiers succès de la géométrie birationnelle italienne. Soient S une surface lisse unirationnelle, et $f : \mathbb{P}^2 \dashrightarrow S$ une application dominante. On montre comme plus haut qu'on a $H^o(S, \Omega_S^1) = H^o(S, \Omega_S^2) = 0$ (noter que f est défini en dehors d'un nombre fini de points, ce qui permet de définir l'image inverse par f d'une forme holomorphe) ; plus généralement, on a $H^o(S, \Omega_S^1)^{\otimes k}) = 0$ pour tout k. Il reste à montrer que cette propriété entraîne que S est rationnelle, et c'est là le résultat essentiel de Castelnuovo :

<u>Théorème</u> : <u>Toute surface lisse S telle que</u> $H^o(S, \Omega_S^1) = H^o(S, (\Omega_S^2)^{\otimes 2}) = 0$ <u>est rationnelle.</u>

3. LE PROBLÈME DE LÜROTH EN DIMENSION TROIS.

Le problème se trouvait dès lors posé en dimension 3 ; Max Nœther avait d'ailleurs déjà observé que l'hypersurface cubique dans $\mathbb{P}^4$ est unirationnelle, et posé la question de sa rationalité. Mais c'est surtout le nom du mathématicien italien G. Fano qui reste attaché au problème de Lüroth en dimension 3. Dès 1908, Fano "prouve" l'irrationalité de la quartique de $\mathbb{P}^4$ et de l'intersection complète d'une quadrique et d'une cubique dans $\mathbb{P}^5$ [F1]. En 1912, Enriques démontre que cette dernière variété est unirationnelle [E], fournissant ainsi en principe un contre-exemple au problème de Lüroth. Malheureusement l'argument de Fano se heurte à des questions délicates sur les points-base des systèmes linéaires, auxquelles les techniques de l'époque ne permettent pas de répondre rigoureusement ; Fano doit faire (implicitement) des hypothèses de position générale, qui nous paraissent aujourd'hui injustifiables. Fano donne en 1915 une autre "démonstration" [F2], mais qui n'échappe pas aux mêmes critiques. Dans les années qui suivent, Fano étudie longuement les variétés de dimension 3 plongées dans $\mathbb{P}^n$ par leur système anticanonique (les deux variétés citées plus haut en constituent les premiers exemples). En 1947, il prétend démontrer que trois autres types de cette série sont irrationnels [F4] ; certaines de ces variétés étant birationnellement équivalentes à une hypersurface cubique de $\mathbb{P}^n$, il répondrait ainsi à la question de Nœther -malheureusement la démonstration n'est pas plus rigoureuse que les précédentes.

Il semble que les résultats de Fano aient été largement acceptés par ses contemporains (cf. par exemple [G]). Des réserves sont apparues plus tard : un exposé critique des travaux de Fano se trouve dans le livre de Roth [R]. Roth conclut qu'aucune des démonstrations de Fano n'est à l'abri de la critique. Il poursuit en exhibant, à son tour, un contre-exemple "rigoureux" au problème de Lüroth : il s'agit du "solide d'Enriques", obtenu en normalisant une sextique dans $\mathbb{P}^n$ astreinte à passer doublement par 6 plans, intersections deux à deux de 4 hyperplans en position générale. Roth démontre que cette variété est unirationnelle, puis que son groupe de Picard admet un élément de torsion, ou, ce qui revient au même, qu'elle n'est pas simplement connexe. Puisque le π_1 est un invariant birationnel, une telle variété est certainement irrationnelle. Malheureusement (pour Roth), Serre démontrait quelques années plus tard qu'une variété unirationnelle est toujours simplement connexe [Se] ! L'erreur de Roth provient de ce que le solide d'Enriques possède des singularités isolées en dehors des 6 plans doubles [Ty].

Il a en fait fallu attendre 1970 pour que le problème de Lüroth soit résolu, et ce par 3 paires de mathématiciens, schématisés dans le diagramme suivant :

auteurs	exemple	méthode
Clemens-Griffiths	cubique dans $\mathbb{P}^4$	jacobienne intermédiaire
Iskovskikh-Manin	quartique dans $\mathbb{P}^4$	automorphismes birationnels
Artin-Mumford	spécial (cf. § 9)	torsion de $H^3(X,\mathbb{Z})$

Le problème de Lüroth est cependant loin d'être entièrement résolu. En effet, le diagramme ci-dessus montre que des variétés d'un type extrêmement simple (hypersurfaces de bas degré dans $\mathbb{P}^4$) sont unirationnelles et non rationnelles. Dès lors, la question se pose de disposer de critères permettant d'affirmer qu'une variété donnée est ou n'est pas rationnelle. Le but de cet exposé est de montrer que les méthodes 1 et 2 permettent dans une certaine mesure de répondre à cette question en dimension 3. Le problème est par contre entièrement ouvert en dimension ≥ 4 (cf. § 10).

4. <u>LES CANDIDATS</u>.

Il s'agit ici de décrire une classe assez générale de variétés lisses X parmi lesquelles figurent d'éventuels contre-exemples. Une condition nécessaire, d'après ce qui précède (§ 2), est que tous les espaces de tenseurs contravariants holomorphes soient nuls, c'est-à-dire

$$H^0(X,(\Omega_X^1)^{\otimes k}) = 0 \quad \text{pour tout } k \geq 1 \quad .$$

Cette condition décrit une classe intéressante de variétés, mais certainement trop large pour qu'on puisse espérer une classification. Il est raisonnable, pour simplifier, d'ajouter la condition $b_2 = 1$ (équivalente à $\text{Pic}(X) = \mathbb{Z}$). On obtient alors les <u>variétés de Fano de première espèce</u>, qui ont été classifiées par Iskovskikh [I 1,2]. Pour énoncer ses résultats, notons H_X le générateur ample de $\text{Pic}(X)$; on a $K_X = -rH_X$, où r est un entier > 0 qu'on appelle l'indice de X. On a alors

Théorème : Soit X une variété de Fano de première espèce, d'indice r.

(i) On a $r \leq 4$; si $r = 4$ (resp. $r = 3$), X est isomorphe à $\mathbb{P}^3$ (resp. à une quadrique lisse de $\mathbb{P}^4$).

(ii) Si $r = 2$, X est isomorphe à l'une des variétés lisses suivantes :

(A_1) hypersurface de degré 6 dans l'espace projectif quasi-homogène $\mathbb{P}(1,1,1,2,3)$;
(A_2) revêtement double de $\mathbb{P}^3$ ramifié le long d'une quartique ;
(A_3) cubique dans $\mathbb{P}^4$;
(A_4) intersection de deux quadriques dans $\mathbb{P}^5$;
(A_5) section linéaire (dans le plongement de Plücker) de la grassmannienne $G(2,5)$.

(iii) Si $r = 1$, X est isomorphe à l'une des variétés lisses suivantes :

(B_2) revêtement double de $\mathbb{P}^3$ ramifié le long d'une sextique ;
(B_4) quartique dans $\mathbb{P}^4$;
(B_4') revêtement double d'une quadrique Q de $\mathbb{P}^4$, ramifié le long d'un diviseur découpé sur Q par une quartique de $\mathbb{P}^4$;
(B_6) intersection d'une quadrique et d'une cubique dans $\mathbb{P}^5$;
(B_8) intersection de trois quadriques dans $\mathbb{P}^6$;
(B_{10}) section quadratique de $G(2,5)$;
(B_{14}) section linéaire de $G(2,6)$;
(B_d) ($d = 12, 16, 18, 22$) une variété de degré d dans $\mathbb{P}^{d/2+2}$.

Les résultats positifs sur la rationalité ou l'unirationalité de ces variétés sont classiques ; ils sont essentiellement dus à Fano :

Proposition : Toutes les variétés de Fano de 1ère espèce sont unirationnelles sauf (peut-être) les types A_1, B_2 et B_4.
Les variétés de type A_4, A_5 ; B_{12}, B_{16}, B_{18} sont rationnelles.

Certaines quartiques sont unirationnelles [Sg] ; on ignore tout quant à l'unirationalité de la quartique générique, ainsi que des variétés de type A_1 et B_2. Certaines variétés de type B_{22} sont rationnelles ; Iskovskikh affirme qu'elles le sont toutes, mais sa démonstration est incomplète.

Ces résultats se démontrent par des méthodes projectives classiques : par exemple, soit X une intersection de deux quadriques dans $\mathbb{P}^5$ (type A_4) ; la projection depuis une droite contenue dans X définit une application birationnelle de X dans $\mathbb{P}^3$. Nous verrons plus loin (§ 6) une démonstration de l'unirationalité des types A_3 et B_8.

5. LA JACOBIENNE INTERMÉDIAIRE.

L'outil utilisé par Clemens-Griffiths [C-G] est la jacobienne intermédiaire. Je me bornerai à la définir dans un cas très particulier, celui d'une variété X de dimension 3 (lisse, projective) vérifiant $H^o(X,\Omega_X^3) = 0$. Dans ce cas la décomposition de Hodge de $H^3(X,\mathbb{C})$ s'écrit simplement

$$H^3(X,\mathbb{C}) = H^{2,1} \oplus H^{1,2} \quad ,$$

où $H^{2,1}$ et $H^{1,2}$ sont deux sous-espaces complexes conjugués de $H^3(X,\mathbb{C})$ (on considère $H^3(X,\mathbb{C})$ comme le complexifié de $H^3(X,\mathbb{R})$, ce qui le munit d'une conjugaison complexe). Cette propriété de conjugaison entraîne que la projection de $H^3(X,\mathbb{Z})$ dans $H^{1,2}$ est un réseau, de sorte que le quotient $H^{1,2}/H^3(X,\mathbb{Z})$ est un tore complexe J(X), appelé <u>jacobienne intermédiaire</u> de X. De plus, la forme $(\alpha,\beta) \mapsto -2i \int_X \bar{\alpha} \wedge \beta$ sur $H^{1,2}$ possède les propriétés suivantes :
- c'est une forme hermitienne positive séparante sur $H^{1,2}$;
- sa partie imaginaire induit sur $H^3(X,\mathbb{Z})$ le cup-produit, c'est-à-dire une forme alternée unimodulaire (à valeurs entières).

Elle définit par conséquent sur J(X) une <u>polarisation principale</u>, autrement dit un <u>diviseur thêta</u>, bien défini à translation près (cf. par exemple [M1]). La jacobienne intermédiaire est donc dans ce cas une <u>variété abélienne principalement polarisée</u>, et c'est toujours ainsi que nous la considèrerons.

La construction précédente est strictement parallèle à celle de la jacobienne d'une courbe ; de fait la jacobienne intermédiaire joue, pour les variétés qui nous occupent, le même rôle fondamental que la jacobienne pour les courbes. Je me contenterai de mentionner ici d'une part qu'elle intervient dans l'étude des cycles de dimension un sur X, et d'autre part qu'on peut espérer qu'elle détermine la variété X (problème de Torelli). Mais son importance dans les questions de rationalité vient du résultat suivant ([C-G], cor. 3.26) :

Proposition : **Si la variété X est rationnelle, J(X) est isomorphe à une jacobienne ou un produit de jacobiennes.**

Par jacobienne on entend la jacobienne d'une courbe ; d'autre part il s'agit bien sûr d'un isomorphisme de variétés abéliennes polarisées.

Esquisse de démonstration : On vérifie d'abord, par un calcul cohomologique facile, que lorsqu'on éclate une courbe lisse B dans X, la jacobienne intermédiaire de la variété éclatée est le produit de J(X) et de J(B). Si X est rationnelle, il existe d'après Hironaka un diagramme

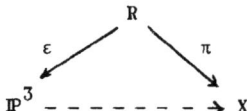

où ε est composé d'un nombre fini d'éclatements de points ou de courbes lisses, et où π est un morphisme birationnel. D'après ce qui précède, J(R) est un produit de jacobiennes, et J(X) en est un facteur direct. Or une variété abélienne principalement polarisée se décompose **de manière unique** en produit de facteurs irréductibles, ces facteurs correspondant aux composantes irréductibles du diviseur thêta. La proposition résulte alors de ce que les jacobiennes sont irréductibles (leur diviseur thêta est irréductible par le théorème de Riemann).

Le problème de Lüroth rejoint ainsi un autre problème classique, le **problème de Schottky** : déterminer, parmi toutes les variétés abéliennes principalement polarisées, celles qui sont des jacobiennes. Notons que les premières dépendent de $\frac{1}{2} g(g+1)$-modules et les secondes de $3g-3$, de sorte que le problème est (hautement !) non trivial dès que $g \geq 4$. Parmi les propriétés géométriques des jacobiennes, une conséquence facile du théorème des singularités de Riemann fournit un critère commode :

- **le lieu singulier du diviseur Θ d'une jacobienne est de codimension ≤ 4 dans la jacobienne.**

Andreotti et Mayer ont prouvé que cette propriété n'est pas très loin de caractériser les jacobiennes, cf. [A-M].

On notera qu'un produit de variétés abéliennes a un diviseur réductible, qui a donc un lieu singulier de codimension 2. Pour prouver que la variété X n'est pas rationnelle, on cherchera donc à montrer que le diviseur Θ de J(X) est peu singulier. Il faut pour cela disposer d'une description déométrique de J(X). On est loin de savoir répondre à cette question en général, ne serait-ce que pour les variétés de Fano. Nous allons décrire maintenant une classe de variétés pour lesquelles une telle description existe.

6. FIBRÉS EN CONIQUES.

Définition : On dit que la variété X est fibrée en coniques s'il existe une surface rationnelle S et un morphisme $f : X \to S$ dont les fibres sont des coniques (éventuellement dégénérées).

Il est facile de voir que toute variété admettant une application rationnelle sur $\mathbb{P}^2$ dont la fibre générique est une courbe rationnelle (ou, en langage classique, une congruence rationnelle du 1er ordre de courbes rationnelles) est birationnellement équivalente à un fibré en coniques. On décrit donc ainsi une classe de variétés qui semblent "assez proches" d'être rationnelles.

Sous les hypothèses de la définition, on vérifie facilement qu'il existe une courbe $C \subset S$ telle que
- si $p \in S - C$, $f^{-1}(p)$ est une courbe rationnelle lisse ;
- si p est un point lisse de C, $f^{-1}(p)$ est réunion de 2 courbes rationnelles se coupant transversalement ;
- si p est un point singulier de C, $f^{-1}(p)$ est une droite double ; p est alors un point double ordinaire de C .

On dit que C est la courbe discriminante du fibré en coniques.
On ignore si tout fibré en coniques est unirationnel (cf. § 10). On a cependant le résultat suivant, dû à Enriques :

Proposition : Soit $f : X \to S$ un fibré en coniques, et soit $R \subset X$ une surface rationnelle telle que la restriction de f à R soit surjective,

de degré d. Alors la variété X <u>est unirationnelle</u> ; <u>plus précisément,
il existe une application rationnelle dominante</u> $\mathbb{P}^3 \dashrightarrow X$ <u>de degré d</u>.

<u>Démonstration</u> : Traitons d'abord le cas $d = 1$. Alors la fibre générique de f est une conique sur le corps $K = \mathbb{C}(x,y)$, qui admet un point rationnel sur K ; elle est donc K-isomorphe à $\mathbb{P}^1_K$ (par projection stéréographique !), ce qui entraîne que X est rationnelle. Le cas général s'en déduit par le changement de base $R \to S$.

<u>Exemples</u> : 1) La cubique de $\mathbb{P}^4$.

Soit X une cubique (lisse) dans $\mathbb{P}^4$, et soit X_ℓ la variété obtenue en éclatant une droite ℓ contenue dans X. La projection de centre ℓ définit un morphisme $f : X_\ell \to \mathbb{P}^2$, qui fait de X_ℓ un fibré en coniques. La courbe discriminante $C \subset \mathbb{P}^2$ est de degré 5.
Notons E le diviseur exceptionnel dans l'éclatement de ℓ ; la restriction de f à E est surjective, de degré 2. Ainsi X <u>est unirationnelle</u>.

2) L'intersection de 3 quadriques dans $\mathbb{P}^6$.

Soit X une variété de ce type. Notons Π le réseau de quadriques de $\mathbb{P}^6$ contenant X ; c'est un plan projectif. Choisissons une droite ℓ contenue dans X ; notons G_ℓ la variété (isomorphe à $\mathbb{P}^4$) des 2-plans de $\mathbb{P}^6$ contenant ℓ. Soit $Q_\ell(X)$ la sous-variété de $\Pi \times G_\ell$ formée des couples (q,π) tels que $\pi \subset q$. La projection $f : Q_\ell(X) \to \Pi$ fait de $Q_\ell(X)$ un fibré en coniques. On vérifie immédiatement que la sous-variété de G_ℓ formée des plans contenus dans une quadrique q est singulière si et seulement si q l'est, de sorte que la courbe C est l'ensemble des quadriques singulières de Π ; elle est définie par l'annulation d'un déterminant symétrique d'ordre 7, à coefficients linéaires, et par suite est de degré 7.

Notons X_ℓ la variété obtenue en éclatant ℓ dans X, et S la sous-variété de $\Pi \times X_\ell$ formée des couples (q,x) tels que $\langle \ell, x \rangle \subset q$ (on désigne par $\langle \ell, x \rangle$ le plan engendré par ℓ et x ; ce symbole a un sens pour tout $x \in X_\ell$). On a un diagramme

avec $a(q,x) = (q, \langle \ell, x \rangle)$.

Il est immédiat que p_2 et a sont des morphismes birationnels, de sorte que X est birationnellement équivalent au fibré en coniques $Q_\ell(X)$.

De nouveau le diviseur exceptionnel dans l'éclatement de ℓ se projette surjectivement sur Π (le degré est 4). Ainsi X est unirationnelle.

3) Autres exemples.

Les variétés de Fano de 1ère espèce ne sont pas des fibrés en coniques ; mais elles le deviennent par des spécialisations convenables si on leur permet d'acquérir des singularités. Citons par exemple (cf. [B1]) :
- le revêtement double de $\mathbb{P}^3$ ramifié le long d'une quartique, lorsque celle-ci admet au moins un point double ordinaire ;
- la quartique avec une droite double ;
- l'intersection dans $\mathbb{P}^5$ d'une quadrique et d'une cubique contenant un plan.

L'importance des fibrés en coniques pour les questions de rationalité vient du théorème suivant :

Théorème : Soit X un fibré en coniques sur $\mathbb{P}^2$. Si le degré de la courbe discriminante C est ≥ 6, J(X) n'est pas isomorphe à une jacobienne (ou un produit de jacobiennes). En particulier, la variété X n'est pas rationnelle.

La démonstration de ce théorème consiste d'abord à décrire géométriquement J(X). On observe que la courbe C est munie naturellement d'un revêtement double $\tilde{C} \to C$, correspondant aux deux composantes des coniques singulières $f^{-1}(p)$ pour $p \in C$. Mumford a prouvé que J(X) est isomorphe à la variété de Prym associée à $(\tilde{C}, C)$ (cf. [B1]) ; il a d'autre part montré que le lieu singulier du diviseur Θ d'une variété de Prym est de codimension ≥ 5 sauf pour un petit nombre d'exceptions ([M2], et [B2] pour le cas où C a des points doubles). Il reste à s'assurer qu'une courbe plane de degré au moins 6 ne rentre pas dans ces exceptions, ce qui est facile.

Corollaire : Les variétés de Fano de type A_3, B_8 et B_{14} sont irrationnelles.

Pour les types A_3 et B_8, cela résulte du théorème et des exemples 1 et 2 (je triche un peu pour la cubique, dont la courbe discriminante est de degré 5 ; une étude plus précise montre que le théorème s'étend à ce cas). Fano a montré qu'une variété de type B_{14} est birationnellement équivalente à une cubique de $\mathbb{P}^4$ [F3].

7. IRRATIONALITÉ GÉNÉRIQUE.

La méthode des jacobiennes intermédiaires permet de régler génériquement les cas encore en suspens :

Théorème [B1] : <u>Une variété de Fano générique de type</u> A_1, A_2, B_2, B_4, B'_4, B_6 <u>ou</u> B_{10} <u>est irrationnelle</u>.

De manière précise, il est clair que l'ensemble des variétés de chaque type considéré peut être paramétré par une variété irréductible T ; il existe une sous-variété Z de T, distincte de T, telle que les variétés paramétrées par T - Z soient irrationnelles.

Nous allons montrer en fait que la jacobienne intermédiaire d'une variété de T - Z n'est pas une jacobienne ou un produit de jacobiennes ; comme cette propriété est <u>ouverte</u>, il suffit de le prouver en un point de T - Z. La démonstration repose alors sur le lemme suivant :

Lemme : <u>Soient S une courbe lisse, $X \to S$ une famille de variétés projectives de dimension trois, o un point de S. On suppose que</u> :
- <u>pour $s \neq o$, la fibre X_s est lisse, et</u> $H^o(X_s, \Omega^3_{X_s}) = 0$;
- X_o <u>n'a que des points doubles ordinaires ; si X'_o désigne la variété lisse obtenue en éclatant ces points doubles, $J(X'_o)$ n'est pas isomorphe à une jacobienne ou un produit de jacobiennes.</u>
<u>Alors il existe un ouvert non vide U de S tel que $J(X_u)$ ne soit pas une jacobienne (ou un produit de jacobiennes) pour tout $u \in U$.</u>

Démonstration : Notons $\mathcal{Q}_g$ l'espace des modules des variétés abéliennes principalement polarisées de dimension g (quotient du demi-espace de Siegel par le groupe modulaire), avec $g = \frac{1}{2} b_3(X_s)$. La famille $J(X_s)_{s \in S-o}$ définit une application classifiante $c : S - o \to \mathcal{Q}_g$; l'hypothèse sur X entraîne que c se prolonge en $\bar{c} : S \to \bar{\mathcal{Q}}_g$, où $\bar{\mathcal{Q}}_g$ désigne la compactification de Satake de $\mathcal{A}_g$. On a ensemblistement

$$\overline{\mathcal{A}}_g = \mathcal{A}_g \cup \mathcal{A}_{g-1} \cup \ldots \cup \mathcal{A}_o \quad,$$

et le point $\overline{c}(o)$ de $\overline{\mathcal{A}}_g$ est la classe de $J(X'_o)$ dans $\mathcal{A}_p \subset \overline{\mathcal{A}}_g$, avec $p = \dim J(X'_o) \leq g$. Soit d'autre part J_g l'ensemble des points de $\mathcal{A}_g$ correspondant aux jacobiennes et aux produits de jacobiennes ; il est fermé dans $\mathcal{A}_g$, et son adhérence dans $\overline{\mathcal{A}}_g$ est

$$\overline{J}_g = J_g \cup J_{g-1} \cup \ldots \cup J_o \quad.$$

Comme $\overline{c}(0) \notin \overline{J}_g$ par hypothèse, il existe un voisinage ouvert U de a dans S tel que $\overline{c}(U) \cap \overline{J}_g = \emptyset$, d'où le lemme.

Pour achever la démonstration du théorème, il reste à mettre en évidence, dans chaque cas, une déformation $X \to S$ possédant les propriétés du lemme, et telle que X'_o soit un fibré en coniques. Je me contenterai d'indiquer quelques exemples. Si l'on projette une intersection de 3 quadriques depuis une droite générale contenue dans la variété, on obtient une quartique X_o avec 17 points doubles ordinaires que l'on peut déformer en une quartique générique. En projetant de nouveau depuis un point double de X_o, on obtient le type B_2 ; en faisant acquérir à l'intersection de 3 quadriques un point double ordinaire et en projetant depuis ce point, le type B_6, etc.

Remarque : La méthode s'applique bien entendu à d'autres variétés que les variétés de Fano.

8. LES TRAVAUX D'ISKOVSKIKH ET MANIN.

Les résultats d'Iskovskikh et Manin sont les suivants :

Théorème : Les variétés de Fano de type B_2, B_4, B'_4, B_6 sont irrationnelles.

Les démonstrations reprennent les idées de Fano (essentiellement celles de [F2]) en les complétant. Elles sont longues et difficiles, et j'avoue ne pas avoir tout lu. Je vais me borner à indiquer les énoncés qui conduisent au théorème.

Pour une variété X de type B_2 ou B_4, Iskovskikh et Manin prouvent dans [I-M] (cf. aussi [I3]) :

- <u>Toute application birationnelle de</u> X <u>dans une variété de Fano de 1ère espèce est un isomorphisme.</u>

En particulier, le groupe des automorphismes birationnels de X est fini, et X n'est pas birationnellement équivalente à $\mathbb{P}^3$.

La situation est plus compliquée pour les types B_4' et B_6, car la variété X admet alors des automorphismes birationnels qui ne sont pas partout définis. La méthode d'Iskovskikh dans [I3] consiste à mettre en évidence certains de ces automorphismes ; notant B(X) le groupe qu'ils engendrent, il obtient l'énoncé suivant :

- <u>Si</u> $\chi : X \to V$ <u>est une application birationnelle de</u> X <u>dans une variété de Fano de 1ère espèce</u>, <u>il existe</u> $\psi \in B(X)$ <u>tel que</u> $\chi \circ \psi$ <u>soit un isomorphisme.</u>

Il en déduit en particulier l'irrationalité de X (prenant $V = \mathbb{P}^3$), ainsi que des renseignements profonds sur la structure du groupe des automorphismes birationnels de X.

Signalons pour terminer que ces méthodes s'appliquent avec succès aux fibrés en coniques :

<u>Théorème</u> [S] : <u>Soit</u> $X \to S$ <u>un fibré en coniques, dont la courbe discriminante</u> C <u>vérifie</u> $|4K_S + C| \neq \emptyset$. <u>Alors tout automorphisme birationnel de</u> X <u>préserve la fibration, et</u> X <u>est irrationnel.</u>

Pour $S = \mathbb{P}^2$, ce résultat est moins fort que le théorème du § 6 ; mais en éclatant des points dans $\mathbb{P}^2$, Sarkisov donne des exemples de <u>fibrés en coniques</u> X <u>irrationnel avec</u> $H^3(X, \mathbb{Z}) = 0$ (la courbe C est alors une courbe elliptique). Il n'est pas clair qu'il existe de tels exemples avec X unirationnel.

9. <u>L'EXEMPLE D'ARTIN-MUMFORD.</u>

La méthode de [Ar-M] sort un peu du cadre de cet exposé, puisqu'elle ne s'applique pas aux variétés de Fano ; elle a cependant l'avantage de fournir un critère d'irrationalité en toute dimension. Elle est basée sur le résultat suivant :

Proposition : <u>Pour une variété projective lisse</u> X, <u>le sous-groupe de torsion de</u> $H^3(X,\mathbb{Z})$ <u>est un invariant birationnel</u>.

La démonstration est analogue à celle de le proposition du § 5, mais plus facile.

Il s'agit donc de construire des variétés unirationnelles X pour lesquelles $H^3(X,\mathbb{Z})$ contienne des éléments de torsion. Voici une description géométrique de l'exemple d'Artin-Mumford. Soit Π un système linéaire de quadriques de $\mathbb{P}^3$, de dimension projective 3, vérifiant les conditions de position générale suivantes :
  (i) Π n'a pas de point-base ;
  (ii) si ℓ est une droite singulière pour une quadrique de Π, les autres quadriques de Π ne contiennent pas ℓ.

Notons G la grassmannienne des droites de $\mathbb{P}^3$; soit R la sous-variété de G formée des droites contenues dans un pinceau de quadriques de Π. On sait que R est une <u>surface d'Enriques</u>, appelée classiquement <u>congruence de Reye</u> (cf. [B3], p. 136). Posons

$$\hat{G} = \{(\ell,q) \in G \times \Pi \mid \ell \subset q\} \ .$$

Il est immédiat que la projection $\hat{G} \to G$ n'est autre que l'éclatement de R dans G. Notons $f : \hat{G} \to \Pi$ la seconde projection ; pour $q \in \Pi$, la fibre $f^{-1}(q)$ s'identifie à l'ensemble des génératrices de q. L'application f se factorise donc en

$$f : \hat{G} \xrightarrow{g} X' \xrightarrow{\pi} \Pi \quad ,$$

où π est un revêtement double, ramifié le long de la quartique de Π correspondant aux quadriques singulières. Celle-ci a dix points doubles ordinaires $s_1,\ldots,s_{10}$, correspondant aux quadriques de rang 2 de Π ; on a $\pi^{-1}(s_i) = \{p_i\}$, où p_i est un point double ordinaire de X'. On note X la variété (lisse) obtenue en éclatant les points $p_1,\ldots,p_{10}$ dans X. On voit facilement que X est unirationnelle (cf. § 6, exemple 3).

Proposition : Le groupe $H^3(X,\mathbb{Z})$ contient un élément d'ordre 2.

Démonstration : Nous noterons Q_i la diviseur exceptionnel de X au-dessus de p_i ; on pose $U = X' - \{p_1,\ldots,p_{10}\}$ et $V = g^{-1}(U)$. Le morphisme $g : V \to U$ est une fibration en droites projectives, tandis que $P_i = g^{-1}(p_i)$ est la réunion de deux plans se coupant en un point. La cohomologie considérée dans ce qui suit est toujours à coefficients entiers.

a) Le groupe $H^4(G)$ contient un élément d'ordre 2 : il est en effet isomorphe à $H^4(G) \oplus H^2(R)$, et $c_1(R)$ est un élément d'ordre 2 dans $H^2(R)$.

b) Il en est de même de $H^4_c(V)$, à cause de la suite exacte

$$\bigoplus_i H^3(P_i) \longrightarrow H^4_c(V) \longrightarrow H^4(G) \longrightarrow \bigoplus_i H^4(P_i)$$

et des relations $H^3(P_i) = 0$, $H^4(P_i) = \mathbb{Z} \oplus \mathbb{Z}$.

c) Le groupe $H^2_c(U)$ est sans torsion ; en effet $H^2(X)$ est sans torsion puisque X est simplement connexe, et $H^2_c(U)$ est un sous-groupe de $H^2(X)$ puisque $H^1(Q_i) = 0$.

d) Le groupe $H^4_c(U)$ contient un élément d'ordre 2. En effet, la suite exacte de Gysin pour la fibration en sphères $g : V \to U$ s'écrit

$$H^1_c(U) \xrightarrow{e} H^4_c(U) \xrightarrow{g^*} H^4_c(V) \xrightarrow{g_*} H^2_c(V) \quad ,$$

où e désigne le cup-produit avec la classe d'Euler du fibré en sphères ; celle-ci est annulée par 2. Si $\text{Im}(e) \neq 0$, l'assertion est claire ; sinon l'élément d'ordre 2 de $H^4_c(V)$ provient (d'après c)) d'un élément d'ordre 2 de $H^4_c(U)$.

e) Puisque $H^3(Q_i) = 0$, le groupe $H^4_c(U)$ est isomorphe à un sous-groupe de $H^4(X)$; celui-ci contient donc un élément d'ordre 2, et il en est de même de $H^3(X)$ par dualité de Poincaré et par la formule des coefficients universels.

Remarques : 1) Pour tout n, la variété $X \times \mathbb{P}^n$ est unirationnelle et non rationnelle (puisque $H^3(X \times \mathbb{P}^n, \mathbb{Z})$ contient un élément d'ordre 2).

2) La variété V est une variété de Severi-Brauer au-dessus de U, ce qui signifie que la fibration $g : V \to U$ n'est pas la fibration projective associée à un fibré vectoriel : en effet dans

le cas contraire, V serait birationnellement équivalente à $X \times \mathbb{P}^1$, donc irrationnelle. Un argument formel montre que V s'étend en une variété de Severi-Brauer au-dessus de X.

10. PROBLÈMES OUVERTS.

1) <u>Compléter les résultats pour les variétés de Fano</u>.

Il est possible qu'une étude détaillée de chaque cas, dans le style de [C-G], fournisse la réponse. Il serait plus intéressant de comprendre la question suivante :

2) <u>Toute déformation (lisse) d'une variété irrationnelle de dimension 3 est-elle irrationnelle</u> ?

Notons qu'une **spécialisation** d'une variété rationnelle de dimension 3 est rationnelle [T].

3) Fano observe souvent que la série des variétés de Fano "s'approche de la rationalité" quand le degré croît. Il en est de même lorsqu'on impose un nombre croissant de points doubles à une variété donnée. Peut-on donner un sens précis à ces assertions expérimentales ?

4) <u>Donner des critères en dimension</u> > 3, applicables aux variétés usuelles. En particulier :

5) <u>Prouver qu'une cubique générique de dimension 4 est irrationnelle</u>. Il existe une famille de codimension un de cubiques de $\mathbb{P}^5$ qui sont rationnelles ; conjecturalement , la réponse à la question 2 devrait donc être négative en dimension 4.

6) Signalons un problème de Zariski :
<u>Si</u> $X \times \mathbb{P}^1$ <u>est rationnelle, la variété</u> X <u>est-elle rationnelle</u> ?

Elle est évidemment unirationnelle. Notons qu'on a construit au § 9 une variété irrationnelle X admettant une variété de Severi-Brauer ("forme tordue" de $X \times \mathbb{P}^1$) rationnelle.

7) On rencontre souvent dans les problèmes de modules des variétés du type V/G, où G est un groupe semi-simple complexe opérant linéairement sur l'espace vectoriel complexe V. Ces variétés sont-elles toutes rationnelles ? Si H est un sous-groupe fermé de G, l'espace homogène G/H est-il rationnel ?

L'espace des modules des courbes de genre g est unirationnel pour g ≤ 10. Est-il rationnel ? J'ignore la réponse dès que g ≥ 3.

Les problèmes d'unirationalité semblent à l'heure actuelle encore plus inaccessibles, vu l'absence totale de méthodes existantes. Voici trois questions classiques, qui sont d'ailleurs liées entre elles.

8) <u>Donner un exemple de variété</u> X <u>non unirationnelle, telle que</u> $H^0(X,(\Omega_X^1)^{\otimes k}) = 0$ <u>pour tout</u> k .

9) <u>Une quartique générique de</u> $\mathbb{P}^4$ <u>est-elle unirationnelle</u> ?

10) <u>Donner un exemple de fibré en coniques non unirationnel</u>.

Un candidat possible, suggéré par Enriques, est l'hypersurface de degré d dans $\mathbb{P}^4$ contenant une droite avec multiplicité (d-2), pour d ≥ 5.

BIBLIOGRAPHIE

[A-M] A. ANDREOTTI et A. MAYER : On period relations for abelian integrals on algebraic curves, Ann. Sc. Norm. Sup. Pisa 21 (1967), 189-238.

[Ar-M] M. ARTIN et D. MUMFORD : Some elementary examples of unirational varieties which are not rational, Proc. London Math. Soc. 25 (1972), 75-95.

[B1] A. BEAUVILLE : Variétés de Prym et jacobiennes intermédiaires, Ann. E.N.S. 10 (1977), 309-391.

[B2] A. BEAUVILLE : Prym varieties and the Schottky problem, Inventiones Math. 41 (1977), 149-196.

[B3] A. BEAUVILLE : Surfaces algébriques complexes, Astérisque n° 54, S.M.F. (1978).

[C3] G. CASTELNUOVO : Sulla razionalità delle involuzioni piane, Math. Annalen 44 (1894).

[C-G] H. CLEMENS et P. GRIFFITHS : The intermediate Jacobian of the cubic threefold, Ann. of Math. 95 (1972), 281-356.

[E] F. ENRIQUES : Sopra una involuzione non razionale dello spazio, Rend. Acc. Lincei, s. 5^a, 31 (1912), 81-83.

[F1] G. FANO : Sopra alcune varietà algebriche a tre dimensioni aventi tutti i generi nulli, Atti Acc. Torino 43 (1908), 973-977.

[F2] G. FANO : Osservazioni sopra alcune varietà non razionali aventi tutti i generi nulli, Atti Acc. Torino 50 (1915), 1067-1072.

[F3] G. FANO : Sulle sezioni spaziali della varietà Grassmanniana delle rette dello spazio a cinque dimensioni, Rend. Acc. Lincei 11 (1930), 329-356.

[F4] G. FANO : Nuove ricerche sulle varietà algebriche a tre dimensioni a curve-sezioni canoniche, Comm. Pont. Ac. Sci. (1947), 635-720.

[G] L. GODEAUX : Questions non résolues de géométrie algébrique, Hermann, Paris (1933).

[I1] V. ISKOVSKIKH : Fano threefolds I, Math. USSR Izvestia 11 (1977), 485-527.

[I2] V. ISKOVSKIKH : Fano threefolds II, Math. USSR Izvestia 12 (1978), 469-506.

[I3] V. ISKOVSKIKH : Birational automorphisms of three-dimensional algebraic varieties, J. Soviet Math. 13 (1980), 815-868.

[I-M] V. ISKOVSKIKH et J. MANIN : Three-dimensional quartics and counterexamples to the Lüroth problem, Math. USSR Sbornik (1971), 141-166.

[L] J. LÜROTH : Beweis eines Satzes über rationale Curven, Math. Annalen 9 (1876), 163-165.

[M1] D. MUMFORD : Abelian varieties, Oxford University Press (1970).

[M2] D. MUMFORD : Prym varieties I. Contributions to analysis, Academic Press, New York (1974).

[R] L. ROTH : Algebraic threefolds, Ergebnisse der Math. 6, Springer-Verlag, Berlin-Heidelberg-New York (1955).

[S] V. SARKISOV : Birational automorphisms of conic bundles, Mat. USSR Izvestia 17 (1981), 177-202.

[Se] J.P. SERRE : On the fundamental group of a unirational variety, J. London Math. Soc. 34 (1959), 481-484.

[Sg] B. SEGRE : Variazione continua ed omotopia in geometria algebrica, Ann. Mat. Pura Appl. 50 (1960), 149-186.

[T] K. TIMMERSCHEIDT : On deformations of threedimensional rational manifolds, Math. Annalen 258 (1982), 267-275.

[Ty] J. TYRELL : The Enriques threefold, Proc. Cambridge Phil. Soc. 57 (1961), 897-898.

CONIC BUNDLES ON NON-RATIONAL SURFACES

by

M. Beltrametti and P. Francia[*]

Contents

Introduction

§1. Notations, definitions, and preliminary results.
§2. The Chow group $A^2(X)$ of a conic bundle.
§3. The algebraic representative of $A^2(X)$.
§4. The classical intermediate Jacobian of a conic bundle.
§5. Some open questions.
References.

INTRODUCTION

In this paper we state some results concerning conic bundles X on nonsingular surfaces S. A reason for this study is that conic bundles occur in the classification of threefolds with negative Kodaira dimension. We mainly describe, up to isogenies, the group $A^2(X)$ of cycles of codimension 2 which are algebraically equivalent to zero and the intermediate Jacobian $J(X)$.

---

[*] This is an expanded version of a talk presented at the Conference "Open Problems in Algebraic Geometry", Ravello (Italy), on June 2, 1982.

In the first section we recall some well known general results on conic bundles: the main reference for this is [B] where the case $S = \mathbb{P}^2$ is studied. Moreover we point out some relations between conic bundles and threefolds with negative Kodaira dimension.

In Section 2 we find a decomposition of the group $A^2(X)$, up to isogeny, as direct sum $A^2(S) \oplus A^1(S) \oplus P_X$, where P_X is the Prym variety associated with X. In particular for a surface S with $q(S) = p_g(S) = 0$ (and not of general type) one has the isomorphism $A^2(X) \simeq P_X$ and the theory runs as well as in [B].

In Section 3 the algebraic representative A_X of $A^2(X)$ is considered. We prove that the decomposition of $A^2(X)$ induces the one of A_X, via canonical morphisms.

Section 4 is devoted to the intermediate Jacobian $J(X)$. First, $J(X)$ is canonically isogenous to A_X. Moreover we prove that there exists an isogeny $\sigma: J(X) \tilde{\sim} \text{Alb}(S) \oplus \text{Pic}^o(S) \oplus P_X$. In the general case σ is not a morphism of principally polarized abelian varieties. However, under the assumption $q(S) = 0$, one sees that σ induces an isomorphism $J(X) \simeq P_X$ of principally polarized abelian varieties.

A list of some open questions is contained in Section 5.

Here we give a rather elementary exposition even though this may result in rather a lengthy paper.

§1. Notations, definitions, and preliminary results

Throughout this paper we fix an algebraically closed field k of characteristic zero. By __variety__ we mean a complete irreducible nonsingular scheme defined over k.

For a coherent sheaf F on a d-dimensional variety V we denote by $h^i(F)$ the dimension of the k-vector space $H^i(V,F)$, $i \geq 0$. We call __irregularity__ of V the integer $q(V) = h^1(\mathcal{O}_V)$. Moreover the __m-plurigenus__ of V is $p_m(V) = h^0(\omega_V^{\otimes m})$, where ω_V is the canonical sheaf. We denote by $\kappa(V)$ the Kodaira dimension of V and by $\chi(\mathcal{O}_V) = \sum_{i=0}^{d} (-1)^i h^i(\mathcal{O}_V)$ the Euler-Poincaré characteristic of V. Finally, for all p, $0 \leq p \leq d$, let $S^r(\Omega_V^p)$ be the r-symmetric tensor of the sheaf of regular p-forms Ω_V^p.

By __threefold__ (resp. __surface__) we mean a nonsingular projective variety of dimension three (resp. two).

DEFINITION 1.1 We say that a threefold X is a __conic bundle__ if there exist a surface S and a morphism $f: X \to S$ such that all the fibres are conics. □

Conic bundles on a nonsingular surface occur in the classification of threefolds with negative Kodaira dimension. This is clear from the recent results of Mori [Mo], and also from the following statements contained in [B-F].

THEOREM 1.2 Let X be a threefold with $\kappa(X) < 0$, $q(X) > 0$. Then we have:

(I) X is birationally equivalent to a conic bundle on a surface S such that $\kappa(S) \geq 0$, or

(II) the image of the Albanese mapping $\alpha: X \to \mathrm{Alb}(X)$ is a nonsingular curve and the general fibre of α is a rational surface.

Furthermore if $\chi(\mathcal{O}_X) \geq 0$ and $h^0(S^{12}(\Omega_X^2)) > 0$ then X belongs to family (I), while, whenever $\chi(\mathcal{O}_X) \leq 0$ and $h^0(S^{12}(\Omega_X^2)) = 0$, then X belongs to family (II).

PROPOSITION 1.3 Let X be a threefold with $\kappa(X) < 0$, $q(X) > 0$. Suppose there exists a morphism $\alpha: X \to C$ such that C is a nonsingular projective curve and the general fibre is a rational surface. Then X is birationally equivalent to one of the following types of threefolds $\tilde{X}$:

(a) $\tilde{X} = C \times \mathbb{P}^2$;

(b) $\tilde{X}$ is a conic bundle on a surface S birationally equivalent to $C \times \mathbb{P}^1$;

(c) There exists a morphism $\tilde{X} \to C$ such that the generic fibre S is a Del Pezzo surface with $\mathrm{Pic}(S) \simeq \mathbb{Z}$ generated by the anticanonical sheaf ω_S^{-1}. Moreover $1 \leq \omega_S^2 \leq 6$.

REMARK 1.4 The case $\omega_S^2 = 5$ does not occur in family (c) of the previous Proposition. In fact, if $\omega_S^2 = 5$ then X is birational to $C \times \mathbb{P}^2$. This is a consequence of a classical construction due to Enriques (see [E], §8 and also [Co], p.473-474). □

We summarize here without proof some well known results about conic bundles on a surface S. They are proved in [B], Ch. I in the case $S = \mathbb{P}^2$. Since all questions involve local properties the arguments contained in [B] can be extended in a standard way to the case of a nonsingular surface.

THEOREM 1.5 Let $f:X \to S$ be a conic bundle. Then we have:

(1) The morphism f is flat.

(2) Let ω_X be the canonical sheaf on X. Then the direct image $\mathcal{E} = f_*\omega_X^{-1}$ is a locally free sheaf of rank 3 and X is defined in $\mathbb{P}(\mathcal{E})$ by a quadratic form.

(3) There exists a curve C in S with at most ordinary double points, such that:

- for every point $s \in S \setminus C$ the fibre $f^{-1}(s)$ is nonsingular;
- if s is a nonsingular point of C then the fibre $f^{-1}(s)$ is isomorphic to two distinct lines;
- if s is an ordinary double point of C then the fibre $f^{-1}(s)$ is isomorphic to a double line.

DEFINITION 1.6 The curve C of statement 1.2 is named discriminant curve of X. Moreover we say that a conic bundle is ordinary if the discriminant curve is irreducible and nonsingular. □

From now on, we only consider ordinary conic bundles. Let $G_1(\mathcal{E})$ be the Grassmannian of the lines in $\mathbb{P}(\mathcal{E})$, and let $\tilde{C}$ be the curve in $G_1(\mathcal{E})$ whose closed points are the lines of the fibres $f^{-1}(s)$, $s \in C$. Then the natural projection $G_1(\mathcal{E}) \to S$ induces a double etale covering $\nu: \tilde{C} \to C$. Let $\tau: \tilde{C} \to \tilde{C}$ be the canonical involution.

DEFINITION 1.7 The Prym variety P_X associated to the conic bundle X is the abelian subvariety of $J(\tilde{C})$ given by
$P_X = \text{Im}(\mathbb{1}_{J(\tilde{C})} - \tau^*)$. □

Moreover P_X has a principal polarization Θ such that 2Θ is algebraically equivalent to $m^*\tilde{\Theta}$ where $m: P_X \to J(\tilde{C})$ is the inclusion and $\tilde{\Theta}$ is the canonical principal polarization on $J(\tilde{C})$. For more details and further properties of P_X, see f.e. [B], [M2].

Examples of ordinary conic bundles can be obtained as follows. Let S be a surface such that $q(S) > 0$. We can choose a nonsingular irreducible curve C such that $C \equiv 2D$ for some divisor D ("≡" means linear equivalence) and an element η of order 2 in $\text{Pic}^o(S)$ whose restriction to $\text{Pic}^o(C)$ is not trivial. If $\{h_\alpha\}_\alpha$ is a system of local equations of C, then the equation

$$h_\alpha x_0^2 + x_1^2 + x_2^2 = 0$$

locally defines in the projective scheme $\mathbb{P}(\mathcal{O}_S(D) \oplus \mathcal{O}_S(\eta) \oplus \mathcal{O}_S)$ a conic bundle X having C as discriminant curve. Moreover the double covering $\nu:\tilde{C} \to C$ doesn't split since η is not zero in $\operatorname{Pic}^o(C)$.

§2. **The Chow group $A^2(X)$ of a conic bundle.**

Let $f: X \to S$ be a conic bundle. We denote by $C^q(X)$ the Chow group of the cycle-classes of codimension q modulo rational equivalence, by $A^q(X) \subset C^q(X)$ the subgroup of the classes algebraically equivalent to zero. Our purpose is to find a decomposition for the group $A^2(X)$ in terms of the Prym variety P_X and $A^2(S) \oplus A^1(S)$.

First of all, choose a very ample sheaf on S such that the restriction $\mathcal{O}_{\mathbb{P}(f_*\omega_X^{-1}\otimes L)}(1) \otimes \mathcal{O}_X$ is a very ample sheaf on X ([EGA], II, 4.4.10) and look at the embedding $X \hookrightarrow \mathbb{P}(f_*\omega_X^{-1}\otimes L) \simeq \mathbb{P}(f_*\omega_X^{-1})$. Therefore we can find an irreducible nonsingular divisor T_X belonging to the complete linear system $|\mathcal{O}_{\mathbb{P}(f_*\omega_X^{-1}\otimes L)}(1) \otimes \mathcal{O}_X|$, by use of Bertini's Theorem. We say that T_X is a **tautological divisor** of X. As it can be seen in the sequel our arguments do not depend on the choice of L (see 2.1.4).

2.1. Let C be the discriminant curve of X and let $Y = f^{-1}(C)$ be the ruled surface of the singular fibres. Put $U = X \setminus Y$. In this section we want to describe the group $A^2(U)$.

We denote by T the restriction of T_X to the open set U, by i the inclusion $T \to U$ and we set $V = S \setminus C$. The morphism $\pi = f \circ i$ is generically finite of degree 2 and we get the commutative base change diagram

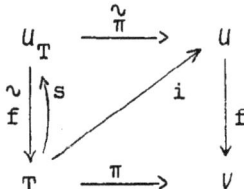

where $s: T \to U_T$ is the section defined by $s(t) = (t, i(t))$.

To begin with, we study the group $A^2(U)$.

LEMMA 2.1.1 <u>The projection $\tilde{f}: U_T \to T$ is birationally equivalent to a fibration in projective lines. Moreover one has a natural isomorphism</u>

$$A^2(U_T) \simeq \tilde{f}^* A^2(T) \oplus s_* A^1(T).$$

<u>Proof.</u> Let L be the invertible sheaf on U_T defined by the section s. The function

$$h^0(t, L) = \dim H^0(\tilde{f}^{-1}(t), L_t) = \dim H^0(\mathbb{P}^1, \mathbb{P}^1(1)) = 2$$

is constant on T. Moreover L is flat on T since $\tilde{f}$ is a flat morphism. Therefore $\tilde{f}_* L$ is a locally free sheaf and rank $(\tilde{f}_* L \otimes k(t)) = h^0(t, L)$ (see [H], p. 288). On the other hand there exists a closed embedding $U_T \to \mathbb{P}(\tilde{f}_*L)$, which is just an isomorphism as U_T and $\mathbb{P}(\tilde{f}_*L)$ are both of dimension three (see [EGA], III, 5.5.4). Then the projection $\tilde{f}$ is birationally equivalent to a fibration in projective lines. Hence we have an isomorphism (see [C], exp. IV)

$$C^q(U_T) \simeq \tilde{f}^*C^q(T) \oplus \tilde{f}^*C^{q-1}(T) \cdot s_*A^0(T), \quad q = 1,2,$$

where "·" means intersection of cycles. In particular we can write for every $\gamma \in A^2(U_T)$:

$$\gamma = \tilde{f}^*\alpha + \tilde{f}^*\beta \cdot s_*(T), \quad \alpha \in C^2(T), \ \beta \in C^1(T).$$

Since $\tilde{f}_*\tilde{f}^*\alpha = 0$ and $\tilde{f}_*s_* = \mathrm{id}$. we obtain
$\tilde{f}_*\gamma = \tilde{f}_*(\tilde{f}^*\beta \cdot s_*(T)) = \beta \cdot \tilde{f}_*s_*(T) = \beta$. It follows $\beta \in A^1(T)$, hence $\tilde{f}^*\alpha \in A^2(U_T)$. Therefore $s^*\tilde{f}^*\alpha = \alpha \in A^2(T)$.
Then one has

$$A^2(U_T) \simeq \tilde{f}^*A^2(T) \oplus \tilde{f}^*A^1(T) \cdot s_*A^0(T) \simeq \tilde{f}^*A^2(T) \oplus s_*A^1(T).$$

q.e.d.

Consider now the morphism

$$\Phi_U : A^2(V) \oplus A^1(V) \to A^2(U)$$

defined by $\phi_U(a,b) = f^*a + f^*b \cdot i_*T$.

PROPOSITION 2.1.2. *The morphism* ϕ_U *is surjective*.

Proof. We have to show that all elements of $A^2(U)$ can be written in the form $f^*a + f^*b \cdot i_*T$, for some $(a,b) \in A^2(V) \oplus A^1(V)$.

Take a cycle $\tilde{\pi}^*\gamma \in A^2(U_T)$, $\gamma \in A^2(U)$. We can write $\tilde{\pi}^*\gamma = \tilde{f}^*\alpha + s_*\beta$, $(\alpha,\beta) \in A^2(T) \oplus A^1(T)$ by Lemma 2.1.1. Then, since $\tilde{f}_*\tilde{f}^*\alpha = 0$ and $\tilde{f}_*s_* = \text{id.}$, one has $\tilde{f}_*\tilde{\pi}^*\gamma = \beta$. Moreover $\tilde{f}_*\tilde{\pi}^* = \pi^*f_*$ (see [F], 2.2). Therefore we find $\beta = \pi^*\beta'$ with $\beta' = f_*\gamma \in A^1(V)$, so we get an embedding

$$\tilde{\pi}^*A^2(U) \to \tilde{f}^*A^2(T) \oplus s_*\pi^*A^1(V).$$

Recalling that the composition $\tilde{\pi}_*\tilde{\pi}^* = \cdot 2$ (multiplication by 2) is surjective, the diagram

$$\tilde{f}^*A^2(T) \oplus s_*\pi^*A^1(V) \xrightarrow{\tilde{\pi}_*} A^2(U) \to 0$$
$$\uparrow \qquad \qquad \nearrow$$
$$\tilde{\pi}^*A^2(U) \qquad \tilde{\pi}_*$$

is commutative with exact row and gives an isomorphism

$$(\cdot) \quad A^2(U) \simeq \tilde{\pi}_*\tilde{f}^*A^2(T) \oplus \tilde{\pi}_*s_*\pi^*A^1(V) \simeq f^*\pi_*A^2(T) \oplus i_*\pi^*A^1(V).$$

Now we compute $i_*\pi^*A^1(V)$. For all $x = \pi^*x' \in \pi^*A^1(V)$, take the cycle $i^*f^*\pi_*x \in A^1(T)$. Since $i^*f^* = \pi^*$ and $\pi_*\pi^* = \cdot 2$ one has $i^*f^*\pi_*x = \pi^*\pi_*\pi^*x' = 2x$. Then in $A^2(U)$:

$$i_*(2x) = i_*(i^*f^*\pi_*x) = f^*\pi_*x \cdot i_*T.$$

The group $\pi^*A^1(V)$ is divisible, so we can assume every element $y \in \pi^*A^1(V)$ to be of the form $y = 2x$, $x \in \pi^*A^1(V)$. Thus we have shown that for all $y \in \pi^*A^1(V)$ there exists an element $x \in \pi^*A^1(V)$ such that

$$i_*y = f^*\pi_*x \cdot i_*T.$$

Recalling $(\cdot)$ and putting $b = \pi_*x$ we are done.

q.e.d.

As done for the morphism Φ_U, we can define

$$\Phi = (f^*, i_*\pi^*) : A^2(S) \oplus A^1(S) \to A^2(X)$$

such that

$$\Phi(a,b) = f^*a + i_*\pi^*b = f^*a + f^*b \cdot i_*T_X$$

where $i : T_X \to X$, $\pi = f \circ i : T_X \to S$. The morphism Φ needs not to be surjective : the cokernel

$$G(X) = A^2(X)/\mathrm{Im}\Phi$$

plays an important role in section 2.2. The following remarks are needed.

REMARKS 2.1.3 1) The quotient $G(X)$ does not depend on the choice of the line bundle L on S which gives the embedding $X \to \tilde{\mathbb{P}}(f_*\omega_X^{-1} \otimes L)$. To see this, let T'_X be the tautological divisor corresponding to some other inclusion $X \to \tilde{\mathbb{P}}(f_*\omega_X^{-1} \otimes L')$, $L' \in \mathrm{Pic}(S)$, and consider the associated map $\Phi' = (f^*, i_*\pi^*)$. It is not difficult to verify that $\mathrm{Im}\Phi = \mathrm{Im}\Phi'$. In fact, reasoning as in Proposition 2.1.2 we find an isomorphism

$$\tilde{\pi}_*\tilde{\pi}^* C^1(X) \simeq f^*C^1(S) \oplus \mathbb{Z},$$

with $\mathbb{Z}$ generated by the class of the cycle i_*T_X. Then, since $\tilde{\pi}_*\tilde{\pi}^* = \cdot 2$, we can write $2(i_*T'_X) = f^*\xi + m i_* T_X$ for suitable $\xi \in C^1(S)$, $m \in \mathbb{Z}$, so that $2\Phi'(a,b) = \Phi(2a+b\cdot\xi, mb) \in \mathrm{Im}\Phi$, for all $(a,b) \in A^2(S) \oplus A^1(S)$. As $\mathrm{Im}\Phi'$ is a divisible group it follows $\mathrm{Im}\Phi' \subset \mathrm{Im}\Phi$; similarly $\mathrm{Im}\Phi' \supset \mathrm{Im}\Phi$.

2) The kernel of Φ is a finite torsion group contained in $A^2(S)_4 \oplus A^1(S)_2$. In fact, suppose $(a,b) \in \ker\Phi$. Then $f_*(f^*a + f^*b \cdot i_*T_X) = 0$. Since $f_*f^*a = 0$ we find

$$f_*(f^*b \cdot i_*T_X) = b \cdot f_* i_* T_X = b \cdot \pi_* T_X = 2b.$$

Therefore $b \in A^1(S)_2$, so that

$$2(f^*a + f^*b \cdot i_*T_X) = f^*2a + f^*2b \cdot i_*T_X = f^*2a = 0.$$

It follows $0 = i^*f^*2a = f^*2a \cdot i_*T_X$. Hence, as before

$$f_*(f^*2a \cdot i_*T_X) = 2a \cdot f_* i_* T_X = 2a \cdot \pi_* T_X = 4a,$$

that is $a \in A^2(S)_4$. As proved in [R] the group $A^2(S)_4 \oplus A^1(S)_2$ is isomorphic to $\mathrm{Alb}(S)_4 \oplus \mathrm{Pic}^o(S)_2$, so it is finite.

2.2. We return to the discriminant curve C of the conic bundle X and the ruled surface $Y = f^{-1}(C)$ of singular fibres. There exists a canonical section $e : C \to Y$ such that $e(s)$ is the singular point of the fibre $X_s = f^{-1}(s)$, for all $s \in C$. We denote by $\varepsilon : X' \to X$ the blowing-up of X along $e(C)$, by E the exceptional divisor and by Y' the proper transform of Y. Let $p':Y' \to C$ be the restriction of $f \circ \varepsilon$ to Y'. The fibre $p'^{-1}(s)$, $s \in C$, is isomorphic to the blowing-up of X_s in its singular point. Furthermore p' factorizes through p,ν

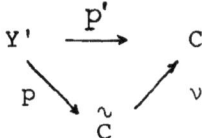

where p is a fibration in projective lines and ν is the double etale covering of C. Moreover the divisors E,Y' are nonsingular with transversal intersection. Their intersection can be identified with the curve $\tilde{C}$ via the projection p. Thus we have a commutative diagram

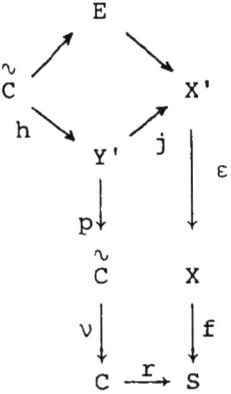

where the upper square is cartesian and the inclusion h is a section of the morphism p, i.e. poh = id.

Write $U = X \setminus Y$, $V = S \setminus C$ and let $j_U: U \to X$, $r_V: V \to S$ be the inclusions.

LEMMA 2.2.1 a) **The sequence**

$$A^1(Y') \xrightarrow{(\varepsilon \circ j)_*} A^2(X) \xrightarrow{j_U^*} A^2(U) \to 0$$

is exact.

b) $r_V^*: A^1(S) \to A^1(V)$ **is an isomorphism.**

Proof. a) Look at $K = \ker j_U^*$ as subgroup of $C^2(X)$ and consider the subgroup $K_0 = (\varepsilon \circ j)_*^{-1} K$ of $C^1(Y')$. Since the sequence

$$C^1(Y') \xrightarrow{(\varepsilon \circ j)_*} C^2(X) \xrightarrow{j_U^*} C^2(U) \to 0$$

is exact and the morphism $j_U^*: A^2(X) \to A^2(U)$ is surjective (see [B], 0.1.2) we get the exactness of

$$K_0 \xrightarrow{(\varepsilon \circ j)_*} A^2(X) \xrightarrow{j_U^*} A^2(U) \to 0.$$

Then it is sufficient to show that $K_0 = A^1(Y')$. The last sequence gives $A^1(Y') \subset K_0$.

To prove the converse, let $\text{Pic}^n(Y')$ be the group of the classes of divisors numerically equivalent to zero. The quotient group $\text{Num}(Y') = \text{Pic}(Y')/\text{Pic}^n(Y')$ is generated by $\tilde{C}_0$ and F where F is a fibre of $p: Y' \to \tilde{C}$ and $\tilde{C}_0$ a section. Choose an ample divisor D on S. Then a direct calculation shows that the divisors associated to the sheaves $(\varepsilon \circ j)^* \mathcal{O}_X(f^*D)$, $(\varepsilon \circ j)^* \mathcal{O}_X(1)$ generate $\text{Num}(Y')$ over $\mathbb{Q}$. Suppose now $\gamma \in K_0$. Then $(\varepsilon \circ j)_* \gamma$ is numerically equivalent to zero as element of $A^2(X)$. Therefore $(\varepsilon \circ j)_* \gamma \cdot \xi = 0$ for all $\xi \in \text{Pic}(X)$. The projection formula gives $\gamma \cdot (\varepsilon \circ j)^* \xi = 0$, so the claim implies $\gamma \cdot \mu = 0$ for all $\mu \in \text{Num}(Y')$. That is γ is numerically equivalent to zero in Y'. Since numerical and algebraic equivalence coincide on the ruled surface Y', it follows $\gamma \in A^1(Y')$.

b) Set $K_0 = \ker r_V^*$ and consider the exact diagram (see [B], 0.1.2)

$$\begin{array}{ccccc} C^0(C) & \xrightarrow{r_*} & C^1(S) & \xrightarrow{r_V^*} & C^1(V) \to 0 \\ \uparrow & & \uparrow & & \uparrow \\ K_0 & \xrightarrow{r_*} & A^1(S) & \xrightarrow{r_V^*} & A^1(V) \to 0 \end{array}$$

Since $C^0(C) \simeq \mathbb{Z}$ generated by the class $\mathbb{1}_C$ all elements γ in $C^0(C)$ can be written in the form $\gamma = a\mathbb{1}_C$, $a \in \mathbb{Z}$. It follows $r_*\gamma = a\mathbb{1}_C$ in $C^1(S)$. Suppose $\gamma \in K_0$ so that $a \mathbb{1}_C$ is algebraically equivalent to zero. This implies $a\mathbb{1}_C$ to be numerically equivalent to zero as cycle of S, hence $a = 0$ ($\mathbb{1}_C$ is an effective cycle). Then $\gamma = 0$ in $C^0(C)$. This means $K_0 = (0)$, and we are done.

q.e.d.

Next define the <u>cylinder map</u>

$$\psi_A = \varepsilon_* j_* p^* : J(\tilde{C}) \to A^2(X)$$

and let $G(X) = A^2(X)/\text{Im}\Phi$ be the cokernel of the map Φ introduced in section 2.1. Denote by

$$\psi : J(\tilde{C}) \to G(X)$$

the composition with the projection $A^2(X) \xrightarrow{q} G(X)$.

<u>PROPOSITION 2.2.2</u>   <u>The morphism</u> $\psi : J(\tilde{C}) \to G(X)$ <u>induced by the cylinder map is surjective.</u>

<u>Proof.</u> It suffices to prove that for all cycles $\gamma \in A^2(X)$ there exists a cycle $\beta \in \ker \{j_U^* : A^2(X) \to A^2(U)\}$ such that $\gamma - \beta \in \text{Im}\Phi$. Therefore by Lemma 2.2.1 all elements of $G(X)$ are classes of a cycle β belonging to $\text{Im}(\varepsilon \circ j)_*$, that is $\beta = \varepsilon_* j_* p^* x$, where $x \in J(\tilde{C})$ and $J(\tilde{C}) \simeq A^1(Y')$ via p^*, so that ψ is surjective.

For simplicity, set $u = j_U$. We can write all cycles $\gamma \in A^2(X)$ in the form $\gamma = \alpha + \delta$ where $\alpha = \overline{u^*\gamma} \in C^2(X)$ is the Zariski closure of $u^*\gamma$ in X and $\delta = \gamma - \alpha$. The cycle $u^*\alpha = u^*\gamma$ belongs to $A^2(U)$ as $\gamma \in A^2(X)$. Then the surjectivity of the morphism $\Phi_U : A^2(V) \oplus A^1(V) \to A^2(U)$ (see 2.1.2) gives

$$u^*\alpha = f^*a + f^*b \cdot i_*T$$

with $(a,b) \in A^2(V) \oplus A^1(V)$, $T = u^*T_X$. Now put $v = r_V$ and consider the commutative square, $q = 1,2$,

$$\begin{array}{ccc} A^q(S) & \xrightarrow{v^*} & A^q(V) \\ f^* \downarrow & & \downarrow f^* \\ A^q(X) & \xrightarrow{u^*} & A^q(U) \end{array}$$

The morphisms v^*, u^* are surjective ([B], 0.1.2(ii)), hence we find $a = v^*\bar{a}$, $b = v^*\bar{b}$ for some $(\bar{a},\bar{b}) \in A^2(S) \oplus A^1(S)$. Thus we have

$$u^*a = f^*v^*\bar{a} + f^*v^*\bar{b} \cdot i_*T ,$$

then

$$u^*a = u^*f^*\bar{a} + u^*f^*\bar{b} \cdot u^*i_*T_X ,$$

that is

$$\alpha - \Phi(\bar{a},\bar{b}) \in \ker\{j_{\tilde{u}}^* : C^2(X) \to C^2(U)\} .$$

By use of the exact sequence (see again [B], 0.1.2)

$$C^1(Y') \xrightarrow{(\varepsilon \circ j)_*} C^2(X) \xrightarrow{u^*} C^2(U) \to 0$$

we find in $C^2(X)$:

$$\alpha = \Phi(\overline{a},\overline{b}) + (\varepsilon \circ j)_*\mu \quad , \quad \mu \in C^1(Y') \; .$$

Hence

$$\gamma = \Phi(\overline{a},\overline{b}) + \delta + (\varepsilon \circ j)_*\mu \; .$$

Putting $\beta = \delta + (\varepsilon \circ j)_*\mu$ one has $\gamma - \beta \in \text{Im}\Phi$. Moreover $j_U^*\beta = 0$ and $\beta \in A^2(X)$, so that $\beta \in \ker\{j_U^*: A^2(X) \to A^2(U)\}$ as required.

q.e.d.

Now we point out some properties of the morphisms ψ_A, ψ. The notations are as in section 2.1.

CLAIM 1. For all $\alpha \in \widetilde{J(C)}$ such that $\psi(\alpha) = 0$ one has

$$\varepsilon_*j_*p^*2\alpha = f^*\alpha', \quad \alpha' \in A^2(S) \; .$$

Proof. Consider the commutative diagram

$$\begin{array}{ccc} A^2(S) \oplus A^1(S) & \xrightarrow{\Phi} & A^2(X) \\ {\scriptstyle r_V^*}\downarrow & & \downarrow{\scriptstyle j_U^*} \\ A^2(V) \oplus A^1(V) & \xrightarrow{\Phi_U} & A^2(U) \end{array}$$

The hypothesis $\psi(\alpha) = 0$ implies

$$\varepsilon_* j_* p^* \alpha = f^* a + f^* b \cdot i_* T_X$$

for some $(a,b) \in A^2(S) \oplus A^1(S)$. Therefore reasoning as in the proof of Proposition 2.2.2 one has

$$j_{\tilde{U}}^* (f^* a + f^* b \cdot i_* T_X) = 0 .$$

Since $j_{\tilde{U}}^* i_* = i_* j_{\tilde{U}}^*$, we see $j_{\tilde{U}}^* i_* T_X = i_* T$, so that the last equality is equivalent to

$$\Phi_U (r_{\tilde{V}}^* (a,b)) = 0 .$$

With the same argument as in Remark 2.1.3,2) we find

$$\ker \Phi_U \subset A^2(V)_4 \oplus A^1(V)_2 .$$

Hence in particular $r_{\tilde{V}}^*(2b) = 0$ which implies $2b = 0$ since $r_{\tilde{V}}^* : A^1(S) \to A^1(V)$ is an isomorphism (see 2.2.1b)). Thus $\varepsilon_* j_* p^* 2\alpha = f^* 2a$.

q.e.d.

Let now $\tau^* : J(\tilde{C}) \to J(\tilde{C})$ be the canonical involution induced by the double etale covering $\nu : \tilde{C} \to C$. Then we have

CLAIM 2. The equality

$$p_* j^* \varepsilon^* \psi_A(\alpha) = \tau^*\alpha - \alpha$$

holds, for all $\alpha \in J(\tilde{C})$.

Proof. It runs as well as in [B], 3.1.4 where the assert is proved in the case $S = \mathbb{P}^2$.

□

Recall the definition of Prym variety $P_X = \text{Im}(\mathbb{1}_{J(\tilde{C})} - \tau^*)$ associated to X. Let $\rho: P_X \to A^2(X)$ be the restriction to P_X of the cylinder map $\psi_A = \varepsilon_* j_* p^* : J(\tilde{C}) \to A^2(X)$, and denote by P_n the set of points of order n of P_X.

LEMMA 2.2.3 Let $\nu: \tilde{C} \to C$ be the double etale covering of the discriminant curve C. Then one has

$$\ker(\mathbb{1}_{J(\tilde{C})} - \tau^*) = \nu^* J(C) .$$

Proof. The relation $\nu^* J(C) \subset \ker(\mathbb{1}_{J(\tilde{C})} - \tau^*)$ is clear since $\tau^* r^* a = \nu^* a$ for all $a \in J(C)$.

To prove the converse, take $a = \mathcal{O}_{\tilde{C}}(D) \in J(\tilde{C})$ such that

$a = \tau^*a$, that is $D \equiv \tau^*D$ ("$\equiv$" means linear equivalence). We can choose a divisor A on C such that $V = H^0(\tilde{C}, \mathcal{O}_{\tilde{C}}(D+\nu^*A))$ has positive dimension. Let i^* be the endomorphism of V induced by τ. Put $E = \text{div}(s)$ where s is an eigenvector of i^*. Then we have $i^*E = E$, hence $E = \nu^*B$ for some divisor B of C. Moreover $E \equiv D + \nu^*A$. It follows $D \equiv \nu^*(B-A)$ that is $a \in \nu^*J(C)$.

q.e.d.

The following Proposition gives the required decomposition for the group $A^2(X)$.

PROPOSITION 2.2.4 i) <u>There exists a surjective morphism</u> $\phi_X : P_X \to G(X)$ <u>whose kernel is contained in</u> P_2 <u>which makes commutative the diagram with exact row</u>

$$0 \to \text{Im}\,\Phi \to A^2(X) \xrightarrow{q} G(X) \to 0$$

$$\rho \uparrow \quad \nearrow \phi_X \circ (\cdot 2) = 2\phi_X$$

$$P_X$$

ii) $A^2(X) = \text{Im}\,\Phi + \rho(P_X)$;

iii) $\rho(\ker 2\phi_X) \supset \text{Im}\,\Phi \cap \rho(P_X)$;

iv) <u>the morphism</u> $\Phi + \rho : A^2(S) \oplus A^1(S) \oplus P_X \to A^2(X)$ <u>is an isogeny</u>.

<u>Proof</u>. i) We have to show that there exists a surjective

morphism with $\ker \phi_X \subset P_2$ such that the diagram

$$\begin{array}{ccccc} P_X & \xrightarrow{m} & J(\tilde{C}) & \xrightarrow{\psi_A} & A^2(X) \\ & \searrow{\cdot 2} & \Big\downarrow{1\!\!1_{J(\tilde{C})} - \tau^*} & & \Big\downarrow{q} \\ & & P_X & \xrightarrow{\phi_X} & G(X) \end{array}$$

commutes. Putting $\psi = q \circ \psi_A$, it is sufficient to prove the relations

$$\ker \psi \supset \ker(1\!\!1_{J(\tilde{C})} - \tau^*) \ , \ 2 \ker \psi \subset \ker(1\!\!1_{J(\tilde{C})} - \tau^*) \ .$$

Suppose $\alpha = \tau^* \alpha$, so that $\alpha = \nu^* \beta$ for a suitable $\beta \in J(C)$ by Lemma 2.2.3. Moreover $\varepsilon_* j_* p^* \nu^* = f^* r_*$ (see also [B], 3.1.5), hence

$$\psi(\alpha) = [f^* r_* \beta] \ , \ \text{where "[ ]" means the class in } G(X).$$

Since $f^* r_* \beta \in \text{Im} \tilde{\Phi}$, it follows $\psi(\alpha) = 0$ in $G(X)$.

Assume now $\psi(\alpha) = 0$, $\alpha \in J(\tilde{C})$. Claims 1,2 give, for some $\alpha' \in A^2(S)$

$$p_* j^* \varepsilon^* (\varepsilon_* j_* p^* 2\alpha) = p_* j^* \varepsilon^* (f^* \alpha') = 2(\tau^* \alpha - \alpha) \ .$$

As $j^* \varepsilon^* f^* = p^* \nu^* r^*$, we find

$$p_*p^*\nu^*r^*\alpha' = 2(\tau^*\alpha-\alpha) .$$

Since $r^*:A^2(S) \to A^1(C)$ is the zero map, then we obtain $2\tau^*\alpha = 2\alpha$, that is $2\alpha \in \ker(\mathbb{1}_{J(\tilde{C})} -\tau^*)$.

ii), iii). They follow immediately by i).

iv). The morphism $\Phi+\rho$ is surjective by ii). Take $(a,b,c) \in \ker(\Phi+\rho)$, so that iii) implies $\Phi(a,b), \rho(c) \in \rho(\ker 2\phi_X)$. Moreover $\ker 2\phi_X$ is a finite group of order d ($d \leq 4$) by i), therefore

$$\Phi(da,db) = \rho(dc) = 0 .$$

Since $\ker \Phi \subset A^2(S)_4 \oplus A^1(S)_2$ we find $(a,b) \in A^2(S)_{4d} \oplus A^1(S)_{2d}$ which is a finite group in view of the isomorphism $A^2(S)_{tors} \simeq \text{Alb}(S)_{tors}$ proved in [R], while $\rho(dc) = 0$ implies $c \in P_{d^2}$. Thus $\ker(\Phi+\rho)$ is a finite group.

q.e.d.

REMARK 2.2.5. Suppose $X \to S$ to be a conic bundle defined over the complex numbers. Then whenever S satisfies the conditions $q(S) = p_g(S) = 0$, and S is not of general type, one has the isomorphism $\phi_X: P_X \simeq A^2(X)$. In fact, if $p_g(S) = 0$ and S is not of general type, the Albanese mapping $A^2(S) \to \text{Alb}(S)$ is injective (see [B-K-L]), so that $A^2(S) = 0$ as $q(S) = 0$. Moreover $q(S) = 0$ also implies $A^1(S) = 0$. Reasoning as in Proposition 2.2.4, we get the isomorphism $\phi_X: P_X \simeq G(X) = A^2(X)$.

§ 3. The algebraic representative of $A^2(X)$.

We show in this section that an analogous result to Proposition 2.2.4 holds true for the algebraic representative A_X of $A^2(X)$. To begin with, we recall some definitions. Here V is any nonsingular threefold.

DEFINITION 3.1 Let W be a nonsingular algebraic variety. We say that a map (of sets) β u:W $\to$ $A^2(V)$ is <u>algebraic</u> if there exists a cycle $Z \in C^2(V \times W)$ such that for all $t \in W$,

$$u(t) = Z \cdot (\{t\} \times V)$$

DEFINITION 3.2 Let B be an abelian variety. We say that a morphism of groups $\beta: A^2(V) \to B$ is <u>regular</u> if for every algebraic map u:W $\to$ $A^2(V)$ the composition $\beta \circ u$ is a morphism of varieties.

DEFINITION 3.3 Let $a_V: A^2(V) \to A_V$ be a regular morphism, A_V abelian variety. We say that the pair (A_V, a_V) is the <u>algebraic representative</u> of $A^2(V)$ if, for all regular morphisms $\beta: A^2(V) \to B$, there exists one and only one morphism h:$A_V \to$ B of abelian varieties which makes the following diagram

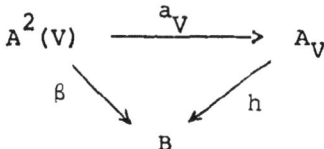

commute.

REMARK 3.4 For every nonsingular variety of dimension $n \geq 2$, analogous definitions are given for the groups $A^q(V)$, $q \leq n$. In particular, it is not difficult to verify that $Alb(V)$ (resp. $Pic^o(V)$) is the algebraic representative of $A^n(V)$ (resp. $A^1(V)$). Moreover if the pair (A_V, a_V) exists then it is unique up to isomorphisms and the morphism a_V is always surjective (see [B], 3.2.4).

The following Proposition states the existence of the algebraic representative of $A^2(V)$ for a large class of threefolds, including conic bundles.

PROPOSITION 3.5 <u>Let Y,V be nonsingular threefolds and let</u> $\phi: Y \dashrightarrow V$ <u>be a rational map. Suppose that there exists the algebraic representative</u> (A_Y, a_Y) <u>of</u> $A^2(Y)$. <u>Then there exists also that one of</u> $A^2(V)$.

Proof. We get a commutative diagram

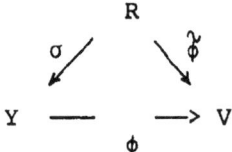

where R is a nonsingular threefold, σ is a sequence of blowing-ups along points or nonsingular curves, and $\tilde{\phi}$ is a morphism.

Since the blowingup along points does not change neither $A^2(Y)$ nor its algebraic representative, we can suppose σ to be a sequence of blowingups along nonsingular curves C_i. Therefore one can easily prove that the direct sum $A_R = A_Y \oplus \prod_i J(C_i)$ is the algebraic representative of the group $A^2(R)$.

Now, let $\beta: A^2(V) \to B$ be a regular morphism, B an abelian variety. It is easy to see that the morphism $\beta \circ \tilde{\phi}_*: A^2(R) \to B$ is regular. Then, by the universal property of the algebraic representative, there exists a morphism g such that the diagram

$$\begin{array}{ccc} A^2(R) & \xrightarrow{\tilde{\phi}_*} & A^2(V) \\ a_R \downarrow & & \downarrow \beta \\ A_R & \xrightarrow{g} & B \end{array}$$

commutes. Since the morphism $\tilde{\phi}_*$ is surjective, hence one has dim Im β = dim. Im g $\leq$ dim A_R = M for all pair (B,β). This is equivalent to the existence of the algebraic representative of $A^2(V)$, as proved in [S], 2.2.

q.e.d.

COROLLARY 3.6. Suppose V to be a uniruled threefold (in particular a conic bundle). Then there exists the algebraic representative A_V of $A^2(V)$.

Proof. As V is uniruled there exist a nonsingular surface Y and a rational map $Y \times \mathbb{P}^1 \dashrightarrow V$. On the other hand the abelian variety $Alb(Y) \oplus Pic^o(Y)$ is the algebraic representative of $A^2(Y \times \mathbb{P}^1)$ via the canonical isomorphism $A^2(Y \times \mathbb{P}^1) \simeq A^2(Y) \oplus A^1(Y)$. Then the claim follows by the previous Proposition.

q.e.d.

We go back to the case of a conic bundle $f: X \to S$. With the notations as in section 2.2, look at the commutative diagram

$$\begin{array}{ccccc} X' & \xrightarrow{\varepsilon} & X & \xrightarrow{f} & S \\ j \uparrow & & & & \uparrow r \\ Y' & \xrightarrow{p} & \tilde{C} & \xrightarrow{\nu} & C \end{array}$$

and consider again the morphism $\Phi: A^2(S) \oplus A^1(S) \to A^2(X)$, and the isogeny $\phi_X: P_X \to G(X) = A^2(X)/\text{Im } \Phi$ (see section 2.1). We recall that $\ker \phi_X$ is contained in P_2, so that there exists an isogeny $\tilde{\phi}_X: G(X) \to P_X$ such that $\tilde{\phi}_X \circ \phi_X = \cdot 2$ is the multiplication by 2 in P_X.

We need the following technical result

LEMMA 3.7(i) <u>Let</u> $q: A^2(X) \to G(X)$ <u>be the canonical projection.</u> <u>Then the composition</u> $\tilde{\phi}_X \circ q: A^2(X) \to P_X$ <u>is a regular morphism.</u>

(ii) <u>Let $u: W \to A^2(S) \oplus A^1(S)$ be an algebraic map</u>(*).
<u>Then $\phi \circ u: W \to A^2(X)$ is also.</u>

<u>Proof</u>: (i) Let $u: W \to A^2(X)$ be an algebraic map defined by a cycle $z \in C^2(X \times W)$ and put $\bar{u} = q \circ u$. We have to show that the composition $\tilde{\phi}_X \circ \bar{u}$ is a morphism of algebraic varieties. Consider the diagram

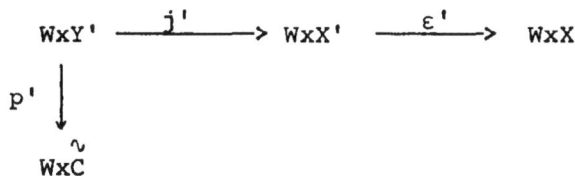

where "$\cdot$'" means $(id_W, \cdot)$. Take the algebraic map $v: W \to A^1(\tilde{C})$ defined by the cycle $d = p'_* j'^* \varepsilon'^* z \in C^1(W \times \tilde{C})$, that is $v(t) = d \cdot (\{t\} \times \tilde{C})$, $t \in W$. Via the isomorphism $A^1(\tilde{C}) \simeq J(\tilde{C})$, the composition $v: W \to J(\tilde{C})$ is just a morphism of algebraic varieties. Moreover the commutativity of the diagram:

(*) We mean that $u = (u_1, u_2)$ with $u_i: W \to A^i(S)$ algebraic maps, $i = 1, 2$.

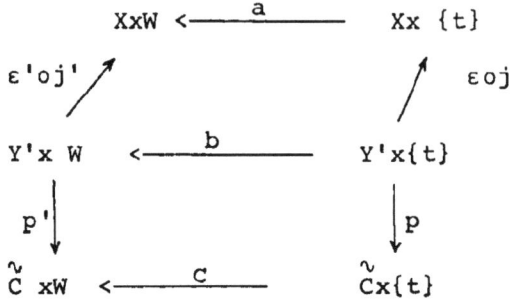

gives

$$j^*\varepsilon^*(z.(\{t\}\times X)) = j^*\varepsilon^* a^* z = b^* j'^* \varepsilon'^* z$$

Put $x = j'^*\varepsilon'^*z$. Then we have (see [F], 2.2),

$$p_* b^* x = c^* p'_* x = p'_* x.(\{t\}\times \tilde{C})$$

that is

(*) $\quad p_* j^* \varepsilon^*(z.(\{t\}\times X)) = p'_* j'^* \varepsilon'^* z.(\{t\}\times \tilde{C}) \,(=v(t))$.

Consider now the diagram

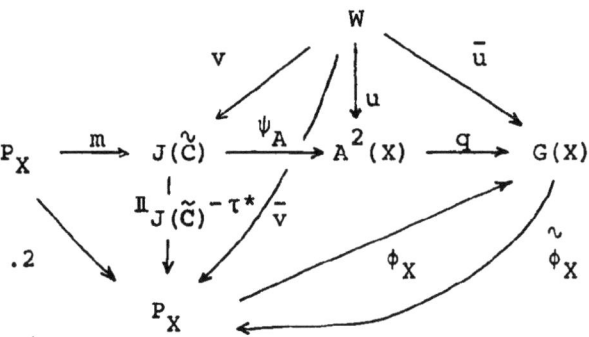

where $\bar{u} = q \circ u$, $\bar{v} = (\mathbf{1}_{J(\tilde{C})} - \tau^*) \circ v$ and $\tilde{\phi}_X \circ \phi_X = \cdot 2$. Since the composition $\psi = q \circ \psi_A$ is a surjective morphism (see Proposition 2.2.2), for all $t \in W$ there exist some $\alpha_t \in J(\tilde{C})$ such that $\bar{u}(t) = \psi(\alpha_t)$, that is

$$[z \cdot (\{t\} \times X)] = [\varepsilon_* j_* p^* \alpha_t],$$

where "$[\cdot]$" means the class modulo $\mathrm{Im}\,\phi$. Then we find

$$z \cdot (\{t\} \times X) = \varepsilon_* j_* p^* \alpha_t + \Phi(\alpha, \beta)$$

for some $(\alpha, \beta) \in A^2(S) \oplus A^1(S)$. Hence by use of (*):

$$v(t) = p_* j^* \varepsilon^* (\varepsilon_* j_* p^* \alpha_t) + p_* j^* \varepsilon^* \tilde{\Phi}(\alpha, \beta).$$

Claim 2 of Section 2.2 gives

$$p_* j^* \varepsilon^* (\varepsilon_* j_* p^* \alpha_t) = \tau^* \alpha_t - \alpha_t.$$

Moreover a standard computation shows (see the proof of Proposition 2.2.4):

$$2(\mathbf{1}_{J(\tilde{C})} - \tau^*)(p_* j^* \varepsilon^* \tilde{\Phi}(\alpha, \beta)) = 0.$$

Therefore we have

$$2\bar{v}(t) = 2(\mathbf{1}_{J(\tilde{C})} - \tau^*)v(t) = 4(\tau^*\alpha_t - \alpha_t).$$

Thus the equalities $\bar{u}(t) = \psi(\alpha_t) = \phi_X \circ (\mathbf{1}_{J(\tilde{C})} - \tau^*)(\alpha_t)$ imply $4\bar{u} = -\phi_X \circ 2\bar{v}$, hence

$$-\tilde{\phi}_X \circ 4\bar{u} = \tilde{\phi}_X \circ \phi_X \circ 2\bar{v}.$$

Since $\bar{v}$, $\tilde{\phi}_X \circ \phi_X = \cdot 2$ are morphisms of varieties and the multiplication by 4,2 are isogenies, also $\tilde{\phi}_X \circ \bar{u}$ is a morphism of varieties.

(ii) Let $u: W \to A^2(S) \oplus A^1(S)$ be the algebraic map defined by the cycles $x \in C^2(S \times W)$, $y \in C^1(S \times W)$, that is $u(t) = (x \cdot (\{t\} \times S), y \cdot (\{t\} \times S))$. A straightforward computation shows that $\Phi \circ u$ is the algebraic map defined by the cycle $z = f'^*x + f'^*y \cdot (T_X \times W) \in C^2(X \times W)$ where $f' = (f, id_W)$.

q.e.d.

With the same notations as in Proposition 2.2.4, we can prove the following

PROPOSITION 3.8 <u>There exist morphisms h_1, h_2 and a commutative diagram with exact row</u>

$$\text{Alb}(S) \oplus \text{Pic}^0(S) \xrightarrow{h_1} A_X \xrightarrow{h_2} P_X \to 0$$

with $a_X \circ \rho$ going up into A_X and diagonal arrow labeled $.2$ into P_X, from P_X below.

Moreover

(i) $A_X = \text{Im } h_1 + a_X \circ \rho(P_X)$;

(ii) $a_X \circ \rho(P_2) \supset \text{Im } h_1 \cap a_X \circ \rho(P_X)$;

(iii) Suppose the surface S has irregularity $q(S) = 0$. Then $h_2: A_X \to P_X$ is an abelian variety isomorphism.

(iv) In the complex case the morphism
$h_1 + a_X \circ \rho: \text{Alb}(S) \oplus \text{Pic}^0(S) \oplus P_X \to A_X$ is an isogeny.

Proof. By use of Lemma 3.7 the universal property of the algebraic representatives a_S, a_X gives abelian variety morphisms h_1, h_2 such that the following diagram

$$\begin{array}{ccccccccc} 0 & \to & \ker \phi & \to & A^2(S) \oplus A^1(S) & \xrightarrow{\phi} & A^2(X) & \xrightarrow{q} & G(X) \to 0 \\ & & & & \downarrow a_S & & \downarrow a_X & & \downarrow \tilde{\phi}_X \\ & & & & \text{Alb}(S) \oplus \text{Pic}^0(S) & \xrightarrow{h_1} & A_X & \xrightarrow{h_2} & P_X \to 0 \end{array}$$

commutes. Note that $a_S = (a_S^2, a_S^1)$ where $a_S^2: A^2(S) \to \text{Alb}(S)$, $a_S^1: A^1(S) \to \text{Pic}^0(S)$ are the algebraic representative maps. The exactness of the first row is proved in Proposition 2.2.4(i), while the exactness of the second one is easily achieved by taking into account that the morphisms a_S, a_X are surjective. Now (i), (ii), (iii) immediately follow, while (iv) will be proved in Section 4, 4.1.2, 4.2.5.

q.e.d.

§4. <u>The classical intermediate Jacobian of a conic bundle.</u>

Through this section we work over the complex field. In the first part we recall some definitions and give some general statements. In the second one we go back to conic bundles and we get a decomposition of the intermediate Jacobian (up to isogenies).

<u>4.1</u>. Let V be a nonsingular threefold defined over the complex field $\mathbb{C}$. Consider the Hodge decompositon $H^3(V,\mathbb{C}) = \bigoplus_{p+q=3} H^{p,q}(V)$. The 2th <u>Griffiths Jacobian</u> of V is the complex torus

$$T^2(V) = H^{1,2}(V) \oplus H^{0,3}(V)/\text{poi } H^3(V,\mathbb{Z})$$

where $p: H^3(V,\mathbb{Z}) \to H^{1,2}(V) \oplus H^{0,3}(V)$ is the projection and $i: H^3(V,\mathbb{Z}) \to H^3(V,\mathbb{C})$ the natural inclusion.

Now we define the <u>Weil morphism</u> (called Abel-Jacobi homomorphism in [G])

$$w: A^2(V) \to T^2(V).$$

Fix closed 3-forms $\phi_1,\ldots,\phi_N$ such that the cohomology classes of the ϕ's give a basis for $H^{2,1}(V) \oplus H^{3,0}(V)$. Let $\{\gamma_j\}_j$ be a basis for $H_3(V,\mathbb{Z})$ and let Γ be the subgroup of $\mathbb{C}^N$ generated by the vectors $\{(\int_{\gamma_j}\phi_1,\ldots,\int_{\gamma_j}\phi_N)\}_j$. Then $T^2(V) \simeq \mathbb{C}^N/\Gamma$ and the Weil mapping is given by

$$x = \partial C \mapsto (\int_C \phi_1,\ldots,\int_C \phi_N) \quad \text{modulo } \Gamma,$$

where C is a 3-chain on V.

The image $w(A^2(V))$ under the morphism w is an abelian subvariety $J(V)$ of $T^2(V)$, which is contained in $H^{1,2}(V)/(H^3(V,\mathbb{Z}) \cap H^{1,2}(V))$ (see [G] n.2 and [L], n.3). We say that $J(V)$ is the <u>intermediate Jacobian</u> of V. Moreover the Weil morphism is <u>regular</u> in the sense of definition 3.2 (see [S], n.5). Therefore, if there exists the algebraic representative (A_V, a_V) of the group $A^2(V)$, then a (surjective) abelian variety morphism is defined such that the diagram

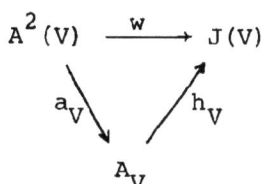

commutes.

REMARK 4.1.1. For every nonsingular variety of dimension $n \geq 2$ analogous definitions give Weil morphisms $w_q : A^q(V) \to T^q(V)$, $q \leq n$. In particular w_n coincides with the Albanese mapping $A^n(V) \to Alb(V)$, while w_1 is the canonical isomorphism $A^1(V) \simeq Pic^o(V)$.

PROPOSITION 4.1.2. <u>Let</u> Y, V <u>be nonsingular threefolds and let</u> $\phi : Y \to V$ <u>be a rational map. Suppose there exists the algebraic representative</u> A_Y <u>of</u> $A^2(Y)$ <u>and assume</u> $J(Y)$ <u>to be isogenous to</u> A_Y <u>via the morphism</u> h_Y. <u>Then there exists the algebraic representative</u> A_V <u>of</u> $A^2(V)$ <u>and</u> $J(V)$ <u>is isogenous to</u> A_V <u>via</u> h_V.

<u>Proof</u>. (essentially due to Murre-Block, see [M-B], 5.3). The existence of (A_V, a_V) has been proved in section 3, Prop. 3.5. Reasoning again as in the proof of Proposition 3.5 we can assume ϕ to be a morphism. Moreover it is not difficult to verify that the morphisms $a_Y \circ \phi^* : A^2(V) \to A_Y$, $a_V \circ \phi_* : A^2(Y) \to A_V$ are regular: note for this that ϕ is a generically finite morphism. Hence the universal property of a_V, a_Y gives the commutative squares

$$\begin{array}{ccc} A^2(V) & \xrightarrow{\phi^*} & A^2(Y) \\ a_V \downarrow & & \downarrow a_Y \\ A_V & \xrightarrow{\phi^*} & A_Y \end{array} \qquad \begin{array}{ccc} A^2(Y) & \xrightarrow{\phi_*} & A^2(V) \\ a_Y \downarrow & & \downarrow a_V \\ A_Y & \xrightarrow{\phi_*} & A_V \end{array}$$

Thus we get the commutative diagram

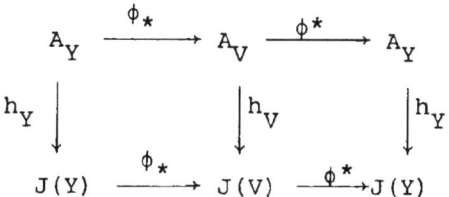

As h_V is surjective, it suffices to show that the kernel ker h_V is a finite group. If $a \in \ker h_V$, $h_Y \circ \phi^*(a) = 0$ that is $\phi^* a \in \ker h_Y$. Since $\ker h_Y$ is a finite group by hypothesis, one has $n\phi^*(a) = \phi^*(na) = 0$ for some $n \in \mathbb{N}$, that is $na \in \ker \phi^*$. On the other hand $\ker \phi^* \subset \ker(\phi_* \circ \phi^*)$, moreover $\phi_* \circ \phi^*$ is the multiplication by $d = \deg \phi$. It follows $nda = 0$, so we are done.

q.e.d.

Let us consider the case of <u>uniruled</u> threefolds (in particular conic bundles).

PROPOSITION 4.1.3. <u>Let</u> V <u>be a uniruled threefold. Then we have</u>:

i) $J(V)$ <u>is isogenous to the algebraic representative</u> A_V <u>of</u> $A^2(V)$.

ii) <u>The Weil morphism</u> $w: A^2(V) \to T^2(V)$ <u>is surjective</u>.

iii) $J(V)$ is equal to $H^{1,2}(V)/\text{poi } H^3(V,\mathbb{Z})$ and it is a principally polarized abelian variety.

Proof. i) It is a consequence of Proposition 4.1.2. In fact, as V is uniruled, there exists a rational map $\phi: Y \times \mathbb{P}^1 \dashrightarrow V$, Y nonsingular projective surface. Then $\text{Alb}(Y) \oplus \text{Pic}^0(Y)$ is the algebraic representative of $A^2(Y \times \mathbb{P}^1)$ via the canonical isomorphism $A^2(Y \times \mathbb{P}^1) \simeq A^2(Y) \oplus A^1(Y)$.

ii) By use of Künneth formulas, we get an isomorphism $T^2(Y \times \mathbb{P}^1) \simeq \text{Alb}(Y) \oplus \text{Pic}^0(Y)$, therefore the Weil morphism $w: A^2(Y \times \mathbb{P}^1) \to T^2(Y \times \mathbb{P}^1)$ is surjective. As usual, we can assume ϕ to be a morphism. Since the Weil morphisms are regular there exists a commutative square

$$\begin{array}{ccc} A^2(Y \times \mathbb{P}^1) & \xrightarrow{\phi_*} & A^2(V) \\ w \downarrow & & \downarrow w \\ T^2(Y \times \mathbb{P}^1) & \xrightarrow{\phi_*} & T^2(V) \end{array}$$

The direct images ϕ_* are surjective, so that $w: A^2(V) \to T^2(V)$ is also.

iii) Recalling that $H^{0,3}(V) = (0)$ as V is uniruled, we get by ii):

$$J(V) = H^{1,2}(V)/poiH^3(V,\mathbb{Z}) = T^2(V).$$

Moreover $J(V)$ has a canonical principal polarization θ. The corresponding hermitian form is defined by

$$H(\alpha,\beta) = 2i\int_V \alpha\wedge\bar{\beta}, \quad \alpha,\beta \in H^{1,2}(V).$$

The fact that the polarization is principal is a consequence of Poincare duality (see also [B], O.2 and [T], Ch.I).

q.e.d.

<u>4.2</u>. Through this section we return to consider a conic bundle $f:X \to S$. Look again at the commutative diagram

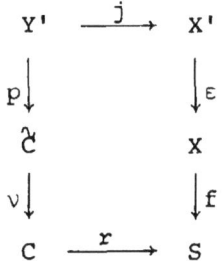

introduced at the beginning of section 2.2. Let g be the genus of the discriminant curve C, Y the surface of singular fibres, $e:C \to Y$ the canonical section, $\varepsilon:X' \to X$ the blowing up of X along $e(C)$, Y' the strict transform of Y. As in section 2.1, let $i: T_X \to X$ be the embedding of the tautological divisor T_X, and put $\pi = f \circ i : T_X \to S$.

We need some preparatory results. Let $\tau: \tilde{C} \to \tilde{C}$ be the canonical involution and define

$$H_1^{\pm}(\tilde{C}, \mathbb{Z}) = \{\gamma \in H_1(\tilde{C}, \mathbb{Z}), \gamma \mp \tau_* \gamma = 0\}.$$

LEMMA 4.2.1 i) <u>There exists a canonical homology basis for</u> $H_1(\tilde{C}, \mathbb{Z})$ <u>of the form</u>

$$\{\sigma_1, \ldots, \sigma_g, \sigma_{g+1}, \ldots, \sigma_{2g}, \tau(\sigma_1), \ldots, \tau(\sigma_{g-1}), \tau(\sigma_{g+1}), \ldots, \tau(\sigma_{2g-1})\}$$

<u>such that</u> σ_g <u>is homologous to</u> $\tau(\sigma_g)$ <u>and</u> $\sigma_{2g} = \tau(\sigma_{2g})$. <u>Moreover the intersection matrix is</u>

$$\begin{bmatrix} 0 & I_g & & \\ & & & 0 \\ -I_g & 0 & & \\ \hline & & 0 & I_{g-1} \\ & 0 & & \\ & & -I_{g-1} & 0 \end{bmatrix}$$

ii) <u>The</u> $\mathbb{Z}$-<u>module</u> $H_1^-(\tilde{C}, \mathbb{Z})$ <u>is generated by the set of cycles</u>

$$\{\sigma_i - \tau(\sigma_i)\}_i, \quad i = 1, \ldots, 2g-1, \quad i \neq g.$$

Proof. i) See f.e. [Fa], II.

ii) It follows by i) since $H_1^-(\tilde{C},\mathbb{Z}) = \ker(\mathbb{1}+\tau*)$.

q.e.d.

For any d-dimensional variety V, let (α,β) be an element of $H_{2d-n_1}(V,\mathbb{Z}) \times H_{2d-n_2}(V,\mathbb{Z})$. As in [G], Ch. 0,4, we denote by $\alpha \cdot \beta \in H_{2d-n_1-n_2}(V,\mathbb{Z})$ the intersection cycle. Consider now the topological maps

$$\psi_t : \varepsilon_* j_* p^* : H_1(\tilde{C},\mathbb{Z}) \to H_3(X,\mathbb{Z}) ,$$

$$\phi_t : (f^*, i_*\pi^*) : H_1(S,\mathbb{Z}) \oplus H_3(S,\mathbb{Z}) \to H_3(X,\mathbb{Z}) ,$$

where the inverse images p^*, f^*, π^* are defined via Poincare duality.

LEMMA 4.2.2 (Beauville) The equality

$$\psi_t(a) \cdot \psi_t(b) = -a \cdot (b - \tau*b)$$

holds, for all $a,b \in H_1(\tilde{C},\mathbb{Z})$ (or $a,b \in H_1(\tilde{C},\mathbb{Q})$) .

Proof. See [B], 2.3.

□

The maps ψ_t, ϕ_t induce morphisms of homology groups with rational coefficients, which we denote again by ψ_t, ϕ_t. Let ψ_t^- be the restriction of ψ_t to $H_1^-(\tilde{C},\mathbb{Q})$.

LEMMA 4.2.3(i) <u>The morphisms ϕ_t, ψ_t^- are injective on</u> $\mathbb{Q}$.

(ii) $H_3(X,\mathbb{Q}) = \phi_t(H_1(S,\mathbb{Q}) \oplus H_3(S,\mathbb{Q})) \oplus \psi_t^-(H_1^-(\tilde{C},\mathbb{Q}))$.

<u>Proof.</u> (i) Take $(a,b) \neq (0,0)$ in $H_1(S,\mathbb{Z}) \oplus H_3(S,\mathbb{Z})$. It is easy to see that there exists a cycle $\gamma \in H_3(X,\mathbb{Q})$ such that $\phi_t(a,b) \cdot \gamma \neq 0$. Then ϕ_t is injective.

Beauville formula gives $\ker \psi_t \subset H_1^+(\tilde{C},\mathbb{Q})$. Moreover $H_1^+(\tilde{C},\mathbb{Q}) \cap H_1^-(\tilde{C},\mathbb{Q}) = (0)$. Therefore $\ker \psi_t^- = (0)$.

(ii) To begin with, we prove

(*) $\phi_t(H_1(S,\mathbb{Q}) \oplus H_3(S,\mathbb{Q})) \cap \psi_t(H_1^-(\tilde{C},\mathbb{Q})) = (0)$.

Since $H_3(X,\mathbb{Q})$ is torsion free, it is sufficient to show that for all cycles $\alpha \in H_1^-(\tilde{C},\mathbb{Q})$, $\gamma \in \phi_t(H_1(S,\mathbb{Q}) \oplus H_3(S,\mathbb{Q}))$ one has $\psi_t(\alpha) \cdot \gamma = 0$.

We are going to compute the intersections $\psi_t(\alpha) \cdot f^*a$, $\psi_t(b) \cdot f^*b \cdot i_* T_X$ where $\gamma = f^*a + f^*b \cdot i_* T_X$, $(a,b) \in H_1(S,\mathbb{Q}) \oplus H_3(S,\mathbb{Q})$. We have

$$\psi_t(\alpha) \cdot f^*a = f_*(\psi_t(\alpha) \cdot f^*a) = f_*\psi_t(\alpha) \cdot a .$$

On the other hand $\psi_t(\alpha) = \varepsilon_* j_* p^* \alpha$ is a cycle whose support is contained in a finite number of fibres of the morphism f. Therefore $f_* \psi_t(\alpha) = 0$, so that $\psi_t(\alpha) \cdot f^* a = 0$.

Moreover, after recalling that $j^* \varepsilon^* f^* = p^* \nu^* r^*$, we get from projection formula the following equalities

$$\psi_t(\alpha) \cdot f^* b = \varepsilon_* j_* p^* \alpha \cdot f^* b = \varepsilon_* j_* (p^* \alpha \cdot j^* \varepsilon^* f^* b) =$$

$$= \varepsilon_* j_* p^* (\alpha \cdot \nu^* r^* b).$$

By hypothesis $\alpha \in H_1^-(\tilde{C}, \mathbb{Q})$, while $\nu^* r^* b \in H_1^+(\tilde{C}, \mathbb{Q})$ by construction. Then $\alpha \cdot \nu^* r^* b = 0$ and (*) is proved.

Propositions 3.8 and 4.1.3 give

$$\dim J(X) \le \dim(\text{Alb}(S) \oplus \text{Pic}^o(S) \oplus P_X)$$

hence

$$\dim H_3(X, \mathbb{Q}) \le \dim H_1(S, \mathbb{Q}) + \dim H_3(S, \mathbb{Q}) + \dim H_1^-(\tilde{C}, \mathbb{Q}).$$

Therefore the injectivity of ϕ_t, ψ_t^- together with relation (*) give the equality in the last formula.

q.e.d.

Consider the morphisms of abelian varieties

$$\Psi_J : J(\tilde{C}) \to J(X),$$

$$\Phi_J : \text{Alb}(S) \oplus \text{Pic}^o(S) \to J(X)$$

induced by the topological maps ψ_t, ϕ_t defined at the beginning of this section. The Weil morphism satisfies functorial properties leading to commutative diagrams (see [G], n.2)

$$
\begin{array}{ccc}
A^1(\tilde{C}) & \xrightarrow{\psi_A} & A^2(X) \\
\downarrow & & \downarrow w \\
J(\tilde{C}) & \xrightarrow{\Psi_J} & J(X)
\end{array}
\qquad (1)
\qquad
\begin{array}{ccc}
A^2(S) \oplus A^1(S) & \xrightarrow{\Phi} & A^2(X) \\
w_S \downarrow & & \downarrow w \\
\text{Alb}(S) \oplus \text{Pic}^o(S) & \xrightarrow{\Phi_J} & J(X)
\end{array}
\qquad (2)
$$

where ψ_A, Φ have been introduced in section 2.2 and w_S is the direct sum of Weil morphisms. Moreover we denote by θ, Θ the <u>principal divisors</u> on $J(X)$, P_X respectively and by P the <u>Poincaré divisor</u> which defines a principal polarization on $\text{Alb}(S) \oplus \text{Pic}^o(S)$. Let $m: P_X \to J(\tilde{C})$ be the inclusion.

THEOREM 4.2.4 (i) <u>The morphisms</u> $\Psi_J \circ m$, Φ_J <u>induce an isogeny</u>

$$\Phi_J + \Psi_J \circ m : \text{Alb}(S) \oplus \text{Pic}^o(S) \oplus P_X \to J(X).$$

(ii) The equalities $(\Psi_J om)^*\theta = 4\Theta$, $\Phi_J^*\theta = 2P$ hold, up to algebraic equivalence.

Proof. (i) Look at the morphism $\rho = \psi_A om : P_X \to A^2(X)$ and the isogeny

$$\Phi + \rho : A^2(S) \oplus A^1(S) \oplus P_X \to A^2(X)$$

stated in section 2, Proposition 2.2.4. The commutativity of diagrams (1), (2) implies that one of the square

$$\begin{array}{ccc} A^2(S) \oplus A^1(S) \oplus P_X & \xrightarrow{\Phi+\rho} & A^2(X) \\ {\scriptstyle (w_S, id.)} \downarrow & & \downarrow w \\ Alb(S) \oplus Pic^o(S) \oplus P_X & \xrightarrow{\sigma} & J(X) \end{array}$$

where $\sigma = \Phi_J + \Psi_J om$. Since w, $(w_S, id.)$ are surjective then σ is also. Moreover, as we have seen in the proof of Lemma 4.2.3

$$\dim J(X) = \dim(Alb(S) \oplus Pic^o(S) \oplus P_X).$$

Therefore $\ker\sigma$ is a finite group, so that σ is an isogeny.

(ii) Recall that the Riemann forms E_Θ, E_θ associated to the principal divisors Θ, θ are given by

$$E_\Theta(a,b) = -\tfrac{1}{2}(a \cdot b) \quad , \quad a,b \in H_1^-(\widetilde{\widehat{C}}, \mathbb{Z})$$

$$E_\theta(\gamma,\gamma') = (\gamma \cdot \gamma') \quad , \quad \gamma, \gamma' \in H_3(X, \mathbb{Z}) \; ,$$

via Poincaré duality. Since for all $a,b \in H_1^-(\widetilde{\widehat{C}}, \mathbb{Z})$

$$(\psi_t \text{om}(a) \cdot \psi_t \text{om}(b)) = -a \cdot (b - \tau^* b) = -2(a \cdot b) \; ,$$

we find $(\Psi_J \text{om})^* \theta = 4\Theta$.

Now, let E_P be the Riemann form associated to the Poincaré divisor P on $\text{Alb}(S) \oplus \text{Pic}^o(S)$. Consider the $\mathbb{R}$-bilinear form

$$E(x,y) = E_P((x,0),(0,y))$$

on $H_1(S,\mathbb{C}) \times H_3(S,\mathbb{C})$. Note that E is $\mathbb{C}$-linear on $H_1(S,\mathbb{C})$ and anti-linear on $H_3(S,\mathbb{C})$. Moreover E_P is determined by E and for all $(x,y) \in H_1(S,\mathbb{Z}) \times H_3(S,\mathbb{Z})$ we have (see [M1], II, n.9)

$$E_P((x,0),(0,y)) = x \cdot y \; .$$

Since

$$\phi_t(x,0) \cdot \phi_t(0,y) = f^*(x \cdot y) \cdot i_* T_X = 2(x \cdot y)$$

we get

$$\phi_t(x,0) \cdot \phi_t(0,y) = 2E_P((x,0),(0,y))$$

and $\phi_J^* \theta = 2P$ in $NS(Alb(S) \oplus Pic^o(S))$.

q.e.d.

To conclude, we consider the particular case when the base surface S is regular i.e. $q(S) = 0$.

PROPOSITION 4.2.5 Let $f:X \to S$ be a conic bundle such that $q(S) = 0$. Then there exists an isomorphism $\xi: P_X \to J(X)$ of principally polarized abelian varieties.

Proof. First we show that for all cycles $\gamma \in H_3(X,\mathbb{Z})$, $a \in H_1^-(\tilde{C},\mathbb{Z})$ one has

$$\gamma \cdot \psi_t(a) \equiv 0 \text{ modulo } 2 .$$

Lemma 4.2.3 gives

$$H_3(X,\mathbb{Z}) = \psi_t(H_1^-(\tilde{C},\mathbb{Q})) \cap H_3(X,\mathbb{Z}) ,$$

and after Lemma 4.2.1 ii) we can write a cycle $\gamma \in \psi_t(H_1^-(\tilde{C},\mathbb{Q})) \cap H_3(X,\mathbb{Z})$ in the form

$$\gamma = \sum_i \lambda_i \psi_t(\sigma_i - \tau(\sigma_i)) , \quad \lambda_i \in \mathbb{Q}, \ i = 1,\ldots,2g-1, \ i \neq g .$$

By using Beauville formula and Lemma 4.2.1 we find $\gamma \cdot \psi_t(\sigma_i) = 2\lambda_i$, $i \neq g, 2g$. Therefore $2\lambda_i \in \mathbb{Z}$ being γ, $\psi_t(\sigma_i)$ in $H_3(X,\mathbb{Z})$. Again Beauville formula gives

$$\psi_t(\sigma_i - \tau(\sigma_i)) \cdot \psi_t(a) \equiv 0 \text{ modulo } 4, \quad i \neq g, 2g .$$

Hence $\gamma \cdot \psi_t(a) \equiv 0$ modulo 2. Putting $u_a(\gamma) = \psi_t(a) \cdot \gamma/2$, then we get a morphism

$$u_a : H_3(X,\mathbb{Z}) \to \mathbb{Z} .$$

Therefore Poincare duality implies that there exists a cycle $a' \in H_3(X,\mathbb{Z})$ such that

$$a' \cdot \gamma = u_a(\gamma) \cdot \gamma$$

for all $\gamma \in H_3(X,\mathbb{Z})$ (see [G-H] p.53). Thus $\psi_t(a) = 2a'$ modulo torsion.

Consider now the composition $\Psi_J \circ m : P_X \to J(X)$, and let P_2 be the set of points of order 2 in P_X. One has the isomorphisms (of groups)

$$P_X \simeq H_1^-(\tilde{C},\mathbb{R})/H_1^-(\tilde{C},\mathbb{Z}) \; , \; J(X) \simeq H_3(X,\mathbb{R})/H_3(X,\mathbb{Z}) \; .$$

Then $P_2 \subset \ker(\Psi_J \circ m)$ since $\psi_t(a) = 2a'$ modulo torsion. Thus there exists a morphism $\xi : P_X \to J(X)$ such that the diagram

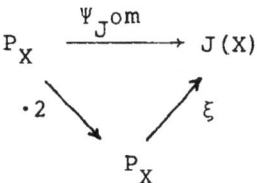

commutes. It follows $2^*\xi^* = (\Psi_J \circ m)^*\theta = 4\theta$ in $NS(P_X)$ by Theorem 4.2.4(ii). Hence $\xi^*\theta = \theta$ as $NS(P_X)$ is torsion free. This achieves the proof after recalling that $J(X)$, P_X have the same dimension by Theorem 4.2.4 (i).

q.e.d.

COROLLARY 4.2.6 Let $X \to S$, $X' \to S'$ be conic bundles and let $\phi : X' \dashrightarrow X$ be a birational map. Suppose $q(S) = 0$.

Then

(i) <u>There exists an embedding of principally polarized abelian varieties</u>

$$P_X \to P_{X'} \oplus \coprod_i J(C_i)$$

<u>with</u> $J(C_i)$ <u>Jacobians of curves.</u>

(ii) <u>Assume</u> X <u>to be ruled. Then</u> P_X <u>is isomorphic, as principally polarized abelian variety, to a product of Jacobians of curves.</u>

Proof. (ii) is an easy consequence of (i) by choosing $X' = S' \times \mathbb{P}^1$; so let us prove (i). According to [C-G] we have a morphism of principally polarized abelian varieties

$$J(X) \to J(X') \oplus \coprod_i J(C_i) ,$$

where C_i are the curves which occur in the resolution of the fundamental locus of ϕ. Morover the hypothesis $q(S) = 0$ implies $q(S') = 0$. Then Proposition 4.2.5 gives an isomorphism of principally polarized abelian varieties $J(X') \simeq P_{X'}$ and we are done.

q.e.d.

§5. Some open questions

The notations are as in previous sections. Let $X \to S$ be a conic bundle.

(1) Consider the morphism of abelian varieties

$$(\Psi_J \text{om}) : (P_X, \Theta) \to (J(X), \theta)$$

where Θ, θ denote the principal polarizations. We proved in 4.2.4(ii) that $(\Psi_J \text{om})^* \theta = 4\Theta$ in the Neron-Severi group $NS(P_X)$. Suppose the map $\Psi_J \text{om}$ factorize through the multiplication by 2 as follows

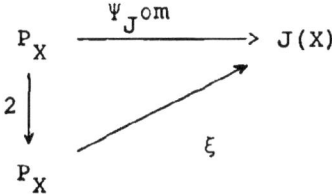

Hence ξ is a morphism of principally polarized abelian varieties, so that we get a decomposition

$$(*) \quad J(X) \simeq P_X \oplus A$$

for some principally polarized abelian variety A isogenous to $\text{Alb}(S) \oplus \text{Pic}^o(S)$. Then a question is to characterize all conic bundles for which the intermediate Jacobian has a decomposition in the weaker form (*) or else in the stronger one

(**) $J(X) \simeq Alb(S) \oplus Pic^o(S) \oplus P_X$.

Recall that (**) holds true by Proposition 4.2.5 whenever $q(S) = 0$.

(2) Consider now the algebraic representative (A_X, a_X) of $A^2(X)$. By universal properties there exists a morphism h_X of abelian varieties such that the diagram

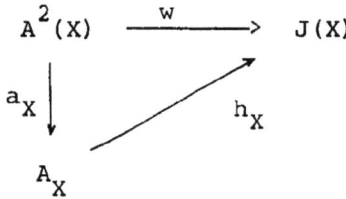

commutes, and h_X is an isogeny by 4.1.2. Again a natural problem is to see if h_X is (or if h_X induces) an isomorphism. To this purpose note that whenever (**) holds then A_X and $J(X)$ are isomorphic (not necessarily via h_X).

(3) A result of Roitman (see [R]) says that the map $a_S: A^2(S) \to Alb(S)$ induces an isomorphism on the points of finite order

$$A^2(S)_{tors.} \simeq Alb(S)_{tors.}$$

A question arising here is to see if an analogous result holds for the group $A^2(X)$, that is if the algebraic representative $a_X : A^2(X) \to A_X$ induces an isomorphism

$$A^2(X)_{\text{tors.}} \overset{?}{\to} (A_X)_{\text{tors.}}$$

Let us consider a particular case. Suppose the surface S satisfy the so called Abel-Jacobi property: this means that $a_S : A^2(S) \to \text{Alb}(S)$ is an isogeny (see [M-B]). It follows that $a_X : A^2(X) \to A_X$ is also an isogeny. Since ker a_X is a divisible group (this property holds true for all uniruled threefolds), then a_X is an isomorphism.

REFERENCES

[B]   Beauville, A. : *Varietés de Prym et Jacobiennes intermédiaires*, Ann. E.N.S. 10 (1977), 309-391.

[B-F]   Beltrametti, M. and Francia, P. : *Threefolds with negative Kodaira dimension and positive irregularity*, Nagoya Math. J. (to appear).

[B-M]   Bloch, S. and Murre, J.P. : *On the Chow group of certain types of Fano threefolds*, Compositio Math. vol. 39 (1979), 47-105.

[C]   Chevalley, C. : *Anneaux de Chow et Applications*, Séminaire Chevalley, Paris (1958).

[C-G]   Clemens, C.H. and Griffiths, P.A. : The intermediate
         Jacobian of the cubic threefold , Annals of Math. vol. 92
         (1970), 281-356.

[Co]   Conforto, F. : Le superficie razionali, Zanichelli (1939).

[E]   Enriques, F. : Sulle irrazionalitá da cui puó farsi
         dipendere la risoluzione di una equazione algebrica
         $f(x,y,z) = 0$ con funzioni razionali di due parametri,
         Math. Ann. Bd. IL (1897), 1-23.

[EGA]   Grothendieck, A. and Dieudonné, J.: Eléments de Géometrie
         Algébrique, Publ. I.H.E.S., II n.8 (1961), IV (3me partie)
         n.28 (1966).

[Fa]   Farkas, H.M. : On the Schottky relation and its
         generalization to arbitrary genus, Annals of Math. vol. 92
         (1970) 57-81.

[F]   Fulton, W. : Rational equivalence on singular varieties,
         (appendix to "Riemann-Roch for Singular Varieties" by
         P. Baum, W. Fulton, R. MacPherson), Publ. I.H.E.S. (1974),
         148-167.

[G]   Griffiths, P.A. : Some trascendental methods in the
         study of algebraic cycles, Several Complex Variables II,
         Maryland, Springer-Verlag, 185 (1970).

[H]   Hartshorne, R. : Algebraic Geometry, GTM, vol. 52
         Springer-Verlag (1977).

[L] Lieberman, D. : Intermediate Jacobians, Algebraic Geometry Oslo 1970, 5th Nordic Summer School in Math., Noordhoff (1972), 125-141.

[M1] Mumford, D. : Abelian Varieties, Oxford Univ. Press (1970).

[M2] Mumford, D. : Prym varieties I, Contribution to Analysis, Academic Press (1974), 325-350.

[Mo] Mori, S. : Threefolds whose canonical bundles are not numerically effective, Annals of Math. 116 (1982), 133-176.

[R] Roitman, A.A. : The torsion of the group of 0-cycles modulo rational equivalence, Annals of Math. 111 (1980) 553-569.

[S] Saito, H. : Abelian varieties attached to cycles of intermediate dimension, Nagoya Math. J., vol. 75 (1979), 95-119.

[T] Tyurin, A.N. : The middle Jacobian of three-dimensional varieties, J. Soviet Math. 13 (1980), 707-814.

[B-K-L] Bloch, S., Kas, A. and Lieberman, D. : Zero cycles on surfaces with $p_g = 0$, Compositio Math. vol. 33 (1976), 135-145.

MODULI OF SURFACES

OF GENERAL TYPE

F. Catanese[*]
Università di Pisa
Dipartimento di Matematica
Via Buonarroti 2, 56100/PISA

Introduction

The present paper follows rather closely the text of the talk given at the Conference, and is therefore rather problem-oriented and of a mostly expository nature.

In the first part we give a very brief survey of the history of the problem of moduli for surfaces and at the very beginning we discuss with some detail a very elementary though important example, namely the deformations of rational ruled surfaces.

Later on we expose some recent results of ours (cfr. [5]) which shed some light on basic questions concerning moduli of surfaces of general type: these results are based on the theory of "bidouble" covers (i.e. Abelian covers with group $(\mathbb{Z}/(2))^2$) and their deformations, and on the application of M. Freedman's recent result on homeomorphisms of 4- manifolds ([7]). We give then a list of some problems, and in the formulation of one of them we are indebted to a private communication of A. Beauville ([1]).

While in [5] the examples we had considered were only bidouble covers of $\mathbb{P}^1 \times \mathbb{P}^1$, we enlarge here in the second part our consideration to bidouble covers of the rational ruled surfaces $\mathbb{F}_{2m}$: on the one hand we can thus explain better the meaning of a certain exact sequence (2.7., 2.18 of [5]), on the other we show how the deformations of the bidouble covers fit together smoothly when the base $\mathbb{P}^1 \times \mathbb{P}^1$ deforms to $\mathbb{F}_{2m}$.

Our notation is as follows:

For a complex space X, Ω_X^i is the sheaf of holomorphic i-forms, O_X is the sheaf of holomorphic functions.

If X is compact, and F is a coherent sheaf of O_X-modules we denote by $H^i(F)$ the finite dimensional $\mathbb{C}$-vector space $H^i(X,F)$, by $h^i(F)$ its dimension, by $\chi(F) = \sum_{i=0}^{\dim X}(-1)^i h^i(F)$.

---

[*] A member of G.N.S.A.G.A. of C.N.R..

For Cartier divisors D,C on X, $O_X(D)$ is the invertible sheaf of sections
of the associated line bundle; $\equiv$ will denote linear equivalence of divisors,
$\sim$ algebraic equivalence, and $|D|$ will be the linear system of effective
divisors linearly equivalent to D; D·C denotes the intersection product.
If X is smooth T_X will denote the sheaf of holomorphic vector fields,
and K_X, when it exists, will denote a canonical divisor, i.e. a divisor
such that $O_X(K_X) \cong \Omega_X^n$, where $n = \dim_{\mathbb{C}} X$.
If X is an algebraic (compact, smooth) surface the geometric genus of
X, p_g, is $h°(\Omega_X^2) = h^2(O_X)$, the irregularity q is $h°(\Omega_X^1) = h^1(O_X)$.
If M is a topological manifold of dimension 4, with a given orientation,
τ is the signature of the quadratic form $q: H^2(M,\mathbb{Z}) \to \mathbb{Z}$ given by Poincaré
duality. As usual $b_i = \dim_{\mathbb{R}} H_i(M,\mathbb{R})$ is the ith Betti number and $e =$
$= \sum_{i=0}^{\dim_{\mathbb{R}} M} (-1)^i b_i$ is the topological Euler-Poincaré characteristic of M.

§ 1. Moduli of surfaces: history and problems.

Let S be an algebraic compact smooth surface, which we assume to be
minimal (i.e. S does not contain curves $E \cong \mathbb{P}^1$ such that $E^2 = -1$).
Like the genus of a curve, the holomorphic invariants K_S^2, $\chi(O_S) = \chi(S)$
depend only on the topology and the orientation of S (this last being
induced by the complex structure).
In fact
$$(1.0) \quad \begin{cases} K^2 = 3\tau + 2e \\ 12\chi = 3\tau + 3e \end{cases}, \text{ as a consequence of}$$
the Hirzebruch - Riemann - Roch theorem (cf. [10]).
Assume that $S \xrightarrow{p} B$ is a connected family of smooth surfaces, i.e.
a) B is a connected complex space
b) p is proper and $S_b = p^{-1}(\{b\})$ is smooth for each $b \in B$.
It is then a classical result that all the S_b's are diffeomorphic to
each other.
According to Mumford ([17]) one has the following definition:
Definition 1.1. The complex space M is said to be a coarse moduli space
for S if there exists a bijection g from the set of isomorphism classes
$\{[S']|$ S' is homeomorphic to S by an orientation preserving homeomorphism$\}$

to M such that for each family $S \xrightarrow{p} B$ the induced mapping $f: B \longrightarrow M$
(such that $f(b) = g([S_b]))$ is holomorphic.

A moduli space does not necessarily exist, as shows the following example, of rational ruled surfaces.

Example 1.2. Consider the rational ruled surfaces $\mathbb{F}_n = \mathbb{P}(O_{\mathbb{P}^1} \oplus O_{\mathbb{P}^1}(n))$, for $n \geq 2$.

Direct computations, which are well known, (cf. e.g. [12] pag. 42) give the result that for $n \geq 2$

$$(1.3) \qquad h^0(T_{\mathbb{F}_n}) = n + 5, \qquad h^1(T_{\mathbb{F}_n}) = n-1.$$

In particular, these surfaces are not biholomorphic to each other, but $\mathbb{F}_n$ and $\mathbb{F}_m$ are diffeomorphic iff $n \equiv m \pmod 2$. In fact, consider the rank 2 - vector bundles V on $\mathbb{P}^1$ which are extensions

$$(1.4) \qquad 0 \longrightarrow O_{\mathbb{P}^1} \longrightarrow V \longrightarrow O_{\mathbb{P}^1}(n) \longrightarrow 0$$

These are classified by the $(n-1)$ - dimensional vector space $B = H^1(O_{\mathbb{P}^1}(-n))$, and we get thus a family $F \xrightarrow{p} B$ of ruled surfaces where $p^{-1}(b) = \mathbb{P}(V_b)$, V_b being the vector bundle corresponding to $b \in B$. By (1.4) any homomorphism $\phi: O_{\mathbb{P}^1}(m) \longrightarrow V$ is trivial if $m > n$, and, if $m = n$, a non zero ϕ gives a splitting of the exact sequence (1.4).

For $b \in B$, let $m(b)$ the maximum m for which there exists a non trivial homomorphism $\phi : O_{\mathbb{P}^1}(m) \longrightarrow V_b$: by the maximality of $m(b)$ such ϕ determines a subline bundle of V_b, moreover $2 m(b) \geq n$ by the Riemann-Roch theorem, hence $V_b \cong O_{\mathbb{P}^1}(m(b)) \oplus O_{\mathbb{P}^1}(n-m(b))$, and $\mathbb{P}(V_b) \cong \mathbb{F}_{2m(b)-n}$. Since $n \leq 2 m(b) \leq 2n$, we get a decreasing filtration B_m of B, for $n/2 \leq m \leq n$ such that $B_m = \{b | \text{ exists a non trivial homomorphism } f: O_{\mathbb{P}^1}(m) \longrightarrow V_b\}$, Clearly $B_m - B_{m+1} = \{b | \mathbb{P}(V_b) \cong \mathbb{F}_{2m-n}\}$, and we have seen that B_n consists of the origin only.

Proposition 1.5. B_m is an algebraic cone of dimension equal to $\min(n-1, 2(n-m))$.

Proof. $\qquad B_m = \{b | H^0(V_b(-m)) \neq 0\}$

By virtue of the exact sequence

$$0 \longrightarrow H^0(V_b(-m)) \longrightarrow H^0(O_{\mathbb{P}^1}(n-m)) \xrightarrow{\beta(b)} H^1(O_{\mathbb{P}^1}(-m))$$

where $\beta(b)$ is given by cup product with $b \in B = H^1(O_{\mathbb{P}^1}(-n))$,

$$B_m = \{ b \mid \beta(b) \text{ is not injective} \}.$$

Fixing two points, 0 and ∞, in $\mathbb{P}^1$ and choosing an affine coordinate z on $\mathbb{P}^1 - \{\infty\}$ such that 0 corresponds to the origin, a basis for $H^1(O_{\mathbb{P}^1}(-n))$ is given by the Čech cocycles

$$z^{-1}, \ldots z^{-n+1} \in H^0(\mathbb{P}^1 - \{0\} - \{\infty\}, O_{\mathbb{P}^1}),$$

whereas a basis for $H^0(O_{\mathbb{P}^1}(n-m))$ is given by $1, z, \ldots z^{n-m}$.

Since $(z^{-i}) \vee z^j = \begin{cases} z^{j-i} & \text{or} \\ 0 & \text{if } j \geq i \text{ or } j - i \leq -m \end{cases}$

$b = \sum_{i=1}^{n-1} b_i z^{-i}$ belongs to B_m

i) if $2m \leq n+1$, or, for $2m \geq n+2$,

ii) if the following matrix $A_m(b)$ has rank strictly less than $(n-m+1)$, where

$$A_m(b) = \begin{vmatrix} b_1 & b_2 & \cdots & b_{n-m+1} \\ b_2 & b_3 & \cdots & b_{n-m+2} \\ \cdots & \cdots & \cdots & \cdots \\ b_{m-1} & b_m & \cdots & b_{n-1} \end{vmatrix}$$

It is immediate now that B_m is an algebraic cone, and we shall prove the assertion on its dimension by considering $V_i = \{b \mid$ the first i columns of $A_m(b)$ are linearly dependent$\}$ and proving, by increasing induction on i, that cod $V_i = m-i$, the case $i=1$ being immediate.

Now $V_i - V_{i-1}$ is covered by open sets where it is a complete intersection of $(m-i)$ hypersurfaces, hence cod $V_i \leq m-i$.
But if it were cod $V_i < m-i$, V_i would intersect the subspace

$P_i = \{b \mid b_j = 0 \text{ for } j \leq i-1 \text{ and } j \geq m\}$ in a locus of dimension at least 1, while it is easily seen that $V_i \cap P_i$ is just the origin.

Q.E.D.

The preceding example shows that the family $F \xrightarrow{p} B$ contains as fibres all the $\mathbf{F}_k$'s with $n \geq k \geq 0$ and $k \equiv n \pmod 2$, and on a Zariski open set of B the fibre is $\cong \mathbf{F}_0$ ($\cong \mathbb{P}^1 \times \mathbb{P}^1$) for n even, and $\cong \mathbf{F}_1$ for n odd.

Now it is easy to see that a moduli space (according to definition 1.1.) cannot exist for these surfaces, because we would have a non costant holomorphic map $f: B \to M$ which should be constant on a dense open set, a contradiction since a complex space is a Haussdorff topological space. The case when S is a surface of general type is a rather lucky one: in this case not only M exists but it is a quasi-projective variety by the theorem of Gieseker (1977, [8]).

The key feature here is given by the properties of the canonical divisor K_S: by the theorem of Bombieri ([2]), when $m \geq 5$ the complete linear system $|m K_S|$ yields a birational morphism $\Phi_m : S \to \mathbb{P}^{P_m-1}$, where

(1.6) $\quad P_m = \dfrac{m(m-1)}{2} K_S^2 + \chi(S)$, and such that

the image X_m of Φ_m is a normal surface enjoying the following properties

a) $X_m \cong X_n$ for $n, m \geq 5$

b) the singularities of $X = X_m$ are R.D.P.'s (Rational Double Points) i.e. biholomorphic (locally) to the hypersurface singularities:

$$A_n = \{z^2 + x^2 + y^{n+1} = 0\}$$
$$D_n = \{z^2 + x(y^2 + x^{n-2}) = 0\} \quad (n \geq 4)$$
$$E_6 = \{z^2 + x^3 + y^4 = 0\}$$
$$E_7 = \{z^2 + x(y^3 + x^2) = 0\}$$
$$E_8 = \{z^2 + x^3 + y^5 = 0\}$$

c) $\Phi = \Phi_m$ is a minimal resolution of singularities of X, i.e. Φ is biholomorphic outside the singular points of X and the inverse image of a singular point is a (connected) union of curves $E \cong \mathbb{P}^1$ with $E^2 = -2$ intersecting transversally, whose structure is described by the Dynkin diagrams

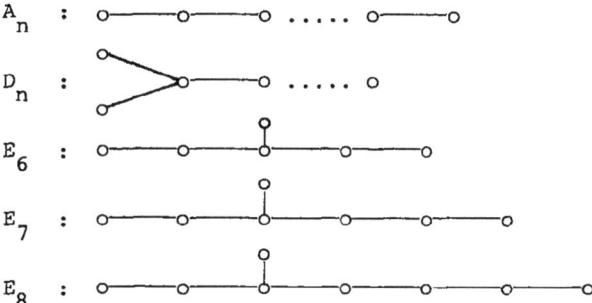

whose vertices correspond to curves, and whose edges correspond to points of intersection of the two curves corresponding to the vertices of the edge. The index (e.g. n in A_n) denoter the number of vertices in the diagram.

The importance of the pluri-canonical model X_m lies in the fact that any morphism $f: S \longrightarrow S'$ induces a projectivity between X_m and X'_m, therefore M appears as a quotient of a locally closed subscheme H of the Hilbert scheme parametrizing surfaces of degree $m^2 K_S^2$ in $\mathbb{P}^{P_m-1}$ by the projective group $PGL(P_m)$. In particular one has:

(1.7) Surfaces of general type with fixed K^2, χ belong to a finite number of families (cf. [3], p. 395).

For later use we show another application of the previous result.

<u>Theorem 1.8.</u> Let G be a finite group and let M^G be the subset corresponding to the isomorphism classes of surfaces S for which G acts faithfully as a group of automorphisms of S. Then M^G is a closed subvariety of M.

<u>Proof.</u> Let $m \geq 5$, and set $N = P_m(S) - 1$. If G acts faithfully on S, G acts linearly on the vector space $H^0(S, O_S(m K_S))$, hence on its dual space, and we get an $(N+1)$-dimensional representation ρ of G inducing a faithful projective representation of G on $\mathbb{P}^N$ leaving X_m invariant. Conversely any faithful representation ρ of G on $\mathbb{P}^n$ leaving X_m invariant, since any automorphism of X_m lifts to an automorphism of S (S being a minimal desingularization), gives an injective homomorphism of G into $Aut(S)$, hence in particular lifts to a linear $(N+1)$- dimensional representation of G. Now there is only a finite number of isomorphism classes of such projective representations ρ, and correspondingly M^G can be expressed as a finite

union of subsets M^ρ. Fix therefore such a (linear) representation ρ inducing a faithful action on $\mathbb{P}^N$, and let H^ρ be the locally closed subscheme of the Hilbert scheme H parametrizing m-canonical images X_m of surfaces S, such that H^ρ is the locus of fixed points for the action of G on H. Clearly M^ρ is the projection to the quotient of the image of H^ρ x PGL(N+1) in H.

To prove that such image is closed we use the valuative criterion of properness (cf. e.g. [9], theorem 4.7.): assume that we have a 1-parameter family $S \xrightarrow{p} B$, and that, setting $B^* = B - \{b_o\}$, $S^* = p^{-1}(B^*)$, we have a faithful action of G on S^*, such that, for $g \in G$, $p \circ g = p$.
By the m^{-th} canonical mapping we get a family $X \xrightarrow{f} B$, with $X \to B \times \mathbb{P}^N$, and it suffices to prove that for every g in G its action extends from X^* to X and in such a way that g does not act as the identity on $X_o = f^{-1}(b_o)$: in fact, X^* being dense, such extension is then unique, hence we get a homomorphism of G into Aut(X_o) which is injective by the second property.

Now, for $g \in G$, we get an invertible matrix $a(t)$, for $t \in B^*$, which we can assume to be given by a regular function on B with $a(b_o) \neq 0$: clearly it suffices to prove that $a(b_o)$ is invertible and is not a multiple of the identity. By continuity, the eigenvalues of $a(b_o)$ are limit of the eigenvalues of $a(t)$, whose ratios are certain fixed roots of unity, therefore, not all the eigenvalues of $a(b_o)$ are equal and if 0 were an eigenvalue of $a(b_o)$, then $a(b_o)$ would be zero, a contradiction.

Q.E.D.

As we heard from D. Mumford's lecture, much is known about the moduli spaces M_g of curves of genus g, the basic fact being that M_g is quasi-projective normal irreducible variety of dimension 3g-3 ($g \geq 2$); on the other side, not many general results are known about the moduli spaces of surfaces of general type, and we shall show here that too optimistic expectations have a negative answer: e.g. these moduli spaces are "in general" highly reducible, with a lot of components of different dimension.

But, in order to explain all this more precisely, let's introduce some notation and let's make same historical remark.

Definition 1.9. Let S be a surface of general type: then the number of

moduli of S, denoted by M(S), is the dimension of the moduli space M
at the point [S] corresponding to S.

M. Noether ([18]) in 1888, under very special hypotheses, postulated for
M a formula which in our terminology reads out as $M = 10\chi - 2K^2$.
This formula is verified quite seldom (especially since M is a positive
integer, whereas the right side can be very negative, even for complete
intersections), but it is the merit of F. Enriques to understand that
$10\chi - 2K^2$ should give a lower bound for M in the case of non ruled
surfaces.

In fact Enriques gave two proofs (see e.g. his book [6], p.204-215, especially the historical note on page 213) which were both incomplete, and in
fact relying on some assumptions which did not hold true. In the first
proof Enriques assumed to have a surface $F \subset \mathbb{P}^3$ with ordinary singularities, of degree n, and with double curve C: he assumed that the characteristic system (cut on the normalization of F by adjoint surfaces of
degree n) should be complete, and this is not true in general as was shown
by Kodaira in 1965 ([11]); similarly in the second proof it was assumed
that the characteristic system of plane curves with cusps and nodes
should be complete, an assertion which was disproven by Wahl in 1974
([22]), relying on the examples of Kodaira (we defer the reader, for a
more thorough discussion, to the appendix to Chapter V of Zariski's book
[25], written by D. Mumford).

A proof finally came in 1963, through the theorem of Kuranishi ([13])
culminating the theory of deformations of complex structures due to
Kodaira and Spencer.

Let $X \xrightarrow{p} B$ be a connected family of smooth manifolds and $b_0 \varepsilon B$: then
the fibres $X_b = p^{-1}(\{b\})$ are said to be deformations of $X_0 = X_{b_0}$.
Any holomorphic map f of a complex space T into B, with $f(t_0)=b_0$, induces
another family of deformations of X_0, namely the fibre product $T \times_B X$.
A family of deformations $(X, X_0) \xrightarrow{p} (B, b_0)$ is said to be semi-universal
if, for every other deformation $(Y, X_0) \xrightarrow{g} (T, t_0)$ the restriction to a
sufficiently small neighbourhood of t_0 is induced by a holomorphic map
f : T->B whose differential at t_0 is uniquely determined; it is said to
be universal if moreover such a f is always unique.

The theorem of Kuranishi asserts that a semiuniversal deformation exists (it is then unique by its defining property), and moreover that its base B is a germ of analytic subset of $(H^1(X_0,T_{X_0}),0)$ defined by $h^2(X_0,T_{X_0})$ equations vanishing of order at least two at the origin.

Later Wavrik ([23]) proved that, if $H^0(T_{X_0})=0$, then the deformation is universal, what implies that if a moduli space M exists for X_0, then the germ of M at $[X_0]$ is biholomorphic to the quotient $B/\mathrm{Aut}(X_0)$ (though, e.g. in the case of Galois covers whose deformations are all Galois covers, the action of $\mathrm{Aut}(X_0)$ on $H^1(T_{X_0})$ need not be effective).

Now, when S is a surface of general type, Aut(S) is a finite group (this is another application of pluricanonical embeddings), hence $H^0(T_S)=0$, being the Lie algebra of a finite group. Deformation theory gives a solution to Enriques' inequality via the Hirzebruch - Riemann - Roch theorem: if a surface X is not ruled, then $H^0(T_X)=0$, and $M = \dim B$, if M exists.

Clearly one has, by the previous remarks on B,

(1.10) $\quad h^1(T_X) - h^2(T_X) \leq \dim B = M \leq h^1(T_X)$

but, since $h^0(T_X)=0$, the left hand side is $-\chi(T_X)$, i.e. $10\chi - 2K^2$ by the Hirzebruch R.-R. theorem, hence (1.10) is exactly Enriques' inequality. One drawback of (1.10) is that the upper bound for M does not depend only on topological invariants: hovever, since by Serre duality

$$h^2(T_X) = h^0(\Omega_X^1 \otimes \Omega_X^2),$$

the right hand side is $10\chi - 2K^2 + h^0(\Omega_X^1 \otimes \Omega_X^2)$, so it is enough to give an upper bound on $h^0(\Omega_X^1 \otimes \Omega_X^2)$, and in the case e.g. of surfaces of general type this can be done via exact sequences, restricting the sheaf $\Omega_S^1 \otimes \Omega_S^2$ to a smooth curve in $|K|$ or $|mK|$.

One gets (theorems B and C of [5]) the upper bounds

(1.11) $\quad M \leq 10\chi + 3K^2 + 108 \qquad$ (in general)

(1.12) $\quad M \leq 10\chi + q + 1 \qquad$ if S contains a smooth canonical curve C.

These extimates appear to be too crude and an interesting question is, roughly speaking:

(1.13) what is asymptotically the best upper bound for M?

I will return later to better bounds for irregular surfaces, for the moment let me remark that, for a surface of general type S, by Castelnuovo's theorem, Noether's inequality, and the inequality of Bogomolov - Miyaoka - Yau, the topological invariants K^2, χ are subject to the following inequalities:

(1.14)
$$\begin{cases} K^2 \geq 1, \chi \geq 1 \\ K^2 \geq 2\chi - 6 \\ K^2 \leq 9\chi. \end{cases}$$

It is possible therefore that, as $K^2, \chi \longrightarrow +\infty$ one may have different best upper bounds according to the limiting value of the ratio K^2/χ between 2 and 9.

One may ask however whether the moduli space is pure-dimensional: we proved recently that this is not true, and that M can attain arbitrarily many different values for orientedly homeomorphic surfaces.

More precisely, we proved ([5] theorem A)

(1.15) for each positive integer n there exist integers $0 < M_1 < M_2 < \ldots M_n$ and homeomorphic simply-connected surfaces of general type $S_1, \ldots S_n$, such that $M(S_i) = M_i$.

An important remark is that the surfaces we consider are such that the canonical map is a biregular embedding, and their invariants K^2, χ are quite "spread" in the region defined by (1.14), so that these examples should be considered the rule rather than the exception.

Let me sketch briefly the idea of the proof, which consists of 3 basic ingredients.

<u>Step I</u>: If S_1, S_2 are simplyconnected, have equal K^2, χ, and $K^2 \neq 9$, they are orientedly homeomorphic if and only if either

  a) $K_{S_i} \in 2\, P_{ic}(S_i)$ (i=1,2)

  b) $K_{S_i} \notin 2\, P_{ic}(S_i)$ (i=1,2)

<u>Step II</u>: Find families of surfaces, with the properties stated in step I, depending on many integral parameters, and compute in terms of those K^2, χ, M.

Step III: Show that one can fix K^2, χ and obtain different values $M_1, \ldots M_n$ for M: this is a number theoretic problem, solved by E. Bombieri (cf. the appendix to [5]), so that I will not talk about this in a Conference on Algebraic Geometry.

Step I was suggested by B. Moishezon and depends almost entirely on the recent deep theorem of M. Freedman (cf. [7]).

(1.16) If S_1, S_2 are simply-connected compact oriented differentiable 4-manifolds with the same intersection form on $H^2(S_i, \mathbb{Z})$, then they are (orientedly) homeomorphic.

and on the theorem of Yau ([24])

(1.17) if $K^2 = 9\chi$ and K is ample, then the universal cover of S is the unit ball in $\mathbb{C}^2$.

In fact, by (1.0), K^2, χ determine the rank and the signature of the unimodular quadratic form $q: H^2(S, \mathbb{Z}) \longrightarrow \mathbb{Z}$, and it is known that indefinite unimodular integral quadratic forms are classified only by the rank, signature and parity (q being even iff for each x, $q(x) \equiv 0$ (mod 2), being odd otherwise).

Therefore Freedman's result applies provided that q is not negative or positive definite.

But q is negative definite only for surfaces of class VII which have $b_1 = 1$, while it is positive definite if and only if $S = \mathbb{P}^2$, as an easy corollary of Yau's theorem (1.17) (cf. [14], [21]).

Step II consists in considering bidouble covers of $\mathbb{P}^1 \times \mathbb{P}^1$ and studying their small deformations: we will return to this point in the second paragraph, where we shall consider, using the results of [5], the more general case of bidouble covers of $\mathbb{F}_{2n}$.

Anyhow 1.15 shows in particular that, fixing the homeomorphism type of S, and varying the complex structure (which necessarily gives a surface of general type if $K^2 \geq 10$), the number of moduli M varies in an interval whose size grows to infinity with K^2, χ.

(1.18) How many irreducible components does $M_{K^2, \chi}$, (the union of the moduli spaces of surfaces with K^2, χ fixed), have at most?

(1.19) Is it true also that the number of connected components of M is unbounded, as $K^2, \chi \longrightarrow \infty$?

We remark here that connected components of M correspond to connected components of the subscheme H of the Hilbert scheme mentioned before (1.7), hence two surfaces S,S' are such that their classes [S], [S'] belong to the same connected component of M if and only if they are deformation of each other: in particular they must be diffeomorphic.
One could have therefore made a different choice for the "moduli space", considering only M_{diff}, i.e. the union of the connected components of M corresponding to diffeomorphic surfaces, and ask, regarding M_{diff}, similar questions to those posed for M, i.e. pure-dimensionality, etc..
As a matter of fact, though bidouble covers can be deformed until the branch locus is a union of lines (one gets then surfaces with only A_1-singularities), even in this last case it is not easy to tell directly whether homeomorphic surfaces are diffeomorphic: we have not pursued this, also there is some hope that Freedman's result can be made stronger as to imply that the two given 4-manifolds should be diffeomorphic.
Let's go back now to the last piece of history: in 1949 G. Castelnuovo ([4]) claimed that

(1.20) For an irregular surface S without irrational pencils, the number M of moduli is $\leq p_g + 2q$.

To explain what this means, we recall the classical theorem of Castelnuovo - De Franchis

(1.21) Assume that η_1, η_2 are independent sections of $H°(\Omega_S^1)$ such that $\eta_1 \wedge \eta_2 \equiv 0$: then there exists a morphism $f: S \to B$, where B is a smooth curve possessing two 1-forms $\xi_1, \xi_2 \in H°(\Omega_B^1)$ such that $\eta_i = f^*(\xi_i)$.

Now, such a map f: $S \to B$ is called a pencil, whose genus is, by definition, the genus of B, and an irrational pencil is just a pencil of genus at least one.
So, if an irregular surface does not have irrational pencils, first of all its image under the Albanese map $\alpha: S \to A = H°(\Omega_S^1)^V / H_1(S, \mathbb{Z})$ is a surface, (hence $q \geq 2!$).

Conversely, it is easy to see that if the Albanese variety A is simple and is not the Jacobian of a curve then S does not have irrational pencils. Unfortunately Castelnuovo's claim is false, as we showed in [5], exhibiting counterexamples where M grows, keeping q fixed, like 4 p_g; anyhow we want to show here (cf. [5], th. D) that it is possible to rescue in some sense his assertion, in view of the Castelnuovo - De Franchis theorem.

Theorem 1.22. If $q \geq 3$ and there exist $\eta_1, \eta_2 \in H^\circ(\Omega_S^1)$ such that C = div $(\eta_1 \wedge \eta_2)$ is a reduced irreducible curve, then $M \leq p_g + 3q - 3$ and $K^2 \geq 6\chi$, this last equality holding if and only if the Albanese map is unramified into an Abelian 3-fold.

Proof. η_1, η_2 define an exact sequence

(1.23) $\quad 0 \longrightarrow O_S^2 \longrightarrow \Omega_S^1 \longrightarrow F \longrightarrow 0$

with supp $(F) = C$, and, since $h^\circ(\Omega_S^1) \geq 3$, we have a non zero section of F, hence a sequence

(1.24) $\quad 0 \longrightarrow O_C \longrightarrow F \longrightarrow \Delta \longrightarrow 0$

where the support of Δ has dimension zero.
By the multiplicativity of global Chern classes with respect to exact sequences, we obtain

$$c_2(\Delta) = - \text{length}(\Delta) = - (c_1^2 - c_2).$$

Hence $c_1^2 \geq c_2$, i.e. $K^2 \geq 6\chi$, and if equality holds $\Delta=0$, $F \cong O_C \Longrightarrow q=3$ and Ω_S^1 is generated by global sections. The assertion about M follows by tensoring (1.23) and (1.24) with Ω_S^2, bounding h° of the middle term with the sum of the h°'s of the two other terms, and $h^\circ(O_C(K))$ with $p_g + q - 1$.

Q.E.D.

Remark 1.25. The hypotheses of 1.22 are verified e.g. if Ω_S^1 is generated by global sections outside a finite set of points.
Castelnuovo's error in fact was based on some wrong results of Severi ([19]): e.g. Severi claimed that for a surface S without irrational pencils of genus q the sections of $H^\circ(\Omega_S^1)$ would have no common zeros,

what is not true (see [5] for a discussion and counterexamples).
In the same paper Severi deduced from these incorrect assertions the
following statement, whose validity we have not checked and we pose then
as a problem

(1.26) Is it true that for an irregular (minimal) surface without irrational pencils $K^2 \geq 4\chi$?

(1.27) Also, it is an interesting question for us whether, under the hypotheses of 1.22, Castelnuovo's inequality $M \leq p_g+2q$ holds: looking at the proof we see that it would be indeed the case if $h^\circ(O_C(K))$ could be bounded by p_g+2. This is not true if S has irrational pencils, and this inequality is related to a question posed by Enriques ([6] page 354):

(1.28) when is the dimension of the paracanonical system $\{K\}$ less than or equal to p_g?

We recall that the paracanonical system can be defined as follows: consider the subscheme $[K]$ of the Hilbert scheme consisting of curves in S algebraically equivalent to a canonical divisor K, and consider the irreducible component $\{K\}$ of $[K]$ which contains the complete linear system $|K|$.

At the conference we posed the problem whether the hypothesis "S without irrational pencils" would imply $\dim\{K\} \leq p_g$, and ideed we asked also more, i.e. whether, under those assumptions, for $\eta \in \text{Pic}^\circ(S)-\{0\}$ it should be $H^1(S,\eta)=0$, a fact which implies $\dim[K] \leq p_g$.

This latter has been answered negatively by A. Beauville ([1]) who gave an example where $[K]$ has dimension bigger than p_g. His example is as follows:

(1.29) Let B, A, be Abelian varieties of respective dimensions g and q, ω an element of $A-\{0\}$ with $2\omega=0$, and let i be the fixed point free involution on $B \times A$ such that $i(b,a) = (-b, a+\omega)$.

Let X be the quotient manifold $B \times A/i$: the direct image of $O_{B \times A}$ splits as $O_X \oplus O_X(\eta)$, where $2\eta \equiv 0$ but η is not a trivial divisor.
It is easily seen that $h^1(O_X) = q$, $h^1(O_X(\eta))=g$, and that A/ω is the

Albanese variety of X.

Taking an embedding of X by a sufficiently very ample linear system, and intersecting X with a general linear subspace of codimension (g+q-2), one gets a surface S whose Albanese variety is just A/ω, and with $h^1(O_S(\eta)) = g$.

But then, if g>q, the dimension of the linear system $|K_S + \eta|$ is $p_g + (g-q)$, > p_g.

Clearly, as we remarked before, if A is not isogenous to a Jacobian and it is simple, S has no irrational pencils.

In this example, the system {K} consists of $|K + \eta|$ and {K}, which has dimension p_g, in fact $H^1(O_S(\varepsilon)) = 0$ for $\varepsilon \in Pic^\tau(S)$, $\varepsilon \neq 0, \eta$, since on an Abelian variety Y the only divisor δ in Pic°(Y) with $H^1(O_Y(\delta)) \neq 0$ is
≡ 0 (cf. [16]).

To end with this first part, let me mention two more problems whose solution I'd like to see.

(1.30) It is known (cf. e.g. [20], page 402 and foll.) that, given any finite group G, one can find, for each n≥2, a variety X of dimension n with $\pi_1(X) = G$. In the case where G is abelian I have proved ([5], Cor. 1.9) the stronger statement that for any simply-connected variety Y of dimension n≥2, there exists an abelian cover of Y with group G^n such that $\pi_1(X) = G$. I guess that something similar could be done for any finite group G, so that, in particular, "every finite group is the fundamental group of infinitely many surfaces".

This last question is a recurrent one when one wants to describe explicitly some particular classes of surfaces.

We recall that the pluricanonical model X of a surface S of general type is isomorphic to S if and only if the canonical bundle of X is ample, i.e. if and only if there are no curves $E \cong \mathbb{P}^1$ with K·E=0 (<=> $E^2 = -2$) (these are the curves coming from the resolution of R.D.P.'s).

It is not clear to me whether these curves can be stable by deformation, i.e..

(1.31) Do there exist irreducible components Z of some moduli space of surfaces of general type such that for each $[S] \in Z$ the canonical bundle K_S is not ample?

R. Klotz has announced the result that $K^2 < 9\chi$ if K is not ample: this result in particular says that there are no discrete cocompact subgroups Γ of automorphisms of the unit ball D in $\mathbb{C}^2$ with D/Γ not smooth and with only R.D.P. as singularities (these subgroups Γ are rigid, by the theorem of Mostow [15]).

§ 2. Bidouble covers of rational ruled surfaces.

Def. 2.1. A bidouble cover $\pi: S \longrightarrow X$ is a Galois finite cover with group $G = (\mathbb{Z}/2)^2$. A bidouble cover is said to be smooth if, moreover, S, X, are smooth varieties.

Let $\pi: S \longrightarrow X$ be a smooth bidouble cover where S,X, are surfaces, and let $\sigma_1, \sigma_2, \sigma_3$ be the 3 non trivial involutions in the group $(\mathbb{Z}/2)^2$. Let $X_i = S/\sigma_i$, and let $\pi_i: X_i \longrightarrow X$ be the induced double cover. The locus $\text{Fix}(\sigma_i)$ of fixed points for σ_i consists of a smooth divisor R_i, and a finite set N_i': it is clear that $R = R_1 + R_2 + R_3$ is the ramification divisor of π, that $\pi(R_i) = D_i$ is a smooth divisor, and $D = D_1 + D_2 + D_3$ is the branch locus of π.

$X_i \xrightarrow{\pi_i} X$ is branched on $D_j + D_k$ ($\{i,j,k\} = \{1,2,3\}$, here and in the following); therefore since the only singularities of X_i are A_1-points (nodes), corresponding to the points in N_i', we have:

(2.2) the divisor D has normal crossings, $N_i' = \pi^{-1}(D_j \cap D_k)$, and there exist divisors L_i on X s.t. $2L_i \equiv D_j + D_k$, so that X_i is the double cover of X in $\mathcal{O}_X(L_i)$ branched on $D_j + D_k$.

In [5] it is proven then that $\pi_* \mathcal{O}_S \cong \mathcal{O}_X \oplus (\bigoplus_{i=1}^{3} \mathcal{O}_X(-L_i))$, and that on X

(2.3) $\quad D_k + L_k \equiv L_i + L_j$.

To describe more explicitly the algebra structure of $\pi_* \mathcal{O}_S$ we use the following notation: x_i is a section of $\mathcal{O}_X(D_i)$ such that $\text{div}(x_i) = D_i$, z_i

is a section of $0_S(R_i)$ with $\operatorname{div}(z_i) = R_i$.
Then $(z_j z_k)^2 = x_j x_k$, and, setting $w_i = z_j z_k$, w_i is a section of $0_S(\pi^* L_i)$ with

$$w_i^2 = x_j x_k$$

and w_i is precisely the square root extracted through the cover π_i. Conversely, in the rank-3 bundle $V = \bigoplus_{i=1}^{3} 0_X(L_i)$ one can consider the bidouble cover described by the equations

(2.4) $$\begin{cases} w_i^2 = x_j x_k \\ x_k w_k = w_i w_j \end{cases}$$

and ([5], prop. 2.3) all smooth bidouble covers arise in this way from divisors D_i, L_i, satisfying (2.2), (2.3).

<u>Def. 2.5.</u> A surface S' is called a natural deformation of S if there exist sections γ_i of $0_X(D_i - L_i)$, x_j' of $0_X(D_j)$ (i,j=1,2,3) such that S' is defined in V by the following equations

(2.6) $$\begin{cases} w_i^2 = (\gamma_j w_j + x_j')(\gamma_k w_k + x_k') \\ w_j w_k = x_i' w_i + \gamma_i w_i^2. \end{cases}$$

Since natural deformations are parametrized by a smooth variety, it is important to know to which subspace of $H^1(T_S)$ they give rise: the answer is given by the following result (thm. 2.19 of [5]).

<u>Theorem 2.7.</u> There exists an exact sequence

$$0 \longrightarrow H^°(T_S) \longrightarrow H^°(\pi^* T_X) \longrightarrow \bigoplus_{i=1}^{3} H^°(0_{D_i}(D_i) \oplus 0_{D_i}(D_i - L_i)(\xrightarrow{\partial}$$

$$\xrightarrow{\partial} H^1(T_S) \longrightarrow H^1(\pi^* T_X)$$

and Im ∂ = Kodaira Spencer image of the natural deformations.

<u>Remark 2.6.</u> In [5] it is also proved that S is simply connected if X is such and the D_i's move in a pencil with transversal base points.

We are going to apply (2.5), in the case when $X = \mathbf{F}_{2m}$. We consider then the family $F \xrightarrow{p} B$ obtained from (1.4) for n=2m. In the exact sequence (1.4) the trivial line subbundle of V_b determines a relative Cartier divisor $S \subset F$ which is a section of the projection of F onto $B \times \mathbb{P}^1$. For each b in $B, S|\mathbf{F}_b = S_b$ is a section of $\mathbf{F}_b$ with normal bundle $\cong \mathcal{O}_{\mathbb{P}^1}(2m)$, hence $S_b^2 = 2m$.

The projection g of F onto $\mathbb{P}^1$ also induces a relative (w.r. to p) Cartier divisor $Y = g^*(\mathcal{O}_{\mathbb{P}^1}(1))$. Clearly $Y_b^2 = 0$, and Pic(F) is a free abelian group with basis given by Y, S: moreover $Y_b \cdot S_b = 1$.

Consider now a divisor $D \equiv a_1 S + a_2 Y$, with $a_1, a_2 \in \mathbb{Z}$: when restricting it to $\mathbf{F}_b \cong \mathbf{F}_{2k}$, $k \leq m$, since $S_b - (m+k)Y_b$ is an effective smooth section of $\mathbf{F}_b$, if the divisor D_b is effective, then either $D_b \cdot Y_b = = a_1 > 0$, or $a_1 = 0$, $a_2 > 0$ and D is a union of fibres of the ruling of $\mathbf{F}_b$ onto $\mathbb{P}^1$. If $a_1 > 0$, also, since $D_b(S_b - (m+k)Y_b) = a_1((2m) - (m+k)) + a_2 = = a_1(m-k) + a_2$, if $(S_b - (m+k)Y_b)$ is not a fixed part of $|D_b|$ then $D_b \equiv a_1 (S_b - (m-k)Y_b) + a_3 Y_b$, with $a_3 \geq 0$. We set for convenience $S_b' = S_b - (m-k)Y_b$: this is a smooth section with $(S_b')^2 = 2k$.

<u>Lemma 2.7.</u> If D_b is an effective divisor on $\mathbf{F}_b \cong \mathbf{F}_{2k}$ $(0 \leq k \leq m)$, and $D_b \equiv a_1 S_b' + a_3 Y_b$, then the linear system $|D_b|$ has no base points if and only if $a_1, a_3 \geq 0$.

<u>Proof.</u> $|S_b'|$ has no base points by the following exact sequence (notice that $S_b' \cong \mathbb{P}^1$)

(2.8) $\quad 0 \longrightarrow H^\circ(\mathcal{O}_{\mathbf{F}_b}) \longrightarrow H^\circ(\mathcal{O}_{\mathbf{F}_b}(S_b')) \longrightarrow H^\circ(\mathcal{O}_{\mathbb{P}^1}(2k)) \longrightarrow 0.$

Moreover, clearly $|Y_b|$ has no base points.

Q.E.D.

In the previous discussion we have also seen that, given $D \equiv a_1 S + a_2 Y$, $H^\circ(\mathbf{F}_b, \mathcal{O}_{\mathbf{F}_b}(D_b)) \geq 1$ if and only if $a_1 \geq 0$, $a_2 \geq -a_1(m+k(b))$, where $k(b) = m(b) - m$ (cf. 1.4. and foll.), $0 \leq k(b) \leq m$ (and $k(b) = m$ only for b=0).

Let K be the relative canonical divisor of $p: F \longrightarrow B$: since $K_b \cdot Y_b = = (K_b + S_b) S_b = -2$,

(2.9) $\quad K \equiv -2S + (2m-2)Y$.

By Serre duality then $H^2(O_{F_b}(D_b))=0$ if D_b is an effective divisor. Moreover, if D_b is effective, by the exact sequence

$$0 \longrightarrow O_{F_b}(-D_b) \longrightarrow O_{F_b} \longrightarrow O_{D_b} \longrightarrow 0,$$

it follows that $H^1(O_{F_b}(-D_b))=0$ if $|D_b|$ has no base points or it has a reduced and connected general member (i.e., in view of 2.7, $a_1=0$, $a_2=1$ or $a_1 > 0$, $a_2 \geq -a_1(m - k(b)) - 2 k(b)$.
Again by Serre duality $H^1(O_{F_b}(D_b)) = H^1(O_{F_b}(-D_b - K_b))) =$

$= H^1(O_{F_b}(-((a_1+2)S_b + (a_2+2-2m)Y_b))$ and is therefore $= 0$ if $a_1 \geq 0$, $a_2 + 2 + a_1(m-k(b)) + \geq 2(m-k(b))$.

Corollary 2.10. Let D be the divisor $a_1S + a_2Y$, and assume $a_1 \geq 0$, $a_2 \geq -2$. Then $R^i p_*(O_F(D))=0$ for $i=1,2$, $p_*(O_F(D))$ is locally free of rank equal to $ma_1(a_1 + 1) + (a_1 + 1)(a_2 + 1)$.

Proof: By the Riemann - Roch theorem and the previous considerations, $h^i(O_{F_b}(D_b))=0$ for $i=1,2$, hence for $i=0$ one obtains $h^° =$

$= \chi(O_{F_b}(D_b)) = 1 + \frac{1}{2}(D_b \cdot (D_b - K_b)) = 1 + \frac{1}{2}(a_1S + a_2Y)((a_1+2)S +$

$+ (a_2 + 2 - 2m)Y) = 1 + \frac{1}{2}[a_1(a_1+2) \cdot 2m + a_2(a_1+2) + a_1(a_2+2-2m)] =$

$= m a_1(a_1+1) + (a_1+1)(a_2+1)$.

The result follows then from the Base change theorems (cf. e.g. [9], chap. III, 12.11, page 290).

Q.E.D.

Let $g: X = F_n \longrightarrow \mathbb{P}^1$ be the canonical projection: then the tangent bundle T_X can be written as an extension of two line bundles, where T_v is the subbundle of vectors tangent to the fibres of g

(2.11) $\qquad 0 \longrightarrow T_v \longrightarrow T_X \longrightarrow g^*(T_{\mathbb{P}^1}) \longrightarrow 0.$

In the case of $X = F_b$, an easy computation gives

$T_v \cong O_{F_b}(2 S_b - 2 m Y_b)$, hence, if $L \equiv d_1 S + d_2 Y$,

then
$$h^i(T_v(-L_b)) = h^i(O_{F_b}((2-d_1)S_b - (d_2+2m)Y_b))$$
is 0, for $i=0,1$, as soon as $d_1 \geq 3$, $d_2 \geq -2m$.
As a consequence we obtain:

(2.12) $H^i(T_X(-L_b)) = 0$ if $d_1 \geq 3$, $d_2 \geq 0$, $i=0,1$.

<u>Proposition 2.13.</u> The family $F \xrightarrow{p} B$ induces the germ of the semi-universal deformation of $\mathbb{F}_n$.

<u>Proof.</u> It suffices to show that the Kodaira - Spencer map $\rho: T_{B,0} \longrightarrow H^1(\mathbb{F}_0, T_{\mathbb{F}_0})$ is an isomorphism. Let V be the vector bundle on $B \times \mathbb{P}^1$ such that $F = \mathbb{P}(V)$ (cf. (1.4)). The relative tangent bundle of p, $T_{F|B}$ fits into an exact sequence

$$0 \longrightarrow T_v \longrightarrow T_{F|B} \longrightarrow g^*(T_{\mathbb{P}^1}) \longrightarrow 0$$

g being the projection on $\mathbb{P}^1$.

Now, in concrete terms, choose an affine coordinate z on $\mathbb{P}^1 - \{\infty\}$, and then on $F|_{\mathbb{P}^1-\{\infty\}}$ we have coordinates

$(y_0, y_1, b_1, \ldots, b_{n-1}, z)$, whereas on $F|_{\mathbb{P}^1-\{0\}}$ we have coordinates $(y_0', y_1', b_1, \ldots b_{n-1}, z')$, with

hence $\quad y_0 = y_0' + y_1' \cdot \sum_{i=1}^{n-1} b_i z^{n-i} \quad$ $\begin{cases} z' = 1/z \\ y_0' = y_0 + y_1 \sum_{i=1}^{n-1} b_i z^{-i} \\ y_1' = y_1 \cdot z^{-n} \end{cases}$

Since $\rho(\frac{\partial}{\partial b_i})$ is the difference of the two liftings of $\frac{\partial}{\partial b_i}$ according to the two given coordinate patches, we obtain

(2.14) $\quad \rho(\frac{\partial}{\partial b_i}) = (y_1' \cdot z^{n-i}) \frac{\partial}{\partial y_0} = (y_1 z^{-i}) \frac{\partial}{\partial y_0}$.

These are $(n-1)$ elements in $H^1(U, T_v)$, U being the cover given by the two open sets above.

An easy computation shows that, for $b=0$, these elements are a basis of $H^1(\mathbb{F}_0, T_{\mathbb{F}_0})$.

Q.E.D.

Let now X be a smooth bidouble cover of $\mathbb{F} = \mathbb{F}_{2m}$ corresponding to the divisors L_1, L_2, L_3 and branched on the divisors D_1, D_2, D_3.
If L_i is $\equiv a_i S + b_i Y$, we shall say that X is of type (a_1, b_1) (a_2, b_2), (a_3, b_3).
Now, if $a_i \geq 3$, $b_i \geq 0$ for each i, by (2.12) $H^1(\pi^* T_{\mathbb{F}}) \cong H^1(\pi_* \pi^* T_{\mathbb{F}})$ is $\cong H^1(T_{\mathbb{F}})$.
Moreover, consider then the family $F \xrightarrow{p} B$ of deformations of $\mathbb{F}$, and, on it, the divisor $L_i \equiv a_i S + b_i Y$, $\mathcal{D}_i \equiv \frac{1}{2}((a_j + a_k)S + (b_j + b_k)Y)$: then the direct images of the associated invertible sheaves in F are locally free (and $R^i p_* = 0$ for $i \geq 1,2$).
On the other hand $p_* O_F(\mathcal{D}_i - L_i)$ is locally free if

(2.15.i) $\begin{cases} a_j + a_k - 2a_i \geq 0 \\ b_j + b_k - 2 b_i \geq 2, \end{cases}$ but also (it is then equal to zero) if

(2.15.ii) $a_j + a_k - 2a_i < 0$.

If, for each i=1,2,3, either (2.15.i) or (2.15.ii) holds, then one can choose a trivialization of $p_* O_F(\mathcal{D}_i)$, $p_* O_F(\mathcal{D}_i - L_i)$ on B.
Then one has a vector space U and, for each $b \in B$, $u \in U$, sections γ_i of $O_F(\mathcal{D}_i - L_i)$, x'_j of $O_F(\mathcal{D}_j)$: according to (2.6) one defines a family of deformations $X \xrightarrow{f} B \times U$ which, restricted to $\{0\} \times U$, gives the natural deformations of X. In view of theorem 2.7 and of proposition 2.13, the associated Kodaira - Spencer map is surjective.
Thus we get the following.

<u>Theorem (2.16)</u> Let X be a smooth bidouble cover of $\mathbb{F} = \mathbb{F}_{2m}$ of type (a_i, b_i) (i=1,2,3) with (2.15) i) or ii) holding for each i. Then the moduli space of X contains only one (unirational) irreducible component passing through X, and its dimension equals $\sum_i h°(O_{\mathbb{F}}(D_i)) + h°(O_{\mathbb{F}}(D_i - L_i)) - 6$.

<u>Proof.</u> In view of the preceding discussion the family $X \xrightarrow{f} B \times U$, which is induced by a morphism h of $B \times U \longrightarrow H^1(T_X)$ from the semi-universal deformation, is such that h is of maximal rank at the origin of the vector space $B \times U$. Therefore the semi-universal deformation has as basis an open neighbourhood of the origin in $H^1(T_X)$, moreover then

B × U dominates an affine neighbourhood of [X] in its moduli space M. The assertion regarding the dimension follows from theorem 2.7, since, by (2.12), $h^1(\pi^* T_{\mathbb{F}}) - h^0(\pi^* T_{\mathbb{F}}) = h^1(T_{\mathbb{F}}) - h^0(T_{\mathbb{F}}) = -6$.

Q.E.D.

References

[1] Beauville, A. : Letter to the Author of September 1982.

[2] Bombieri, E. : "Canonical Models of Surfaces of General Type", Publ. Scient. I.H.E.S. 42(1973), 171-219.

[3] Bombieri, E. - Husemoller, D. : "Classification and Embeddings Surfaces", Alg. Geom. Arcata 1974, Proc. Symp. Pure Math. 23, A.M.S. Providence R.I. (1975), 329-420.

[4] Castelnuovo, G. : "Sul Numero dei Moduli di una Superficie Irregolare" I, II, Rend. Acc. Lincei VII (1949), 3-7, 8-11.

[5] Catanese, F. : "On the Moduli Spaces of Surfaces of General Type", to appear in Jour. of Diff. Geom.

[6] Enriques, F. : "Le Superficie Algebriche", Zanichelli, Bologna (1949).

[7] Freedman, M. : "The Topology of Four-Dimensional Manifolds", Jour. Diff. Geom. 17,3 (1982), 357-453.

[8] Gieseker, D. : "Global Moduli for Surfaces of General Type", Inv. Math. 43 (1977), 233-282.

[9] Hartshorne, R. : "Algebraic Geometry", Springer G.T.M. 52, (1977).

[10] Hirzebruch, F. : "Topological methods in Algebraic Geometry", Grundlehren 131, Springer-Verlag, Heidelberg (3rd ed. 1966).

[11] Kodaira, K. : "On Characteristic Systems of Families of Surfaces with Ordinary Singularities in a Projective Space", Amer. J. Math., 87 (1965), 227-256.

[12] Kodaira, K. - Morrow, J. : "Complex Manifolds", Holt-Rinehart-Winston, New York (1971).

[13] Kuranishi, M. "New Proof for the Existence of Locally Complete Families of Complex Structures", Proc. Conf. Compl. Analysis, Minneapolis, Springer (1965), 142-154.

[14] Miyaoka, Y. : "On the Chern Numbers of Surfaces of General Type", Inv. Math. 42 (1977) 225-237.

[15] Mostow, G.D. : "Strong Rigidity of Locally Symmetric Spaces", Ann. of Math. Studies 78, Princeton Univ. Press, (1973).

[16] Murakami, S. : "A Note on Cohomology Groups of Holomorphic Line Bundles over a Complex Torus", in "Manifolds and Lie Groups", Progress in Mathematics, Birkhäuser, Boston (1981), 301-314.

[17] Mumford, D. : "Geometric Invariant Theory", Ergebnisse 34, Springer-Verlag, Heidelberg (1965).

[18] Noether, M. : "Anzahl der Moduln einer Classe Algebraischer Flächen", Sitzungsberichte der Akademie, Berlin, (1888).

[19] Severi, F. : "La Serie Canonica e la Teoria delle Serie Principali dei Gruppi di Punti sopra una Superficie Algebrica", Comm. Math. Helv. 4(1932), 268-326.

[20] Shafarevitch, I.R. : "Basic Algebraic Geometry", Grundlehren 213, Springer-Verlag, Heidelberg (1974).

[21] Van de Ven, A. : "Some Recent Results on Surfaces of General Type", Sem. Bourbaki 500, (Feb. 1977), 1-12.

[22] Wahl, J. : "Deformations of Plane Curves with Nodes and Cusps", Am. J. of Math., 96, (1974), 529-577.

[23] Wavrik, J.J. : "Obstructions to the Existence of a Space of Moduli", Global Analysis, Princeton Math. Series 29 (1969), 403-414.

[24] Yau, S.T. : "On the Ricci Curvature of a Compact Kähler Manifold and the Complex Monge-Ampère Equations, I", Comm. Pure and Appl. Math. 31 (1978), 339-411.

[25] Zariski, O. : "Algebraic Surfaces", Ergebnisse 61, (2nd suppl. ed.), Springer-Verlag, Heidelberg (1971).

ON A PROOF OF TORELLI'S THEOREM

Ciro Ciliberto[(*)]
Istituto Matematico "R. Caccioppoli"
Università di Napoli
80100 Napoli
Italia

INTRODUCTION

A well known theorem of Torelli states that any compact Riemann surface is determined by its principally polarized jacobian variety. Torelli's original proof (see [T]) has been revisited and adapted to any characteristic of the base field by Matsusaka (see [M]). This Author slightly modified Torelli's argument avoiding the use of some Schubert's enumerative formulas, rigourosly proved only later over $\mathbb{C}$ by Macdonald (see [MC]), and over any algebraically closed field by Ghione (unpublished). On the other hand Matsusaka's proof uses a suitable version of Castelnuovo-Humbert's theorem and a weak form of Riemann's singularity theorem.

A second proof of Torelli's theorem was given in 1952 by Andreotti (see [A1]) and adapted to any characteristic by Weil and Andreotti himself (see [A2], [W]).

There is however a third proof due to Comessatti (see [C]), which seems to have been ignored in the current literature. It is, in our opinion, a very elegant proof, and we believe it useful to give an account of it in this note. Comessatti's argument (see § 3) basically relies upon two results (theorems (1.8) and (2.5)), both somewhat hidden in the, rather obscure, original exposition. We have worked on an algebraically closed field of any characteristic. This required some care which, in spite of the plainess of the involved ideas, has made the exposition longer than we originally believed. Anyhow the proof can be made strictkingly simple in characteristic zero, avoiding the use of Castelnuovo-Humbert's theorem, present in Comessatti's argument and in our theorem (1.8). This simplification (see § 3) has been achieved by virtue of a result (theorem (9.1)) very easy to prove, but only in characteristic zero.

(*) The Author is a member of G.N.S.A.G.A. of C.N.R.

§ 1. - SOME SUBVARIETIES OF THE SYMMETRIC PRODUCT OF A CURVE RELATED TO LINEAR SERIES

Let C be a complete, non-singular curve over an algebraically closed field k. For any positive integer d, we shall denote by C(d) the d-th symmetric product of C.

Let L be a line bundle on C with first Chern class of degree n. Any $(r+1)$-dimensional vector subspace of $H^o(C,L)$ corresponds to a linear series g_n^r contained in the complete one, denoted by $|L|$. For any positive integer $d \leq n$, we shall indicate by $C(d,g_n^r)$ the closed subset of C(d) formed by all divisors $D \in C(d)$ contained in some divisor of the g_n^r. If $d \leq r$, it is $C(d,g_n^r) = C(d)$; thus in order to avoid trivial cases, from now on we shall assume $d > r$. We shall also write $C(d,L)$ instead of $C(d,|L|)$.

(1.1) Lemma. $C(d,g_n^r)$ *is a pure r-dimensional closed subset of* $C(d)$.

Proof. Let $V \subseteq H^o(C,L)$ be the vector space corresponding to the g_n^r and let $s_0,\ldots,s_r$ be a basis of V. A divisor $D = P_1 + \ldots + P_d$ belongs to $C(d,g_n^r)$ if and only if

$$\text{rank} \|s_i(P_j)\|_{\substack{i=0,\ldots,r \\ j=1,\ldots,d}} \leq r \tag{1.2}$$

where, the question being local, $s_i(P_j)$ can be viewed as elements of the field k. From (1.2) it follows that any component of $C(d,g_n^r)$ has at least dimension r. On the other hand, it is clear that any component of $C(d,g_n^r)$ has at most dimension r. Whence the assertion.

(1.3) Remark. The proof of lemma (1.1) shows that $C(d,g_n^r)$ has a natural structure of scheme, but we shall not be interested in it in what follows.

Let g_n^r be a linear series without base points on C, and let

$$\psi: C \to \Gamma \subset \mathbf{P}^r$$

be a morphism determined by g_n^r, $\Gamma = \psi(C)$ being a complete, irreducible, non-degenerate curve in $\mathbf{P}^r$. If $\pi: \Delta \to \Gamma$ is the normalization of Γ, let $\varphi: C \to \Delta$ be the unique morphism such that $\psi = \pi \circ \varphi$. If φ is not separable, there is a commutative diagram, uniquely determined up to isomorphisms

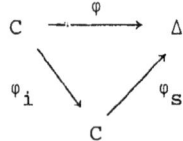

where φ_i is purely inseparable, hence a composition of Frobenius morphisms, and φ_s is inseparable. The morphism

$$\varphi_s = \pi \circ \varphi_s : C \to \Gamma \subseteq \mathbb{P}^r$$

corresponds to a linear series of dimension r on C, denoted by $(g_n^r)_s$. Any divisor of g_n^r is a multiple of a divisor of $(g_n^r)s$. If φ is separable, we put $\varphi = \varphi_s$, $g_n^r = (g_n^r)_s$, and we shall say that g_n^r is *separable*.

(1.4) Remark. If g_n^r is not separable, it is not complete. In fact g_n^r is strictly contained in a multiple of $(g_n^r)_s$.

Let g_n^r be a separable linear series. It is said to be *composite with an involution* of degree m > 1 if there exists a complete, non-singular curve C' and a separable, dominant morphism

$$f: C \to C'$$

of degree m such that g_n^r is the pull-back, via f, of a linear series g_a^r on C'. Any g_n^r, without base points, will be said to be composite with an involution if $(g_n^r)_s$ is so.

With the above notation, the triple (C,C',f) is said to be an *involution* of degree m > 1 on C. Remark that if C' $\simeq \mathbb{P}^1$ it is in fact a separable g_m^1. Given an involution (C,C',f) of degree m, a natural morphism

$$F: C' \to C(m)$$

arises, associating to each point $Q \in C'$ the geometric fibre of f over Q. F maps C' birationally onto a curve denoted by $\gamma_m^1(C,C',f)$, or simply by γ_m^1 if there is no confusion. It is fairly obvious that:
(i) γ_m^1 is not contained in the diagonal of C(m);
(ii) for any point $P \in C$ the divisor

$$C(m,P) = \{D \in C(m) : D \geq P\}$$

cuts γ_m^1 in one single point $\gamma_m^1(P)$ (namely F(f(P))).

(1.5) Remark. If C' $\subseteq$ C(m) is a curve enjoying the properties (i), (ii), then for any $P \in C$, C(m,P) and C' have multiplicity of intersection one at their common point. In fact let $D = P_1 + \ldots + P_m$ be a generic point of C', so that all P_i's are distinct. Then all C(m,P) cut C' at D, and one, at least, transversally, since all $C(m,P_i)$ are transversal at D. The assertion follows observing that $\{C(m,P)\}_{P \in C}$ is a system of algebraically equivalent divisors on C(m). Hence C' is non-singular and the dominant morphism of degree m

$$f: P \in C \to C(m,P) \cap C' \in C'$$

gives rise to an involution (C,C',f) such that $C' = \gamma_m^1(C,C',f)$. In particular F is an isomorphism between C' and γ_m^1, and γ_m^1 completely determines the involution.

Given an involution γ_m^1 on C and a positive integer $d \leq m$, we denote by $C(d,\gamma_m^1)$ the closed subset of $C(d)$ consisting of all divisors $D \in C(d)$ such that $D \in \gamma_m^1$. Clearly all irreducible components of $C(d,\gamma_m^1)$ have dimension one.

If (C,C',f), (C,C'',g) are distinct involutions on C, the first one is said to be *composite* with the latter if there is an involution (C'',C',h) such that $f = h \circ g$.

(1.6) Lemma. *Let (C,C',f), (C,C'',g) be two distinct involutions γ_n^1 and γ_m^1 an C. If there is an integer $\ell > 1$ such that $C(\ell,\gamma_n^1)$ and $C(\ell,\gamma_m^1)$ have a common component, then the two involution an both composite with a third one of degree $d \geq \ell$.*

Proof. Let d be the maximum integer such that $C(d,\gamma_n^1)$ and $C(d,\gamma_m^1)$ have a common component and let γ be this component. If P is the generic point of C, by the definition of d the greatest common divisor of $\gamma_n^1(P)$ and $\gamma_m^1(P)$ is a divisor of degree d on C which exactly coincides with the unique point cut out by $C(d,P)$ on γ. Whence γ is a γ_d^1 on C (see remark (1.5)). It is easy to see that both γ_n^1 and γ_m^1 are composite with this γ_d^1.

(1.7) Remark. We explicitly point out that the common component γ of $C(d,\gamma_n^1)$ and $C(d,\gamma_m^1)$ completely determines the involution γ_d^1 (see remark (1.5)). Further, if γ_n^1 and γ_m^1 are two involutions not composite with a third one, then $C(2,\gamma_n^1)$ and $C(2,\gamma_m^1)$ have no common component. This implies that, for P generic on C, the greatest common divisor of $\gamma_n^1(P)$ and $\gamma_m^1(P)$ is just P.

We are now able to prove the

(1.8) Theorem. *Let $g_{n_1}^r$, $g_{n_2}^r$ be two distinct complete linear series without base points on C. If there is an integer $d \geq r+1$ such that $C(d,g_{n_1}^r)$ and $C(d,g_{n_2}^r)$ have a common component, then the two series are both composite with the same involution.*

Proof. If $r = 1$ the statement is a particular case of lemma (1.6). Let now $r \geq 2$. We have two separable morphisms (see remark (1.4))

$$\psi_i: C \to \Gamma_i \subseteq \mathbb{P}^r, \quad i = 1,2$$

determined by $g_{n_i}^r$, $i = 1,2$; let

$$\pi_i: \Delta_i \to \Gamma_i, \quad i = 1,2$$

be the normalizations of Γ_i, $i = 1,2$ and let

$$\varphi_i: C \to \Delta_i$$

be the morphisms such that $\psi_i = \pi_i \circ \varphi_i$, $i = 1,2$. Let P be any point of C and let us consider the linear series

$$g_{n_i}^r(P) = \{D \in C(n_i-1): D+P \in g_{n_i}^r\}, \quad i = 1,2$$

These are linear series of dimension $r-1 \geq 1$ and degree $n_i - 1$, with, may be, fixed divisors $D_i(P)$, $i = 1,2$. Clearly $P + D_i(P)$ is just the geometric fibre of φ_i containing P, $i = 1,2$. Thus if P is generic on C and $D_1(P)$, $D_2(P)$ have a non-zero greatest common divisor, by lemma (1.6) and remark (1.7), the two involutions (C,Δ_i,φ_i), $i = 1,2$, are both composite with a third one. With this involution are also composite the two series $g_{n_i}^r$, $i = 1,2$, and we get the assertion. So let us examine the case $D_i(P)$, $i = 1,2$, are disjoint divisors, for P in a not empty Zariski open subset of C. Then, eliminating base points from $g_{n_i}^r(P)$ we get two complete linear series $g_{n_i}^{r-1}(P)$, $i = 1,2$. By induction we may assume they are both composite with the same involution $\gamma_\ell^1(P)$. Assume $\gamma_\ell^1(P)$ is rational for infinitely many points P on C. Since $g_{n_i}^{r-1}(P)$ are complete series, we have $g_{n_i}^{r-1}(P) = |(r-1)\gamma_\ell^1(P)|$, $i = 1,2$. This is impossible if φ_1, φ_2 are both birational, since $g_{n_1}^r$, $g_{n_2}^r$ are distinct. So we may assume (C,Δ_1,φ_1) to be a γ_t^1, with $t > 1$. Of course $g_{h_2}^1(P) = g_{h_1}^1(P)$ is composite with this γ_t^1 for infinitely many points P on C. If (C,Δ_i,φ_i), $i = 1,2$, were not composite with the same involution, we could chose two distinct points A,B on C such that $\varphi_1(A) = \varphi_1(B)$ and $A' = \psi_2(A) \neq \psi_2(B) = B'$. Then on the line A'B' in the $\mathbb{P}^r$ containing Γ_2 would lie infinitely many points of Γ_2, which is impossible. Then (C,Δ_i,φ_i), $i = 1,2$, are composite with the same involution, and we again get the assertion. Let us finally assume that for any point P in a not empty Zariski open subset of C, $\gamma_\ell^1(P)$ is not rational. Then it is an easy consequence of the Castelnuovo-Humbert theorem (see [M], Appendix, lemma 3) that for infinitely many points P on C, $\gamma_\ell^1(P)$ is independent of P. In this case, reasoning like above, we get a contradiction unless both (C,Δ_i,φ_i), $i = 1,2$, are composite with this γ_ℓ^1.

(1.9) Remark. In case the characteristic of k is zero it is not difficul to remove the hypothesis of completeness for the two series $g_{n_i}^r$ in theorem (1.8), and is possible to simplify the proof in some points.

§ 2. - ELEMENTARY PROPERTIES OF THE JACOBIAN VARIETY OF A CURVE

We assume, from now on, C of genus $g \geq 1$. Let $J(C)$ be the jacobian variety of C. For any positive integer d, let

$$\psi_d: C(d) \to J(C)$$

be the Abel-Jacobi map, defined with respect to a fixed base point on C. We put, as usual

$$W_d = \psi_d(C(d)), \quad \Theta = W_{g-1}$$

$$\psi_{2g-2}(K_C) = k$$

where K_C is an effective canonical divisor on C. For any $v \in J(C)$ we shall denote by T_v the *translation*

$$u \in J(c) \to u + v \in J(C)$$

and by T_v^* the *reflection*

$$u \in J(C) \to -u + v \in J(C)$$

Since clearly

$$\Theta = T_k^*(\Theta)$$

it is

$$T_v^*(\Theta) = T_{v-k}(\Theta) \tag{2.1}$$

for any $v \in J(C)$. Any divisor on $J(C)$ of the type $T_v(\Theta)$, $v \in J(C)$ is called a *theta divisor*. We also put $SC(g) = C(g, |K_C|)$: this is the locus where ψ_g, which is a birational morphism, fails to be injective. If

$$W = \psi_g(SC(g))$$

we have

$$W = T_k^*(W_{g-2}) \tag{2.2}$$

Hence W, as well as W_{g-2}, is irreducible.

Let now L be any line bundle on C with first Chern class of degree $2g - 1$. Clearly it is $C(g,L) \neq SC(g)$. Further we have the

(2.3) Lemma. (i) *If $|L|$ has no base points, then $C(g,L)$ is an irreducible divisor on $C(g)$;*
(ii) *if $|L|$ has a base point P, then $C(g,L) = SC(g) \cup C(g,P)$;*
(iii) *the image $W(L)$ via ψ_g of the irreducible component of $C(g,L)$ not*

contained in SC(g) *is a theta divisor;*

(iv) *any theta divisor is a* W(L).

Proof. If $E \in C(g-1)$ is general enough, there is only one $D \in C(g)$ such that $E + D \in |L|$. Accordingly there exists a natural birational map

$$f: C(g-1) \dashrightarrow C(g,L)$$

which is dominant onto a component of $C(g,L)$. Further, if $D \in C(g,L)$ and $D \notin SC(g)$, let $E \in C(g-1)$ be such that $D + E \in |L|$. Then D is the unique divisor such that $D + E \in |L|$, namely f is defined at E and $f(E) = D$. Thus, if there is any other component of $C(g,L)$, it has to be a component of $SC(g)$ too. This implies, by theorem (1.8), that $|L|$ has a base Point P. If this happens it is $|L(-P)| = |K_c|$; hence (i) and (ii) follow. Take now any divisor $D \in |L|$ and put $v = \psi_{2g-1}(D)$. Clearly it is

$$W(L) = T_v^*(\Theta) \qquad (2.4)$$

Since ψ_{2g-1} is surjective, v is any point of $J(C)$; thus (2.1) and (2.4) imply both (iii) and (iv).

Consider now the algebraic family of theta divisors on C, which is parametrized by an irreducible component H of the Hilbert scheme of divisors of $J(C)$; H is isomorphic to $J(C)$. Let $\Theta(W)$ be the closed subset of H formed by points of H corresponding to theta divisors containing W. We have the

(2.5) Theorem. W(L) *contains* W *if and only if* $|L|$ *has a base point. Thus* $\Theta(W)$ *is isomorphic to* C.

Proof. If $D \in C(g,L)$, then $|D| = \psi_g^{-1}(\psi_g(D)) \subseteq C(g,L)$. Hence $W(L) \supseteq W$ implies

$$C(g,L) \supseteq SC(g)$$

and, by lemma (2.3), $|L|$ has a base point. Conversely, if $|L|$ has a base point P, again by lemma (2.3), it is

$$W(L) = \psi_g(C(g,P)) \supseteq W$$

Finally the isomorphism between $\Theta(W)$ and C follows by virtue of (iii) and (iv) of lemma (2.3).

§ 3. - COMESSATTI'S PROOF OF TORELLI'S THEOREM

Let C' be another complete, non-singular curve of genus g over k. We shall denote anything concerning C' by an upper "prime".

Assume there exists an isomorphism

$$F: J(C) \to J(C')$$

taking Θ onto Θ'. Let $P \in C$, $P' \in C'$ and set $L = \mathcal{O}_c(K_c+P)$, $L' = \mathcal{O}_c(K_{c'}+P')$. We may as well assume, after having used translations, that

$$F(W(L)) = W'(L') \tag{3.1}$$

By theorem (2.5) Torelli's theorem is a consequence of the following

(3.2) Theorem. *There is either a translation or a reflection in* $J(C')$ *taking* $F(W)$ *onto* W'.

Proof. The assertion is trivially true for $g = 1,2$; thus we shall assume from now on $g \geq 3$. Suppose the theorem is not true. Then, by (2.2), it is

$$F(W) \neq T_{v'}(W') \tag{3.3}$$

$$F(W) \neq T_{v'}(W'_{g-2}) \tag{3.4}$$

for any $v' \in J(C')$. By (3.3), there exists an irreducible, (g-2)-dimensional, closed subset A of $C'(g)$, which dominates $F(W)$ via ψ'_g, and such that

$$A \not\subseteq SC'(g) \tag{3.5}$$

Lemma (2.3), (3.1) and (3.5) imply that

$$A \subseteq C'(g,P') \tag{3.6}$$

Assume now another theta divisor $W'(L'_1)$ containing $F(W)$ corresponds to a linear series with a base point $P'_1 \neq P'$. Then, like above, we get $A \subseteq C'(g,P'_1)$. Thus, by (3.7), it is

$$A \subseteq C'(g,P') \cap C'(g,P'_1) \tag{3.7}$$

Since A is irreducible and the right hand side is isomorphic to $C'(g-2)$, in (3.7) the equality holds. This is easily seen to contradict (3.4). Hence we may suppose $|L'_1|$ without base points. We put

$$B = \{D \in C'(g-1): D + P' \in A\}$$

B is a closed subset of $C'(g-1)$, isomorphic to A. Moreover it is

$$F(W) = T_{p'}(\psi'_{g-1}(B)) \tag{3.8}$$

where $p' = \psi_1(P')$. Now (3.6) implies

$$B \subseteq C'(g-1, L'_1(-P'))$$

If $|L'_1(-P')| = g^{g-2}_{2g-2}$ has an effective divisor M of degree $h > 0$ of base points, there are two cases to be considered:

(i) there is a point Q' in M such that

$$B \subseteq C'(g-1, Q')$$

(ii) $B \subseteq C'(g-1, L'_1(-P'-M))$.

Case (i) leads to a contradiction, argueing like above and taking into account (3.8). Case (ii) is also impossible. In fact $|L'_1(-P-M)| = g^{g-2}_{2g-2-h}$ is a special linear series. Therefore a point R' can be chosen in a divisor of $|0_{C'}(K_{C'}) \otimes L'_1(-P-M)^*|$. So if $v' = \psi_1(R')$, it would be

$$T_{v'}(\psi'_{g-1}(B)) \subseteq W$$

contradicting, by (3.8), (3.3). Thus $|L'_1(-P')|$ has no base points. Take now any L'_1, different from L', such that $W'(L'_i) \supseteq F(W)$, $i = 1,2$, with $L'_1 \neq L'_2$. It is

$$B \subseteq C'(g-1, L'_1(-P')) \cap C'(g-1, L'_2(-P')) \tag{3.9}$$

In force of theorem (1.8), $|L'_i(-P)| = g^{g-2}_{2g-2}(i)$, $i = 1,2$, are composite with the same involution γ^1_m, $m > 1$. Let (C,C'',f) be this involution, and $g^{g-2}_a(i)$ the linear series on C'' of which $g^{g-2}_{2g-2}(i)$ is pull-back via f, $i = 1,2$. It is

$$am = 2g - 2$$

and, since $a \geq g-2$, we get

$$\frac{m}{2} \leq \frac{g-1}{g-2} \leq 2$$

whence the only possible values for m are 2,3,4. But $m = 4$ implies $a = 1$ and $L'_1 = L'_2$, a contradiction; similarly $m = 3$ implies $g = 4$, $a = 2$, and $L'_1 = L'_2$. If $m = 2$, we should have $a = g-1$ and C'' would be elliptic. Moreover, for any $L'' \neq L'$, such that $\bar{W}'(L'') \supseteq F(W)$, $|L''(-P')|$ would be composite with this elliptic γ^1_2: for $g = 3$ this follows by (3.9) and remark (1.7), for $g > 3$ it is obvious. This also leads to a contradiciton, because it would imply that Θ(W) is birational to C", against theorem (2.5).

§ 4. - REMARKS ABOUT THE CASE char k = 0

In the proof of theorem (3.2) we use theorem (1.8), in which, in turn, we made use of Matsusaka's version of Castelnuovo-Humbert's theorem. This can be avoided if char k = 0. We want to briefly sketch here in which way it can be done.

First one proves the following weaker version of theorem (1.8).

(4.1) Theorem. *If g_n^r, $r \geq 2$, is a linear series on C, without base points, not composite with an involution, then for any integer $d \geq r+1$, $C(d,g_n^r)$ is irreducible.*

Proof. The assertion follows from the fact that the monodromy acts as the full symmetric group on the generic divisor of the g_n^r.

Theorem (1.8) can be throughly replaced by theorem (4.1) and lemma (1.6) in § 2, 3.

For instance, the proof of (i) of lemma (2.3), in which theorem (1.8) was employed, follows by theorem (4.1), since, if $|L| = g_{2g-1}^{g-1}$ has no base points, it is not composite with an involution: this can be checked by easy computations.

Only the last part of the proof of theorem (3.2) deserves some comments. We keep all notation introduced there. If $L" \neq L'$ is such that $W(L") \supseteq F(W)$, then $|L"(-P')| = g_{2g-2}^{g-2}$ has no base points and

$$B \subseteq C'(g-1, L"(-P')) \tag{4.2}$$

Since clearly $C'(g-1, L"(-P'))$ completely determines $|L"(-P')|$, in (4.2) the equality does not hold for any $L"$, and B is a common component of all $C'(g-1, L"(-P'))$. By theorem (4.1), or lemma (1.6) for $g = 3$, all linear series $|L"(-P')|$ are composite with an involution γ_m^1. One cheeks again that $m = 2,3,4$, but that for generic $L"$ only the case $m = 2$, γ_2^1 being elliptic, is a priori possible. This γ_2^1 does not depend on $L"$, since C' has finitely many automorphism. Then the conclusion is like in the proof of theorem (3.2).

All the above argument applies to the case char k > 0, provided one proves theorem (4.1). Note that one has to assume at least the g_n^r separable.

REFERENCES

[A1] A. ANDREOTTI: Recherches sur les surfaces algébriques irréguliéres, Acad. Roy. de Belgique, Cl. de Sci., 27 (1952), 3-36.

[A2] A. ANDREOTTI: On a theorem of Torelli, Am. J. of Math., 80 (1958), 801-821.

[C] A. COMESSATTI: Sulle trasformazioni hermitiane della varietà di Jacobi, Atti R. Accad. Sci. Torino, 50 (1914-15), 439-455.

[M] T. MATSUSAKA: On a theorem of Torelli, Am. J. of Math., 80 (1958), 784-800.

[MC] I.G. MACDONALD: Symmetric products of an algebraic curve, Topology, 1 (1962), 319-343.

[T] R. TORELLI: Sulle varietà di Jacobi, Rend. R. Acad. Lincei, 22 (5) (1913), 98-103.

[W] A. WEIL, Zum Beweis des Torellischen Satz, Nachrichten der Akademie der Wissenschaften in Göttingen, Math.-Phis. Klasse, 2 (1957), 33-53.

TWO EXAMPLES OF ALGEBRAIC THREEFOLDS
WHOSE HYPERPLANE SECTIONS
ARE ENRIQUES SURFACES

by

ALBERTO CONTE

Dipartimento di Matematica, Università di Torino

Algebraic three-dimensional varieties whose hyperplane sections are Enriques surfaces were studied extensively in thirties by GODEAUX and FANO. Recently, J. P. MURRE and myself have taken up this subject giving a proof according to modern standards of FANO's main theorem on this class of threefolds:

MAIN THEOREM (see [F] and [C-M]).- Let $W \subseteq P_C^N$ be a projectively normal algebraic threefold such that, for a sufficiently general hyperplane H, the hyperplane section $F = W \cdot H$ is a smooth Enriques surface. Let us denote by p the genus of the curve $C = W \cdot H \cdot H'$, where H' is a hyperplane different from H. Then, $N = p$, $\deg W = 2p - 2$ and $W = W_3^{2p-2}$ is bound to have a finite number of singular points $P_1, \ldots, P_n$.

Assume that W is not a cone, that $p \geq 6$ and that the points P_i are similar, i.e. that they all have the same properties. Then, $n = 8$ and each of the points $P_1, \ldots, P_8$ is a quadruple point with tangent cone the cone over a Veronese surface. Moreover, W carries a linear system $|\varphi|$ of Weil divisors the general member of which is a K3 surface. This system has dimension $p - 1$ and the base points of it are the P_i's, each of which is a rationale double point on a general φ. Let:

$$\lambda_{|\varphi|} : W \dashrightarrow M \subseteq P^{p-1}$$

be the rational map defined by $|\varphi|$. Then, $\lambda_{|\varphi|}$ is always birational and $M = \lambda_{|\varphi|}(W)$ spans a P^{p-1}, has degree $2p - 6$ and has K3 surfaces as (general) hyperplane sections (i.e. is a "Fano threefold" in the classical sense). Furthermore, M contains 8 planes $\pi_1, \ldots, \pi_8$ which are the images of the points $P_1, \ldots, P_8$ and has at most isolated singularities in the points common to two of the π_i's.

We will assume that no three of the π_i's meet in the same point. One can show that π_i and π_j meet if and only if the line $\overline{P_i P_j}$ lies on W. In this case the points P_i and P_j will be called "associated".

FANO claims that such threefolds exist only for p = 6, 7, 9, 13 (plus the exceptional case p = 4, giving rise to the so-called "Enriques threefold). It is however possible that a few more cases exist. In this paper we will give a geometric description of the two most interesting cases, corresponding to the values p = 6 and p = 9. For more details and complete proofs, see the second part of [C-M] and [C-V].

1.- The case p = 6

Here the variety $W = W_3^{10}$ is of degree 10 and lies in P^6. By projecting W from the plane of any three of the P_i's onto P^3, one sees that the P_i's must be two by two associated, so that the planes π_i's will meet two by two in one point.

The corresponding $M = M_3^6$ will be of degree 6, will lie in P^5 and will have canonical curve sections of genus 4, so that, form the classification of Fano threefolds, $M = Q \cdot C$ will be the intersection of a quadric and a cubic hypersurfaces of P^5 containing 8 planes two by two incident.

One can show that Q is a non-singular quadric, so that, in order to describe M and its geometry, it is convenient to identify Q with the Grassmannian $G = G(1,3)$ of lines of P^3 (so that M will be what is classically called a <u>cubic complex</u> of lines).

Remember now that a base for the integral homology ring of G is given by the Schubert cycles (where p_o, l_o, h_o are respectively a fixed point, line and plane of P^3):

$$\sigma_1(l_o) = \{x \in G \mid l_x \cap l_o \neq \emptyset\}$$
$$\sigma_2(p_o) = \{x \in G \mid l_x \ni p_o\}$$
$$\sigma_{1,1}(h_o) = \{x \in G \mid l_x \subseteq h_o\}$$
$$\sigma_{2,1}(p_o,h_o) = \{x \in G \mid p_o \in l_x \subseteq h_o\}$$

with the following intersection relations:

$$\sigma_1^2 = \sigma_2 + \sigma_{1,1}, \quad \sigma_1 \cdot \sigma_2 = \sigma_1 \cdot \sigma_{1,1} = \sigma_{2,1},$$
$$\sigma_2^2 = \sigma_{1,1}^2 = \sigma_1 \cdot \sigma_{2,1} = 1, \quad \sigma_2 \cdot \sigma_{1,1} = 0.$$

Let now R be a <u>net</u> (i.e. a 2-dimensional linear system) of quadrics of P^3. It is well known that:

$$V_o(R) = \{x \in G \mid l_x \text{ lies on some quadric } Q \text{ of } R\}$$

is a cubic complex. On the other hand, M too, as we have seen, is a cubic complex and the 8 planes on it, meeting two by two, will be of the kind $\sigma_2(p_1), \ldots, \sigma_2(p_8)$, where the p_i's are points in P^3. Moreover, through the seven points $p_1, \ldots, p_7$ there will certainly pass a net of quadrics R and, if Q is a quadric of R not containing any of the lines $\overline{p_i p_j}$, then, remembering that:

$$V(Q) = \{x \in G \mid l_x \, Q\} = 4 \, \sigma_{2,1} = c_1 + c_2 \text{ (two conics, corresponding to the two rulings of Q)},$$

one should have, on one hand:

$$M \cdot c_1 = 6,$$

whilst, on the other hand:

$M \cdot c_1 \geq 7$ (the seven lines of the ruling of Q going through $p_1, \ldots, p_7$),

so that $M = V_o(R)$ and R goes also through p_8.

We have therefore proved the following:

PROPOSITION 1.- <u>For any $M = M_3^6 = Q \cdot C \subseteq P^5$ containing 8 planes with the configuration considered, there exists a net of quadrics R of P^3 such that $M = V_o(R)$.</u>

PROPOSITION 2.- <u>M is rational</u>.

<u>Proof</u>.- Let $R = P^2$ and let D be the plane quartic curve (Hesse curve) corresponding to the cones of R. Let S be the rational double plane branched over D. Let's define a rational map:

$$\psi : M \dashrightarrow S$$

by sending the lines of $c_{1,Q}$ and $c_{2,Q}$ respectively into the 2 points of S corresponding to Q S. D is the branch locus because the two rulings coincide for the cones. The fibres of ψ are the conics $c_{i,Q}$, so that M is a conic bundle over the rational surface S. Moreover, any of the planes $\sigma_2(p_j)$ is a rational surface such that:

$$\sigma_2(p_j) \cdot c_{i,Q} = 1 \text{ point (the line of } c_{i,Q} \text{ going through } p_j \in Q).$$

Therefore, M is rational by the Enriques criterion of rationality.

Let now P be a pencil of quadrics contained into R. The base locus of P - the inetrsection of two quadrics spanning P - is a quartic elliptic space curve and:

$$V_o(P) = \{x \in G \mid l_x \text{ lies on some quadric of } P\}$$

is nothing else than the (2,6)-congruence of the chords of C, i. e.:

$V_o(P) \sim 2\sigma_2 + 6\sigma_{1,1}$.

Let now R' be any net other than R going through P. Then:

$V_o(R) \cdot V_o(R') \sim 9\sigma_1^2 = 9\sigma_2 + 9\sigma_{1,1} = (2\sigma_2 + 6\sigma_{1,1}) + (7\sigma_2 + 3\sigma_{1,1}) =$
$= V_o(P) + F$,

where $F \sim 7\sigma_2 + 3\sigma_{1,1}$.

PROPOSITION 3.- F is an Enriques surface.

Proof.- The two nets R and R', having a common pencil, will span a web S and:

$F = \{x \in G \mid 1_x \text{ lies in all quadrics of a pencil } P' \subseteq S\}$

(since 1_x lies on one quadric of R and one of R'). Therefore, F is an Enriques surface, called classically "Reye congruence" (see [B], p. 136).

We have deg F = 7 + 3 = 10 and dim $|F|$ = 6 (since there are ∞^6 nets R' ≠ R going through P) so that $|F|$ will define a rational map:

$$\lambda_{|F|} : M_3^6 \dashrightarrow W_3^{10} \subseteq P^6.$$

To identify $W = W_3^{10}$, let $\check{P}^9$ be the projective space parametrising the dual quadrics of P^3. Inside $\check{P}^9$ there is a well known filtration given by the rank:

$$\check{P}^9$$
$$\cup I$$
$$\Delta_8^4 = \{\text{quadrics of rank} \leq 3\}$$
$$\cup I$$
$$V_6^{10} = \{\text{couples of points of } P^3\}$$
$$\cup I$$
$$V_3^8 = \{\text{points of } P^3 \text{ counted twice}\}$$

Moreover, V_3^8 lies inside V_6^{10} with multiplicity 4.

Let us now remember the following classical:

DEFINITION.- A quadric of equation $\sum a_{ij} x_i x_j = 0$ and a dual quadric of equation $\sum a_{ij} u_i u_j = 0$ are siad to be apolar if:

$\sum a_{ij} \alpha_{ij} = 0$.

It is easy to see that, given a linear system S of quadrics of dimension h, the dual quadrics which are apolar to all quadrics of S make up a linear system S of dimension 8 - h. Moreover, the "couples of points" belonging to S are exactly the couples of points of P^3 which are conjugate (i. e. belong each to the polar plane of the other) with respect to all quadrics of S.

Given now $M = M_3^6 = V_0(R)$, let $\check{R}$ be the 6-dimensional linear system of dual quadrics which are apolar to all quadrics of R.

PROPOSITION 4.- $W_3^{10} = \lambda_{|F|}(M_3^6) = V_6^{10} \cdot \check{R}$.

Proof.- It is enough to remark that, by a well known property of the Reye congruence, a line $l \in F$ if and only if $l = \overline{xy}$, where x and y are points of P^3 which are conjugate with respect to all quadrics of the web $S \subseteq R$, so that $(x,y) \in \check{S} \subseteq \check{R}$ and $\check{S} \cdot W_3^{10}$ is a hyperplane section of W_3^{10}, since dim $\check{S} = 5$.

Note that W_3^{10} has 8 quadruple points in the intersection $V_3^8 \cdot \check{R}$.

2.- The case p = 9

Here the variety $W = W_3^{16}$ has degree 16 and lies in P^9. The corresponding $M = M_3^{12}$ has degree 12 and lies in P^8. As to the configuration of the 8 planes lying on M one can show the following:

PROPOSITION 4.- **The eight planes lying on M can be divided in two disjoint sets** $a_1, a_2, a_3, a_4, \beta_1, \beta_2, \beta_3, \beta_4$ **such that:**
$$a_i \cap a_j = \beta_i \cap \beta_j = \emptyset, \quad a_i \cap \beta_j = 1 \text{ point.}$$

Let now $G = G(2,8)$ be the grassmannian of 2-planes in P^8 and:
$$\xi_i = \{x \in G \mid \dim(\pi_x \cap a_i) \geq 0\}, \quad i = 1, 2, 3, 4.$$
Each ξ_i is an irreducible Schubert cycle of codimension 4 in G (dim G = 18), so that:
$$\tau_a = \bigcap_{i=1}^{4} \xi_i$$
is a 2-dimensional cycle in G.

PROPOSITION 5.-

(i) **If no three of the a_i's are included in a hyperplane of P^8, then** $\tau_a \simeq P^2$.

(ii) **Let:**
$$V_a = \bigcup_{x \in \tau_a} \pi_x \subseteq P^8.$$
Then, V_a **is an irreducible non-singular algebraic variety of dimension 4 and degree 6.**

(iii) **Under the above hypotheses:**
$$V_a = V_\beta = \sigma(P^2 \times P^2),$$
where $\sigma: P^2 \times P^2 \longrightarrow P^8$ **is the Segre embedding.**

Let now $M = M_3^{12}$ be a Fano variety containing 8 planes having the above configuration. Then:

PROPOSITION 6.-

(i) For every $\pi_x \subseteq V_\alpha$, $M \cap \pi_x$ is a curve.

(ii) There exists a quadric hypersurface $Q \subseteq P^8$ such that $M = Q \cap V$.

Proof.- The proof of (i) is rather long and complicated and relies upon properties of the K3 surfaces which are the hyperplane sections of M.

From (i) it follows that $M \subseteq V_\alpha = V$, so that, in particular, M must be of type (p,q) on $P^2 \times P^2$. Since:

$12 = \deg M = \frac{1}{2}(p + q) \deg V = 3(p + q)$,

the only possibilities for (p,q) are:

(0,4), (1,3), (2,2).

The first is ruled out because M intersects both families of planes on V, the second because M intersects them in points which are not all on a line, so that M is of type (2,2), i. e. $M = Q \cap V$.

It is moreover possible to construct two quadrics of P^5 with equations of type:

$Q_1: A_1(y_0,y_1,y_2) + B_1(y_3,y_4,y_5) = 0$
$Q_2: A_2(y_0,y_1,y_2) + B_2(y_3,y_4,y_5) = 0$

such that, if $Y = Q_1 \cdot Q_2$ and $X = Y/i$, where i is the involution defined by:

$i[(y_0, \ldots, y_5)] = (-y_0,-y_1,-y_2,y_3,y_4,y_5)$, there are two suitable embeddings:

$$\begin{array}{c} X \\ \swarrow \quad \searrow \\ P^9 \supseteq W_3^{16} \dashrightarrow M_3^{12} \subseteq P^8 \end{array}$$

Here the hyperplane sections of W_3^{16} are Enriques surfaces (see [B], pp. 135-36) obtained modulo i from the intersection $Q_1 \cdot Q_2 \cdot Q_3$, where Q_3 is a third quadric of P^5 of equation:

$Q_3: A_3(y_0,y_1,y_2) + B_3(y_3,y_4,y_5)$.

REFERENCES

[B]   A. BEAUVILLE, Surfaces algébriques complexes, Astérisque n. 54, Société Mathématique de France 1978.

[C-M] A. CONTE and J. P. MURRE, Three-dimensional algebraic varieties whose hyperplane sections are Enriques surfaces, Institut Mittag-Leffler, Report n. 10, 1981.

[C-V] A. CONTE and A. VERRA, Varietà algebriche tridimensionali immerse in P^9 le cui sezioni iperpiane sono superficie di Enriques, to appear.

[F] G. FANO, Sulle varietà algebriche a tre dimensioni le cui sezioni iperpiane sono superficie di genere zero e bigenere uno, Mem. Soc. It. XL, s. 3^a, t. XXIV (1938), 41-66.

On the Brill-Noether Theorem

D. Eisenbud and J. Harris

The purpose of this note is to give a short, self-contained proof of the Brill-Noether theorem:

Theorem (1): Let C be a general curve of genus g, and suppose that C possesses a linear system of degree d and dimension r. Then

$$\rho = g - (r+1)(g-d+r) \geq 0 \ .$$

This was originally proved in [G-H], and more recently in [E-H]; the converse was established earlier in [K-L I], [K-L II] and [K].

As with all existing proofs, the approach here will be to study the behavior of linear series on a family of curves degenerating to a singular and/or reducible curve. We introduce our family here:

(2) <u>Notational conventions</u>: For the remainder of this paper, $\mathcal{O}$ will be a discrete valuation ring with parameter t, $T = \text{Spec } \mathcal{O}$ its spectrum, and 0 and η the closed and generic points of T respectively. $\pi : X \longrightarrow T$ will be a flat, projective family with total space X smooth, and central fiber $X_0 = \pi^{-1}(0)$ the reduced curve pictured in fig. 1.

Our object is to prove theorem (1) specifically for the geometric general fiber $X_{\bar{\eta}} = X \times_T \text{Spec } \overline{k(\eta)}$ of X; since families X exist for all genera g (see [W]), and since the non-existence of linear series of given degree and dimension is an open condition among smooth curves, this will suffice to prove Theorem (1). We first observe that any line bundle L on $X_{\bar{\eta}}$ is defined over some finite extension of $k(\eta)$. But if we make any finite base change $T' \longrightarrow T$ and minimally

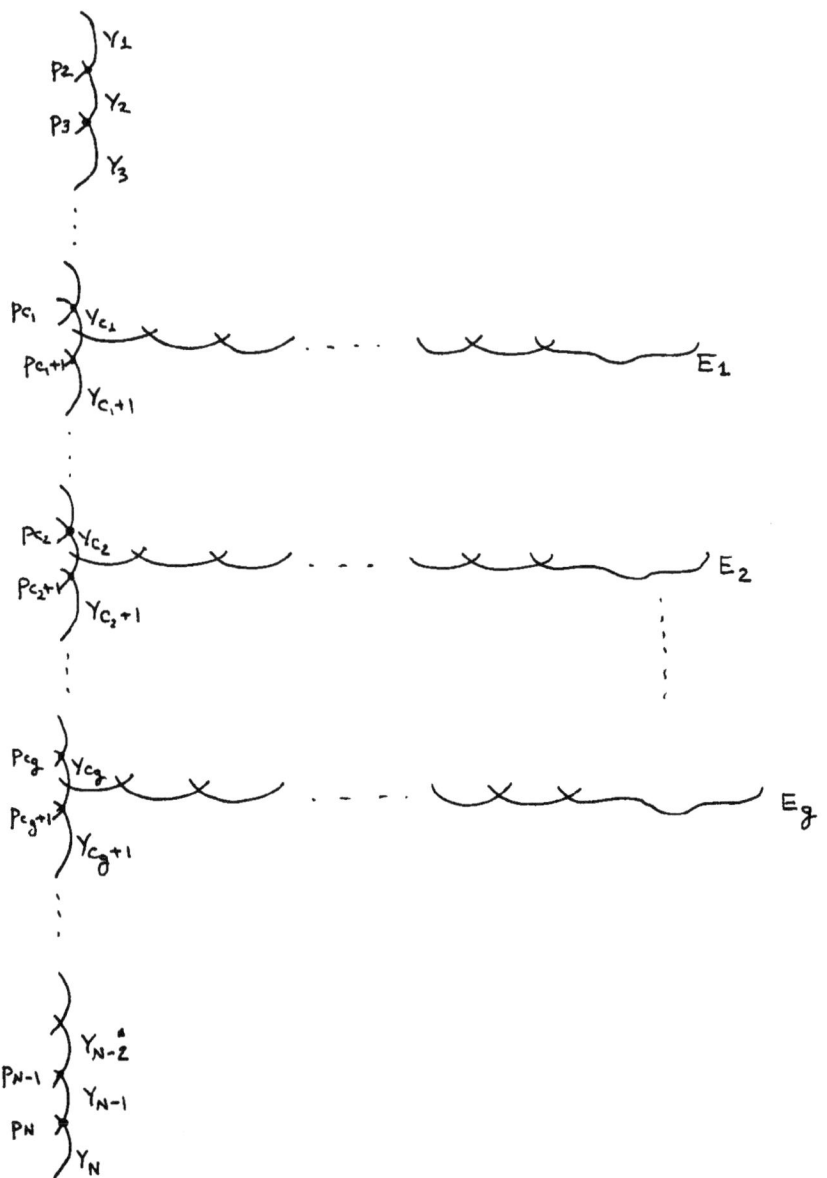

Fig. 1

Components are smooth, and intersect transversally as shown. The E_i are elliptic; all others are rational.

resolve the singularities of $X' = X \times_T T'$, we find that $X' \longrightarrow T'$ is again a family of the same form as X; thus we may assume L is defined over $k(\eta)$. Moreover, since the total space of X is smooth, any line bundle on X_η extends to one on X. Thus, Theorem (1) will follow once we establish

<u>Theorem (3)</u>: Let $X \longrightarrow T$ be as in (2), and let L be any line bundle on X; let d be the relative degree of L and $r+1 = \text{rank}(\pi_* L)$. Then
$\rho = g - (r+1)(g-d+r) \geq 0$.

To prove (3) we consider the limiting behavior of a linear series on X as follows: since the intersection pairing among components of X_0 is unimodular, for each component Y of X_0 there exists a unique line bundle L_Y on X agreeing with L on X_η and such that L_Y has degree d on Y, 0 on all other components of X_0. We define V_Y to be the linear series

$$V_Y = (\pi_* L_Y) \otimes k(0) \subset H^0(X_0, L_Y) \subset H^0(Y, L_Y),$$

the last inclusion coming from the fact that any section of L_Y vanishing on Y vanishes on X_0. Since the V_Y are all limits of the same linear series $(\pi_* L_Y) \otimes k(\eta)$, it is reasonable to expect that they satisfy some compatibility conditions; and indeed, once we establish those conditions Theorem 3') will follow immediately. These conditions may be expressed as follows: for any point $p \in Y$, we define the <u>vanishing sequence</u> $a_0(V_Y, p) < \ldots < a_r(V_Y, p)$ of V_Y at p to be the $(r+1)$ distinct orders of vanishing of sections $\sigma \in V_Y$ at p; in particular, for each $\ell = 2, \ldots, N$ we let $a_0^\ell < \ldots < a_r^\ell$ be the vanishing sequence of the series V_{Y_ℓ} at the point p_ℓ (cf. fig. 1). Our basic condition is then

(4) (i) For all ℓ and i,

$$a_i^{\ell+1} \geq a_i^{\ell} \; ; \text{ and}$$

(ii) If $\ell = c_j$ for some j, then for all but at most one value of i,

$$a_i^{\ell+1} > a_i^{\ell} .$$

We note that Theorem (3) follows immediately from (4): trivially, we have, for any ℓ, $i \leq a_i^{\ell} \leq d-r+i$, so that altogether

$$(r+1)(d-r) \geq \sum_i a_i^N - a_i^2$$

$$= \sum_{i,\ell} a_i^{\ell+1} - a_i^{\ell}$$

$$\geq rg$$

and hence $\rho = (r+1)(d-r) - rg \geq 0$.

We begin the proof of (4) with two lemmas. Both refer to a pair of components Y, Z of X_0, meeting at a point p with p' another point of Y, as in Fig. 2. In this situation, $X_0 - \{p\}$ has two connected components; we will

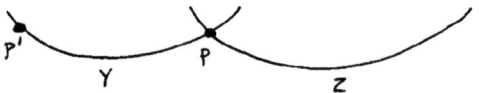

Fig 2

denote by E the divisor on X consisting of the sum of the curves in X_0 in the connected component containing Z. In particular, we have then

$$L_Z \cong L_Y(-dE) \ ;$$

we accordingly regard L_Z as a subsheaf of L_Y and π_*L_Z as a submodule of π_*L_Y. Finally, for any element $\sigma \in \pi_*L_Y$ we will write $\mathrm{ord}_{p,Y}(\sigma)$ for the order of vanishing of the corresponding section of L_Y along Y. With these conventions, then, we have

<u>Lemma 5.</u> There exists a basis $\sigma_0, \ldots, \sigma_r$ of π_*L_Y such that

i) for suitable integers $\alpha_i \geq 0$ the set $t^{\alpha_i}\sigma_i$ is a basis for π_*L_Z;

and ii) the orders $\mathrm{ord}_{p',Y}(\sigma_i)$ are all distinct.

<u>Proof:</u> The matrix expressing the inclusion of free $\mathcal{O}$-modules $\pi_*L_Z \longrightarrow \pi_*L_Y$ may be diagonalized over $\mathcal{O}$ by applying Gaussian elimination to its rows and columns; this procedure yields a basis $\sigma_0, \ldots, \sigma_r$ of π_*L_Y satisfying (i). Now, if $g \in \mathcal{O}$ and $\alpha_i \geq \alpha_j$, then i) will still hold if we replace σ_i by $\sigma_i + g\sigma_j$; these transformations suffice for passing to a basis satisfying (ii) as well.

<u>Lemma 6.</u> If $\sigma \in \pi_*L_Y - t \cdot \pi_*L_Y$ and $\tau = \tau^\alpha \cdot \sigma \in \pi_*L_Z - t\pi_*L_Z$, then we have

$$\mathrm{ord}_{p',Y}(\sigma) \leq d - \mathrm{ord}_{p,Y}(\sigma) \leq \alpha \leq \mathrm{ord}_{p,Z}(\tau)$$

<u>Proof:</u> The first inequality is trivial (but is the key to (7)(iii) below). For the second inequality, observe that since $\tau^\alpha \sigma \in \pi_*L_Z$, the divisor

$$\alpha X_0 + (\sigma) = (t^\alpha \sigma) \geq dE \ ;$$

thus $(\sigma) \geq (d-\alpha)E$ and correspondingly $\mathrm{ord}_{p,Y}(\sigma) \geq d-\alpha$. Likewise, for the last inequality we see that since $t^{-\alpha}\tau \in \pi_*L_Y$,

$$-\alpha X_0 + (\tau) = (t^{-\alpha}\tau) \geq -dE$$

so $(\tau) \geq \alpha(X_0 - E)$ and hence $\text{ord}_{p,Z}(\tau) \geq \alpha$.

Combining lemmas 5 and 6, we have

Lemma 7. With , ,p and p' as in Fig. 2.,

i) $a_i(V_Y, p) + a_{r-i}(V_Z, p) \geq d$

ii) $a_i(V_Z, p) \geq a_i(V_Y, p')$; and

iii) $a_i(V_Z, p) = a_i(V_Y, p')$ for more than one value of i only if there are two or more independent sections of V_Y vanishing only at p and p'.

(In fact, we conclude from Lemmas 5 and 6 that $a_i(V_Z, p) \geq a_{\rho(i)}(V_Y, p')$ for some permutation ρ of $\{0, \ldots, r\}$, and hence that $a_i(V_Z, p) \geq a_i(V_Y, p')$; and similarly for parts (i) and (iii)).

Part (i) of (4) follows immediately from (7)(ii), applied to $Y = Y_\ell$, $Z = Y_{\ell+1}$, $p = p_{\ell+1}$ and $p' = p_\ell$. Part (ii) of (4), and thereby Theorem (3), will follow similarly from 7(iii) once we establish

Lemma 8. If $\ell = c_m$, there is at most one section $\sigma \in V_{Y_\ell}$ non-zero on $Y_\ell - \{p_\ell, p_{\ell+1}\}$.

Proof: Label the components of X_0 between $Y = Y_\ell$ and $E = E_m$ as in Fig. 3:

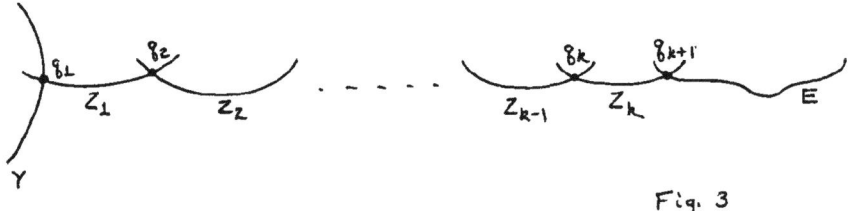

Fig. 3

Suppose there are two independent sections $\sigma, \tau \in V$ vanishing only at p_ℓ and $p_{\ell+1}$. The pencil they span will be totally ramified at p_ℓ and $p_{\ell+1}$ and unramified elsewhere; in particular, there will exist sections $\sigma^\circ, \tau^\bullet \in V$ vanishing to orders exactly 0 and 1 at q_1. Applying (7)(i) once and (7)(ii) k times, then, we have

$$a_0(V_Y, q_1) = 0, \ a_1(V_Y, q_1) = 1$$
$$\Rightarrow a_r(V_{Z_1}, q_1) = d, \ a_{r-1}(V_{Z_1}, q_1) = d-1$$
$$\Rightarrow a_r(V_{Z_2}, q_2) = d, \ a_{r-1}(V_{Z_2}, q_2) = d-1$$
$$\vdots$$
$$\Rightarrow a_r(V_{Z_k}, q_k) = d, \ a_{r-1}(V_{Z_k}, q_k) = d-1$$
$$\Rightarrow a_r(V_E, q_{k+1}) = d, \ a_{r-1}(V_E, q_{k+1}) = d-1 \ .$$

But this is absurd; a pencil of degree d on an elliptic curve can't have d-1 base points.

References

[E-H]: D. Eisenbud and J. Harris: Divisors on general curves and cuspidal rational curves, to appear. Invent. Math.

[G-H]: P. Griffiths and J. Harris: On the variety of special linear systems on an algebraic curve. Duke Math J. 47(1980) p. 233-272.

[K]: G. Kempf: Schubert methods with an application to algebraic curves, Publ. Math. Centrum, Amsterdam, 1971.

[K-L I]: S. Kleiman and D. Laksov: On the existence of special divisors. Amer. J. Math 94(1972) p. 431-436.

[K-L II]: S. Kleiman and D. Laksov: Another proof of the existence of special divisors. Acta Math. 132(1974) p. 163-176.

[W]: G. Winters: On the existence of certain families of curves. Amer. J. Math 96(1972) p. 215-228.

Properties of Arakelov's Intersection Product

by Gerd Faltings

1. Introduction

In Izv. Akad. Nauk. SSSR <u>38</u> (1974), Arakelov introduced an intersection product on the divisors of an arithmetic surface. More precisely, he deals with the following situation: K is a number-field, A its ring of integers,

$$\pi: X \to \text{Spec}(A)$$

a semistable family of curves. An Arakelov-divisor on X is a sum $D = D_f + D_\infty$, where D_f is usual divisor on X, D_∞ is a formal sum $D_\infty = \sum_{v \in S_\infty} r_v F_v$, ($r_v \in \mathbb{R}$, S_∞ = infinite places of K). F_v stands for "the fibre of X at the infinite place v." For a meromorphic function $0 \neq f \in K(X)$, a divisor

$$(f) = (f)_f + (f)_\infty$$

can be defined as follows:

$(f)_f$ = usual divisor of f on X

$$(f)_\infty = \sum_{v \in S_\infty} r_v F_v ,$$

with

$$r_v = -\int_{X(\bar{K}_v)} \log \|f\|_v \, d\mu$$

where:

K_v = local field at v = $\mathbb{R}$ or $\mathbb{C}$.

$X(\bar{K}_v)$ = Riemann-surface defined by $X(\bar{K}_v = \mathbb{C})$.

$$\|f\|_v = \begin{cases} |f|, & K_v = \mathbb{R} \\ |f|^2, & K_v = \mathbb{C} \end{cases}$$

$$d\mu = -\frac{1}{2\pi i} \sum_{k=1}^{g} \omega_k \wedge \bar{\omega}_k,$$

with $\omega_1, \ldots, \omega_g$ an ON-basis of $\Gamma(X(\bar{K}_v), \Omega_X^1)$.

(scalar product $\langle \omega, \omega' \rangle = -\frac{1}{2\pi i} \int_{X(\bar{K}_v)} \omega \wedge \bar{\omega}'$,

g = genus (X), for g=0 the definition has to be modified).

Then an intersection-product can be defined on the Arakelov-divisors, such that a divisor (f) is perpendicular to any other: It is clear how to define the intersection of a fibre of π with a divisor, and what is essentially left is the intersection-product $\langle D, E \rangle$, where D and E are two different sections of $X(K)$. This is given by

$$\langle D, E \rangle = \sum_v \langle D, E \rangle_v \quad \text{(sum over all places of } K\text{)}$$

where for a finite place $\langle D, E \rangle_v$ measures the usual intersection-multiplicity of D and E in the points of the fibre of π over v, and for $v \in S_\infty$ $\langle D, E \rangle_v$ is given by

$$\langle D, E \rangle_v = -\log G(P, Q),$$

where $P, Q \in X(\bar{K}_v)$ are the points defined by D, E, and $G(P,Q)$ is a "Green's-function" on $X(\bar{K}_v)$. $G(P,Q)$ can be used to define a hermitian metric on $O(P)$ (functions with poles in P) by

$$G(P,Q) = \|1\|^2 (Q) \ .$$

In this way we can define a hermitian metric on $O(D)$, for any divisor D on a Riemann-surface Y, and for an arithmetic X as above an Arakelov-divisor $D = D_f + D_\infty$ defines a line-bundle $O(D)$, which is just $O(D_f)$ with a hermitian metric at the infinite places.

We can show that the Hodge-index theorem is true for Arakelov's intersection product, and for this we need a Riemann-Roch theorem. They will be stated in the next chapter, and the last one contains sketches of the proofs.

2. Summary of results

We start with a Riemann-surface Y. For each divisor D on Y, we have defined a hermitian metric on $\mathcal{O}(D)$. For a complex vector-space V we define

$$\lambda(V) = \wedge^{\dim(V)} V \ .$$

A hermitian metric on $\lambda(V)$ is nothing else than a volume-form on V. For a divisor D on Y, we let.

$$\lambda(\mathbb{R}\Gamma(Y,\mathcal{O}(D))) = \lambda(H^0(Y,\mathcal{O}(D))) \otimes \lambda(H^1(Y,\mathcal{O}(D)))^{-1}$$

Theorem 1:

There are unique (up to scalar factors) metrics on the $\lambda(\mathbb{R}\Gamma(Y,\mathcal{O}(D)))$, such that
a) any isometry between the $\mathcal{O}(D)$'s induces an isometry on the $\lambda(\mathbb{R}\Gamma)$'s

b) If $E = D-P$, then the sequence

$$0 \to H^0(Y,\mathcal{O}(E)) \to H^0(Y,\mathcal{O}(D)) \to \mathcal{O}(D)[P] \to H^1(Y,\mathcal{O}(E)) \to H^1(Y,\mathcal{O}(D)) \to$$

is "volume-exact"

($\mathcal{O}(D)[P]$ = fibre of $\mathcal{O}(D)$ at P. This has a hermitian metric).

Remark: D. Quillen can define another volume via "analytic torsion", which hopefully will generalize our volume.

If now $X/\text{Spec}(A)$ is again an arithmetic surface, and D an Arakelov-divisor on X, we have defined a volume-form on

$$\mathbb{R}\Gamma(X,\mathcal{O}(D)) \otimes_{\mathbb{Z}} \mathbb{R} = \prod_{v \in S_\infty} \mathbb{R}\Gamma(X(\bar{K}_v),\mathcal{O}(D)) ,$$

and we define

$$\chi(D) = \chi(H^0(X,\mathcal{O}(D))) - \chi(H^1(X,\mathcal{O}(D))) ,$$

where for any A-module M with a volume form on $M \otimes_{\mathbb{Z}} \mathbb{R}, \chi(M)$ is defined as

$$\chi(M) = -\log[\text{vol}(M \otimes_{\mathbb{Z}} \mathbb{R}/M)/\#(\text{torsion of } M)]$$

Theorem 2: (Riemann-Roch)

$$\chi(D) = \frac{D \cdot (D-K)}{2} + \text{constant} \quad (K=\text{canonical class})$$

Theorem 3: (Hodge index-theorem)

The intersection form has signature $+,-,-,-,\ldots,-$, where the number of $-$ -signs is

(rank of the Mordell-Weil group $\text{Jac}_X(K)$)

$+ \sum_v (\#\text{components of } F_v - 1)$

We can furthermore compare our volume with mor geometric volumes: For this let Y be a Riemann surface.

Theorem 4:

For any $\epsilon > 0$ there exists a d_0, such that for a divisor D of degree $d \geq d_0$,

$$\text{vol}(\{f \in \Gamma(Y, \mathcal{O}(D))) \mid \int_Y \|f\|^2 \, d\mu \leq 1\}) \geq e^{-\epsilon d^2}$$

For arithmetic X, this leads to

Theorem 5:

If D is an Arakelov-divisor with $D^2 > 0$ and $D \cdot F > 0$ (F=fibre), then $n \cdot D$ is effective for $n \gg 0$.

As a corollary, we obtain that for $g > 1$ the relative dualizing sheaf ω_X satisfies $\omega_X^2 \geq 0$, and $\omega_X \cdot D \geq 0$ for any effective divisor D.
If we could bound ω_X^2 in terms of K, g, and the set of bad places, we could derive the Mordell-conjecture

3. Sketches of proofs

Theorem 1:

Property b) tells us how to define the volume. For property a) we can reduce to $\deg(D) = g-1$. If Div_{g-1} denotes the divisors of degree $g-1$ on Y (or more precisely a big subfamily), we have a natural map

$$\text{Div}_{g-1} \longrightarrow \text{Jac}(Y)$$

into the Jacobian of Y.

The $\lambda(\mathbb{R}\Gamma(Y, \mathcal{O}(D)))$'s form a line-bundle on Div_{g-1}, which is the pullback of $\mathcal{O}(-\theta)$ on $\text{Jac}(Y)$ (θ = theta-divisor). We have defined a metric on this bundle, and we want to show that it is the pullback of a metric on $\mathcal{O}(-\theta)$. This is done by calculating its curvature, which indeed comes from $\text{Jac}(Y)$.

Theorem 2:

Just follow the usual proof of R.R..

Theorem 3:

If F denotes a fibre of π, we show that $D \cdot F = 0$ implies $D^2 \leq 0$. It is standard that this is true if D consists of components of fibres of π, and so we may assume that D is perpendicular to all such components, so that $\mathcal{O}(D) \in \text{Pic}^o(X)$.

After replacing D by an integral multiple, we can find a divisor E with $(E \cdot F) = g-1$, such that $H^0(\mathcal{O}(E+nD)) \otimes_A K = 0$ for $n \in \mathbb{Z}$. This means that $H^0(\mathcal{O}(E+nD))$ vanishes, and $H^1(\mathcal{O}(E+nD))$ consists only of torsion. Thus $\lambda(\mathbb{R}\Gamma(X,\mathcal{O}(E+nD))) = \mathbb{C}$, and the volume on this vectorspace is just a number. As this comes from a hermitian metric on $\mathcal{O}(-\theta)$, this number is bounded below. Thus

$$\chi((E+nD)) = -\log(\mathrm{vol}(\lambda(\mathbb{R}\Gamma(X,\mathcal{O}(E+nD))))) - \log(\#\text{torsion of } H^1(X,\mathcal{O}(E+nD)))$$

is bounded above. By Riemann-Roch $D^2 \leq 0$. More precisely, it can be derived that up to a factor D^2 is given by the negative of the Néron-Tate height of D in the Mordell-Weil-group.

Theorem 4:

If vol_D denotes the left side, we get an estimate

$$\frac{1}{\mathrm{vol}_D} \leq \text{constant} \int_{X^d} \prod_{i \neq j} G(P_i, P_j) \cdot d\mu(P_1) \ldots d\mu(P_d)$$

It can be shown that the integrand is bounded by $\text{constant} \cdot e^{\varepsilon d^2}$. For this we use that $\log G(P,Q)$ is a Green's function for the $\partial\bar{\partial}$-operator.

Theorem 5:

Apply Minkowski's theorem to $\Gamma(X, \mathcal{O}(n \cdot D))$ to obtain an element

$$0 \neq f \in \Gamma(X, \mathcal{O}(n \cdot D))$$

with

$$\int_{X(\bar{K}_v)} \|f\|^2 \, d\mu \leq 1$$

for $v \in S_\infty$,

and hence also

$$\int_{X(\bar{K}_v)} \log \|f\| \, d\mu \leq 0 \ .$$

Then $nD + (f) \geq 0$.

The corollaries are obtained just as for algebraic surfaces.

Prof. Dr. G. Faltings

Universität Gesamthochschule Wuppertal

Fachbereich 7 - Mathematik -

Gaußstr. 20

5600 Wuppertal 1

W-Germany

ON NODAL CURVES

William Fulton*

The celebrated Severi problem on the variety of irreducible nodal curves remains a major open question in algebraic geometry. At this conference Arbarello has discussed some recent progress on this problem [3]. Our main purpose is to discuss its relation to other problems, some of which have been solved.

§1. Severi's problem

The plane curves of given degree n form a projective space $\mathbb{P}^N$ of dimension $N = n(n+3)/2$. For any $0 \leq d \leq n(n-1)/2$, let $V(n,d)$ be the set of plane curves of degree n with precisely d nodes and no other singularities. The following facts can be proved by methods outlined by Severi [16], cf. Van der Waerden [18], Alibert - Maltsiniotis [2], Tannenbaum [17], or Nobile [13]:

(1) $V(n,d)$ is a locally closed submanifold of $\mathbb{P}^N$ of codimension d.

(2) The (imbedded) tangent space to $V(n,d)$ at a point $[C]$ corresponding to a curve C is canonically isomorphic to the space of curves of degree n which pass through the nodes of C.

(3) If $d > e$, then $V(n,d)$ is contained in the closure of $V(n,e)$. The branches of $\overline{V(n,e)}$ through a point $[C]$ in $V(n,d)$ are non-singular and correspond to choices of e "assigned" points among the d nodes of C. The other $d-e$ "virtual" nodes disappear when one deforms from $[C]$ to the corresponding branch of

---

* Research partially supported by the Sloan Foundation and the National Science Foundation.

$V(n,e)$.

(4) <u>If</u> $[C] \in V(n,d)$, <u>and</u> S <u>is a set of</u> e <u>nodes of</u> C, <u>the curves</u> C' <u>with</u> $[C']$ <u>in the corresponding branch of</u> $\overline{V(n,e)}$ <u>have k irreducible components, where</u> k <u>is the number of connected components of the complement of</u> S <u>in</u> C.

Locally at $[C] \in V(n,d)$ the varieties $\overline{V(n,e)}$ for $e < d$ form the familiar normal crossing picture of coordinate subspaces of $\mathbb{C}^N$ containing a given coordinate $(N-d)$-plane. The essential point for (1) - (3) is the fact that the nodes of C put independent conditions on the curves of degree n, which follows from Riemann - Roch. For (4) one applies the Enriques - Zariski connectedness theorem to the blow-ups of the curves in a deformation along the deformations of the e "assigned" nodes.

In particular, the existence of irreducible nodal curves of degree n with d nodes, for any $d \leq (n-1)(n-2)/2$, follows from (4), by assigning d nodes appropriately to n lines in general position.

For plane quartics, the possible nodal curves are indicated in the following diagram. Each irreducible component of each $V(4,d)$ is represented by a corresponding nodal curve. The circled "virtual" nodes indicate deformations from $V(4,d)$ into $V(4,d-1)$.

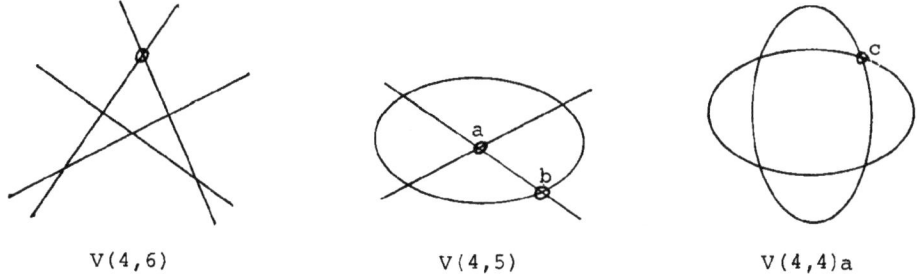

V(4,6)　　　　　　　　V(4,5)　　　　　　　　V(4,4)a

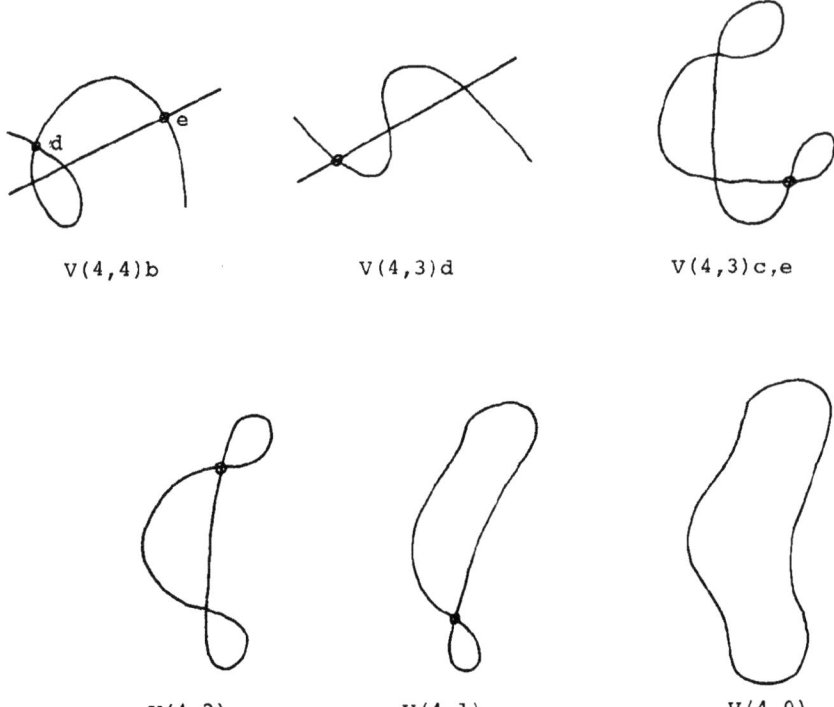

V(4,4)b   V(4,3)d   V(4,3)c,e

V(4,2)   V(4,1)   V(4,0)

Let $V(n,d)^* = \{[C] \in V(n,d) \mid C$ is irreducible$\}$. The <u>Severi problem</u> is to show that $V(n,d)^*$ is connected (irreducible). By analyzing the variety $V(n,n(n-1)/2)$ of lines in general position, in combination with (1) - (4), one sees that this problem is equivalent to showing that the closure of any component of $V(n,d)^*$ must meet $V(n,n(n+1)/2)$. By induction on d it would suffice to show that any irreducible nodal curve of degree n can be degenerated to a nodal curve with more nodes.

In fact, however, no one has proved that plane nodal curves must have <u>any</u> non-trivial degeneration, i.e. a degeneration which is not shared by all other plane curves. The extent of our ignorance may be underscored as follows. Let $g - (n-1)(n-2)/2 - d$, and let $\tau : V(n,d)^* \longrightarrow M_g$ be the canonical map to the moduli space M_g of non-singular curves of genus g.

Challenge. If V is an irreducible component of $V(n,d)^*$, show that $\tau(V)$ is not contained in a complete (compact) subvariety of M_g.

Although M_g can contain complete subvarieties, one does not know how large their dimension can be.

§2. Irreducibility of M_g

One of Severi's reasons for proving $V(n,d)^*$ irreducible was to give a proof of the irreducibility of M_g using methods of algebraic geometry. Indeed, for large n, τ maps $V(n,d)^*$ onto M_g. (Conversely, the irreducibility of M_g has been used in proofs that $V(n,d)^*$ is irreducible for sufficiently large n.)

Deligne and Mumford [5] constructed a compactification $\bar{M}_g$ of M_g, whose boundary points correspond to stable curves whose irreducible components have geometric genus less than g. They show that the irreducibility of M_g is equivalent to the assertion that any non-singular curve of genus g can be degenerated to a stable curve. Until recently even this simple consequence of Severi's assertion had no proof within algebraic geometry - although of course there are well-known topological and analytic proofs of the irreducibility of M_g.

Thanks to a beautiful construction of Harris and Mumford [10], one can now give a simple proof that such degenerations always exist. In place of the space of node curves, one may consider the Hurwitz space $H_{n,b}$ of n-sheeted coverings of $\mathbb{P}^1$, with $b = 2g + 2n - 2$ simple branch points, which is an étale covering space of the space of b-tuples of points in $\mathbb{P}^1$ (modulo automorphisms of $\mathbb{P}^1$). Harris and Mumford construct a compactification of $H_{n,b}$ whose points consist of stable curves C, together with branched coverings $C \longrightarrow D$, as above, where each component of D has at most three points over which ramification may occur, and at least one of these ramification points must be simple. It follows from the Riemann - Hurwitz formula that each component of C must be rational, which concludes the proof. For details we refer to [8].

It is tantalizing to compare this proof with Severi's attempt to prove $V(n,d)*$ irreducible by degenerating a curve to n lines through a point. In fact, if one naively lets the branch points of a branched covering of $\mathbb{P}^1$ come together, one also arrives at a configuration of n copies of $\mathbb{P}^1$ meeting at a point. One may hope that a clever compactification of the spaces $V(n,d)*$ may similarly be used to rescue Severi's argument.

§3. Zariski's problem and multiple planes.

Zariski [9] deduced that the fundamental group $\pi_1(\mathbb{P}^2 - C)$ is abelian for any plane nodal curve C, from Severi's assertion that C can be degenerated to lines in general position. Special cases of this were proved, without appealing to Severi's assertion, by Abhyankar [1] and Prill [15]. Zariski's assertion was proved for the algebraic fundamental group in [7] and in general by Deligne [4]. What was needed beyond methods of [1] and [15] were new "connectedness theorems", which generalize classical results of Bertini; for this we refer to the original papers and [9]. Another proof of Nori [14] is discussed in §4.

When C is an irreducible node curve of degree n defined by an equation $f(x,y) = 0$, the assertion that $\pi_1(\mathbb{P}^2 - C)$ is abelian means that $\pi_1(\mathbb{P}^2 - C)$ is cyclic of order n, or that the complement of the ramification curve in the surface $z^n = f(x,y)$ is simply connected.

For any plane curve C with equation $f(x,y) = 0$, and any integer $k \geq 2$, let $Y_k \longrightarrow \mathbb{P}^2$ be the finite cyclic k-sheeted covering corresponding to the equation $z^k = f(x,y)$; Y_k is taken to be normal, and the ramification occurs over C and perhaps over the line at infinity. Let $Z_k \longrightarrow Y_k$ be a resolution of singularities of Y_k.

Zariski [20] used the first betti number of Z_k as a subtle invariant of C. He showed that this betti number could distinguish families of curves of the same degree with the same number of equivalent singularities. He also used this idea to show that certain invariants allowed by Plücker's equations cannot be realizable by any plane curve. Esnault [6] and Libgober [12] have studied these betti numbers using Hodge theory and Alexander polynomials, respectively.

As Zariski points out, the first betti number of Z_k is intimately related to the fundamental group of $\mathbb{P}^2 - C$. Suppose C is irreducible and meets the line at infinity L transversally. If $\pi_1(\mathbb{P}^2 - C)$ is abelian, Zariski [20] showed that $\pi_1(Z_k) = 0$ for all k. (Indeed, the hypotheses imply that $\pi_1(\mathbb{P}^2 - (C \cup L))$ is infinite cyclic. If U_k is the complement of the inverse image of $C \cup L$ in Z_k, then $\pi_1(U_k)$ is also cyclic, and the surjection $\pi_1(U_k) \longrightarrow \pi_1(Z_k)$ takes a generator to zero.)

Lazarsfeld has pointed out that for <u>any</u> curve C, and <u>any</u> k, the normal variety Y_k is simply connected. Thus the fundamental group and first betti number of Z_k all "come from" the exceptional divisors in the resolution of singularities $Z_k \longrightarrow Y_k$. In fact one has the following:

<u>Theorem</u> (Lazarsfeld). <u>Let X be a normal, projective n-dimensional variety</u>, $f : X \longrightarrow \mathbb{P}^n$ <u>a finite morphism. Suppose there is a positive dimensional locally closed subvariety</u> V <u>of</u> $\mathbb{P}^n$ <u>such that the restriction</u> $f^{-1}(V) \longrightarrow V$ <u>is (set-theoretically) one-to-one. Then</u> $\pi_1(X) = 0$.

<u>Proof</u>. One may reduce to the case where V is a curve, by taking generic hyperplane sections. Let C be a non-singular projective model of V, and $g : C \longrightarrow \mathbb{P}^n$ a morphism that maps C

birationally onto the closure of V in $\mathbb{P}^n$. Let $C' = (C \times_{\mathbb{P}^n} X)_{red}$. The hypothesis implies that the projection from C' to C is finite and generically one-to-one; since C is non-singular, C' maps homeomorphically onto C.

Now apply Deligne's version of the connectedness theorem ([4], [9]) to the product morphism $F : C \times X \longrightarrow \mathbb{P}^n \times \mathbb{P}^n$, which yields the surjectivity of

$$\pi_1(F^{-1}(\Delta_{\mathbb{P}^n})) \longrightarrow \pi_1(C \times X) ,$$

i.e. of the diagonal map $\pi_1(C') \longrightarrow \pi_1(C) \times \pi_1(X)$. But since $\pi_1(C') \xrightarrow{\sim} \pi_1(C)$, $\pi_1(X)$ must be zero. □

§4. Nori's theorem.

M. Nori [14] has proved vast generalizations of the solution to Zariski's problem. He not only gives two different proofs for the original Zariski problem, but he also gives analogues for nodal curves on surfaces other than $\mathbb{P}^2$, as sought for and proved in special cases by Abhyankar [1] and Prill [15].

For a Cartier divisor D on a normal surface, denote by D^2 the self-intersection number, i.e. $\int c_1(\mathcal{O}(D))^2$. For a nodal curve D, let $r(D)$ denote the number of nodes of D.

Theorem (Nori [14]). <u>Let</u> X <u>be a non-singular projective surface,</u> C <u>a nodal curve on</u> X. <u>Assume that for each irreducible component</u> D <u>of</u> C,

$$D^2 > 2 r(D) .$$

<u>Then the kernel of the homomorphism</u>

$$\pi_1(X - C) \longrightarrow \pi_1(X)$$

is abelian.

In particular, if X is simply connected, then $\pi_1(X-C)$ is abelian. Note that when $X = \mathbb{P}^2$, and $m = \deg(D)$, then $D^2 = m^2$ and $r(D) \le (m-1)(m-2)/2$, so Zariski's assertion follows from Nori's theorem. We refer to Nori's paper for the general proof, as well as generalizations to non-compact X. However, we cannot resist presenting his beautiful proof for the algebraic fundamental group. As noted by Abhyankar, the essential point is following lemma, which is of independent interest.

<u>Lemma</u> (Nori [14]). <u>Let $f : Y \longrightarrow X$ be a finite morphism of irreducible projective surfaces, with X non-singular, and let $C \subset X$ be the (reduced) branch curve. Consider the following conditions on an irreducible curve A on Y</u>:

(a) <u>Any singularity of $f(A) \cup C$ at a point of $f(A)$ is a node</u>.
(b) $f(A)^2 > 2\,r(f(A))$.

<u>Then any two curves on Y which satisfy</u> (a) <u>and</u> (b) <u>must intersect</u>.

Proof. One may replace Y by its normalization in the field of a Galois extension of the function field of X, so we may assume Y is normal and $f : Y \longrightarrow X$ is Galois, with Galois group G. For any irreducible subvariety V of Y, set

$$G(V) = \{\sigma \in G \mid \sigma(V) = V\},$$

$$I(V) = \{\sigma \in G(V) \mid \sigma|_V = \mathrm{id}_V\}.$$

Let A be a curve satisfying (a) and (b), and let P be a point of A. An analysis of local fundamental groups ([1], [7], [9]) shows that $I(A)$ is cyclic, $G(P)$ is abelian, and

(1) $I(A) \subset G(P) \subset G(A)$.

Factor f into $h \circ g$:

$$Y \xrightarrow{g} Y/G(A) \xrightarrow{h} X .$$

Let $B = g(A)$, $D = h(B)$. By general ramification theory,

(2) $$A = g^{-1}(B) ,$$

(3)   h <u>maps</u> B <u>birationally onto</u> D ,

(4)   B <u>is not in the ramification locus of</u> h .

In fact, from (1), one has

(5)   h <u>is étale at each point of</u> B .

From (3) and (5) it follows that

(6) $$B^2 - 2\,r(B) = D^2 - 2\,r(D) .$$

Indeed, if $\tilde{B}$ and $\tilde{D}$ are the non-singular models of B and D, the two sides of (6) are the degrees of the normal bundles of the immersions $\tilde{B} \longrightarrow Y/G(A)$ and $\tilde{D} \longrightarrow X$; and $\tilde{B} = \tilde{D}$ by (3), while the normal bundles agree by (5).

In particular,

$$(g^*B)^2 = B^2 \geq D^2 - 2\,r(D) > 0 .$$

Note that, by (2), g^*B is supported on A. Thus any A satisfying (a) and (b) supports a Cartier divisor $\tilde{A}$ with $\tilde{A}^2 > 0$. But if A_1 and A_2 were two such curves which were disjoint, the corresponding Cartier divisors $\tilde{A}_1$ and $\tilde{A}_2$ would satisfy

$$\tilde{A}_1^2 > 0 , \quad \tilde{A}_2 \cdot \tilde{A}_1 = 0 , \quad \tilde{A}_2^2 > 0 ,$$

which contradicts the Hodge index theorem. (By pulling the divisors back to a resolution of singularities of Y, one may use the index theorem on non-singular surfaces; or cf. [11]). □

To prove the theorem for the algebraic fundamental group, let
$f : Y \longrightarrow X$ be a Galois covering, and let H be the subgroup of the
Galois group G generated by the groups $I(A)$, as A runs over the
irreducible components of $f^{-1}(C)$. Then H is a normal subgroup of
G, and $Y/H \longrightarrow X$ is unramified. It suffices to show that H is
abelian. If two components A_1 and A_2 intersect at P, then
$I(A_1)$ and $I(A_2)$ commute by (1). The lemma then shows that all the
cyclic groups $I(A)$ must commute with each other, which makes H
abelian. □

Remark. In characteristic p, the lemma and its proof are valid provided the ramification is tame. The theorem is true for the tame fundamental group. Lazarsfeld's theorem, as proved in §3, also holds for the algebraic fundamental group in arbitrary characteristic.

References

1. Abhyankar, S., Tame coverings and fundamental groups of algebraic varieties, I,II, Amer. J. Math. 81(1959), 46-94; 82(1960), 120-178.

2. Alibert, D. and Maltsiniotis, G., Groupe fondamental du complémentaire d'une courbe à points doubles ordinaires, Bull. Soc. Math. France 102(1974), 335-351.

3. Arborello, E. and Cornalba, M., A few remarks about the variety of irreducible plane curves of given degree and genus, preprint.

4. Deligne, P., Le groupe fondamental du complément d'une courbe plane n'ayant que des points doubles ordinaires est abélien, Séminaire Bourbaki n° 543, Springer Lecture Notes 842(1981), 1-10.

5. Deligne, P. and Mumford, D., The irreducibility of the space of curves of given genus, Publ. Math. I.H.E.S. 36(1969), 75-100.

6. Esnault, H., Fibre de Milnor d'une cône sur une courbe plane singulière, Invent. Math. 68(1982), 477-496.

7. Fulton, W., On the fundamental group of the complement of a node curve, Annals of Math. 111(1980), 407-409.

8. Fulton, W., On the irreducibility of the moduli space of curves, Invent. Math. 67(1982), 87-88.

9. Fulton, W. and Lazarsfeld, R., Connectivity and its applications in algebraic geometry, Springer Lecture Notes 862(1981), 26-92.

"Anzi che trarre delle conclusioni
di priorità il cui interesse per la
scienza è sempre scarso, o dar troppo
peso a qualche debolezza ed agli errori
di grandi scienziati, è meglio rilevare
le nuove conferme di quel fatto che
continuamente si presenta nella storia
della matematica: che alla conescenza
completa, generale, dell'ente o del
risultato esatto si è giunti non in un
sol tratto e per opera di un solo, ma
per opera alternata o simultanea di
vari, passando per più gradi sì di
generalità che di rigore!"

Corrado Segre [1892], concluding paragraph.

ABOUT THE ENUMERATION OF CONTACTS

by

William FULTON, Steven KLEIMAN[1] and Robert MacPHERSON

1. Introduction. Preliminaries.

The enumeration of contacts is a fascinating subject. The number of varieties in a p-parameter family touching p given varieties usually is large and unobvious. For example, over the complex numbers, the number of conics touching 5 given conics is 3264 just as Chasles found in 1864. It is not 7776, that is 6^5, as Steiner asserted in 1848 and Bishoff reaffirmed in 1859. This example has been considered and reconsidered many times over the years with ever increasing degrees of clarity,

---

[1] This report was prepared while the author was on sabbatical leave from the Massachusetts Institute of Technology and a visiting professor at the University of Copenhagen, Denmark. It is a pleasure for him now to thank the Mathematics Institute of the University of Copenhagen for its hospitality and Dita Andersen for her fine typing.

rigor and completeness, but the notable step of showing that, when the given conics are in general position, then there are finitely many (so 3264) conics touching them, each nondegenerate and counted with multiplicity 1, was not taken until 1974. A second step is the number of quadrics touching 9 given quadrics. This number is 666, 841, 088, just as Schubert found in 1870. However, it was not established rigorously until after 1980. A third example is the number of twisted cubics touching 12 given quadrics. It is most likely, 5, 819, 539, 783, 680, just as Schubert found in 1874. However, it has not been veriffied as of this writing.
in 1874.

The first step in the determination of the above numbers and of others like them has been to reduce the problem to the determination of the numbers in the related cases, in which the given varieties are linear spaces - points, lines, planes, etc. - in every combination. Such a reduction, it turns out, is always possible! The reduction in the general case is the major theme of the present work.

The current section continues with preliminaries. Section 2 states and proves the main results, which include several statements that qualify the significance of the final number itself. Section 3 discusses some other points of view and methods of proof. The section closes with a proposition, which generalizes the lemma of Section 2, and which is of independent interest. Finally, Section 4 discusses various open problems.

From now on, the ground field will be algebraically closed and of arbitrary characteristic. Varieties will be reduced algebraic schemes. Subvarieties will be closed and proper (not empty nor the whole ambient variety), but they may be reducible and of impure dimension. A p-parameter family of varieties X will be parametrized by a p-dimensional variety T, but the total space need not be flat over T. The phrase "almost all X" will mean all X parametrized by an open subset of T whose complement is at most (p-1)-dimensional.

Fix a subvariety V of the N-dimensional projective space $\mathbb{P}^N$. The (projective) <u>conormal variety</u> CV is defined as the closure of the set of pairs (P,H) such that P is a simple point of V and H is a hyperplane containing the (embedded) tangent space $T_P V$. It sits in $\mathbb{P}^N \times \check{\mathbb{P}}^N$, where $\check{\mathbb{P}}^N$ is the dual projective N-space, the space of hyperplanes of $\mathbb{P}^N$. Note that CV is of pure dimension N-1.

For a hyperplane H, the H-<u>contact locus</u> V_H is defined as the scheme-theoretic fiber over H of the second projection $pr_2: CV \to \check{\mathbb{P}}^N$, embedded in V via the first projection $pr_1: CV \to \mathbb{P}^N$. It may be thought of as the locus of points P of V at which H touches P. It is not hard to prove that V_H is equal to the singular locus of V∩H endowed with the scheme structure of the Jacobian ideal.

The <u>dual variety</u> (or <u>reciprocal</u>) V* of V is defined as the subvariety $pr_2 CV$ of $\check{\mathbb{P}}^N$. For example, if V is a d-plane, then V* is the "orthogonal" (N-1-d)-plane. In general, for any simple point P of V, obviously V* contains $(T_P V)^*$, and the two are equal if and only if V is linear. So

(1.1) $\qquad \dim(V) \geq N-1-\dim(V^*)$,

and equality holds if and only if V is linear. Furthermore, clearly

(1.2) $\qquad \dim_H(V^*) + \dim(V_H) = N-1$

for all H in a certain dense, open subset of V*.

The conormal variety CV* also lies in $\mathbb{P}^N \times \check{\mathbb{P}}^N$, and if CV* coincides with CV, then V will be called <u>reflexive</u>. It is evident that, if V is reflexive, then V** = V and, for any simple point H of V*, the contact locus V_H is equal to the linear space $T_H(V)^*$. There is a useful criterion for reflexiveness, due to C. Segre and Wallace (see Kleiman [1980]); namely, V is reflexive if and only if the map $pr_2: CV \to V^*$ is separable (smooth on a dense open subset of CV). In characteristic zero, V is, therefore, always reflexive.

For $i = 0, \cdots, N-1$ the very important nonnegative integer

(1.3) $$r_i(V) = \int pr_1^* c_1(\mathcal{O}(1))^i \; pr_2^* c_1(\mathcal{O}(1))^{N-1-i} [CV]$$

will be called the i-<u>th</u> <u>rank</u> of V. (It is equal, when V is irreducible of dimension d, to the (d-i)-th class of V.) It is not hard to see that $r_i(V)$ is just the degree of the polar locus of dimension i; this locus is defined as the closure in V of the scheme of simple points P of V such that, if A is a general (N-2-i)-plane, then $(T_P V) + A$ is exactly a hyperplane.

It is evident that $r_i(V)$ is additive in V; that is, if $V_1, \cdots, V_m$ are the irreducible components of V, then

(1.4) $$r_i(V) = r_i(V_1) + \cdots + r_i(V_m) \; .$$

It is very easy to see using the projection formula that

(1.5) $r_i(V) = 0$ for $i > \dim(V)$,

(1.6) $r_i(V) = \deg(V)$ for $i = \dim(V)$ and

(1.7) $r_i(V) = 0$ for $i < N-1-\dim(V^*)$.

It is not hard to prove (see Hefez-Kleiman [1983]), that if V is irreducible, then

(1.8) $r_i(V) \neq 0$ for $N-1-\dim(V^*) \leq i \leq \dim(V)$.

It is evident that, if V is reflexive, then

(1.9) $r_i(V) = r_{N-1-i}(V^*)$ for all i .

(This lovely observation contains the essence of (3.6), Piene [1978] and (3.3), Urabe [1981].) It is not hard to prove (compare Piene [1978], (4.1) and (4.2)) that, if $\dim(V) \leq N-2$ and if $p: V \to \mathbb{P}^{N-1}$ is a general central projection, then

(1.10) $$r_i(V) = r_i(pV) \quad \text{for } i = 0, \cdots, N-2,$$

and that, if H is a general hyperplane, then

(1.11) $$r_i(V) = r_{i-1}(V \cap H) \quad \text{for } i = 1, \cdots, N-1.$$

Finally, if V is a smooth hypersurface of degree m, then

(1.12) $$r_i(V) = m(m-1)^{N-1-i} \quad \text{for } i = 0, \cdots, N-1,$$

because $CV = \mathbb{P}(\mathcal{O}_V(m-1))$, as is easy to check.

Let $\lambda_0, \cdots, \lambda_{N-1}$ be indeterminates. The expression

(1.13) $$r(V) = r_0(V)\lambda_0 + \cdots + r_{N-1}(V)\lambda_{N-1}$$

will be called the <u>module</u> of V. (More properly, it should be called the module of the condition to touch V, following Chasles, who introduced the term in connection with arbitrary conditions on conics in 1864.)

A second subvariety X of $\mathbb{P}^N$ will be said to <u>touch</u> V, or to make a <u>contact</u> with V, if the two conormal varieties CX and CV have a common pair (P,H), and then (P,H) will be called the <u>element of contact</u>. If, moreover, P is a simple point of X and of V and if H is equal to the span of the tangent spaces $T_P X$ and $T_P V$,

(1.14) $$H = (T_P X) + (T_P V),$$

then the contact will be called <u>proper</u>. Note that, if

(1.15) $$\dim(X) + \dim(V) \leq N-1,$$

then X and V touch just if and only if they meet. Note also that, if both X and V are reflexive, then X touches V if and only if X* touches V*; in fact, then the elements of contact of X and V are the same as those of X* and V*.

Consider a p-parameter family of subvarieties X of $\mathbb{P}^N$. For each sequence of nonnegative integers $(j_0,\ldots,j_{N-1})$ such that $\Sigma j_k = p$, the number $\lambda(j_0,\ldots,j_{N-1})$ of X simultaneously touching j_k general k-planes for $j = 0,\ldots,N-1$ will be called the $(j_0,\ldots,j_{N-1})$-th <u>characteristic number</u>, <u>characteristic</u>, or <u>elementary number</u>. (The terms are primarily due to Chasles, 1864. The word "characteristic" reflects the fact that these numbers suffice to characterize the family in enumerations of contacts. The word "elementary" reflects the fact that these numbers are the basic ones, out of which the others can be formed.) These numbers are finite and well-defined (that is, independent of the choice of the general k-planes). Any subfamily containing almost all X has the same characteristic numbers. If almost all X are reflexive, then the $(j_0,\ldots,j_{N-1})$-th characteristic number is equal to the $(j_{N-1},\ldots,j_0)$-th characteristic number of the family of dual varieties X^*. These facts and others about the numbers result immediately from the theorem below, applied with the k-planes as $V_1,\ldots,V_p$.

2. The Main Results.

<u>Theorem</u>. Given a p-parameter family of subvarieties X of $\mathbb{P}^N$ and p other subvarieties $V_1,\ldots,V_p$, there exists a nonempty open subset U of the p-fold self-product of the automorphism group of $\mathbb{P}^N$ such that, for any p-tuple $(g_1,\ldots,g_p)$ in U, after $V_1,\ldots,V_p$ have been replaced by their translates $g_1 V_1,\ldots,g_p V_p$, the following statements will hold:

(a) (i) The number of X, each X counted with its natural multiplicity, simultaneously touching $V_1,\ldots,V_p$ is finite. (ii) The number is given by the product of the modules of the V's,

(2.1) $$r(V_1)\cdots r(V_p).$$

The product is evaluated by formally multiplying it out and then replacing the monomial $\lambda_0^{j_0}\cdots\lambda_{N-1}^{j_{N-1}}$ by the corresponding characteristic number $\lambda(j_0,\cdots,j_{N-1})$ of the family. (iii) The number is 0 if, for some i and almost all X, either

(2.2) $\qquad\qquad \dim(X) + \dim(V_i) \leq N-2$ or

(2.3) $\qquad\qquad \dim(X^*) + \dim(V_i^*) \leq N-2$.

(b) The expression in (a,ii) for the number of X touching $V_1,\cdots,V_p$ is self-dual; that is, it coincides, up to the order of the terms, with the corresponding expression for the number of X^* touching $V_1^*,\cdots,V_p^*$, when almost all X and $V_1,\cdots,V_p$ are reflexive.

(c) If an X touches $V_1,\cdots,V_p$, then (i) each of the p contacts is proper, (ii) for each i the element of contact (P_i,H_i) of X and V_i lies in any dense, open subset of CX that varies continuously with X and that was given at the outset, and similarly, (P_i,H_i) lies in any given dense, open subset of CV_i, (iii) X itself belongs to any given subfamily that contains almost all X, (iv) for no i does X have a second contact with V_i with element (Q_i,L_i) distinct from (P_i,H_i), when the characteristic is 0, nor, more generally, when almost all X and $V_1,\cdots,V_p$ are reflexive and the characteristic is not 2.

(d) Suppose that the parameter space T is irreducible and that almost all X appear m times in the family (that is, the rational map from T into the Hilbert scheme of $\mathbb{P}^N$ is of degree m); however, set m = 0 if almost all X appear infinitely often. Then each X appears in the count in (a) with multiplicity m, when characteristic is 0. When the characteristic is positive and almost all X and $V_1,\cdots,V_p$ are irreducible, then each X appears in the count with the same multiplicity mcq, where q is a power of the characteristic and c is the number of distinct simultaneous contacts of X with $V_1,\cdots,V_p$.

Remark. In (d), usually $q = 1$ and $c = 1$. In fact, obviously $q = 1$ when the characteristic does not divide the number given by the product of the modules. Moreover, $c = 1$ when almost all X and $V_1, \cdots, V_p$ are reflexive and the characteristic is not 2 by virtue of (c,iv).

Corollary. In $\mathbb{P}^N$ there is always at least one and at most a finite number of smooth hypersurfaces of degree m touching $p = \binom{N+m}{N} - 1$ subvarieties $V_1, \cdots, V_p$ in general position if, for each i, V_i is either a point, a curve of degree ≥ 2, a smooth surface of degree ≥ 2, a smooth complete intersection of degree ≥ 2, or a smooth hypersurface of degree ≥ 2.

Proof of the corollary. The hypotheses on V_i imply $r_0(V_i) \neq 0$ (see Kleiman [1977], IV, D, pp.359-65). On the other hand, it is evident that there is exactly 1 hypersurface passing through p points. Therefore the corollary follows from (a,i), (a,ii) and (c,III) of the Theorem.

Proof of the theorem (plus 2 parentheses, a remark and an example). The theorem is a reflection of the geometry of the incidence correspondence,

(2.4) $\qquad I = \{(P,H) \in \mathbb{P}^N \times \mathbb{P}^{\vee N} | P \in H\}$.

Note that I is a homogeneous space under the induced action of the automorphism group $\text{Aut}(\mathbb{P}^N)$. Note that I is of dimension $2N-1$ and that I contains CV, which is of dimension $N-1$, for every subvariety V of $\mathbb{P}^N$. Note that $gCV = CgV$ for all g in $\text{Aut}(\mathbb{P}^N)$. Recall that, by definition, a second subvariety X touches V if and only if CX meets CV. These observations form the backbone of the proof.

(Remark. One theme of the proof is illustrated in the following short derivation of Bertini's theorem. This derivation is a version of the one in Hodge-Pedoe ([1952] Thm. I, p.153) and in Kleiman ([1974] (10)), and it is worth considering in passing. In the setup above, by dimensional transversality, there exists a nonempty open subset U of Aut($\mathbb{P}^N$) such that CX does not meet gCV = CgV for all g in U; hence, the smooth loci of X and gV meet (differentiably) transversally This conclusion is Bertini's theorem for subvarieties of $\mathbb{P}^N$, and it is valid in any characteristic.)

Proceeding with the proof of the theorem, let T_0 be any open subset of the parameter space of the family whose complement has dimension at most p-1 , and let T be any compactification of T_0. In the product of p+1 factors $I^{\times p} \times T$, form the closure F^p of the union $\cup (CX)^{\times p}$ where X varies in the subfamily parametrized by T_0. Obviously, F^p is of dimension pN .

Since the natural action on $I^{\times p}$ of Aut($\mathbb{P}^N)^{\times p}$ is transitive, by general transversality results (see Kleiman [1974] and Vainsencher [1978] (7.2)), after $V_1, \cdots, V_p$ have been replaced by general translates, the set

(2.5) $$S^p = (pr_1^{-1} CV_1) \cap \cdots \cap (pr_p^{-1} CV_p) \cap F^p$$

becomes finite, where pr_i is the i-th projection. Moreover, S^p lies in any given dense, open subset of each "intersectand". Furthermore, the intersectands meet (differentiably) transversally in characteristic 0 ; in positive characteristic, each point in S^p appears with the same multiplicity, which is a power q of the characteristic.

By using the appropriate open subsets above, it may be assumed that each point of S^p represents an X in the subfamily parametrized by T_0 together with, for each i , an element of contact (P_i, H_i) of X with V_i. Moreover, it may be assumed that P_i is a simple point of X and of V_i and that H_i is a simple point of X* and of V_i^*.

The contact of X with V_i is proper. Indeed, since

(2.6) $$M = (T_{P_i} X) + (T_{P_i} V_i)$$

lies in H_i, if M and H_i were unequal, then there would be infinitely many hyperplanes H containing M. Replacing H_j by H would yield another point of S^p. Thus S^p would be infinite. So $M = H_i$.

Since M is therefore a hyperplane, obviously

(2.7) $$\dim(X) + \dim(V_i) \geq N-1 .$$

If X and V_i are reflexive, then by symmetry also

(2.8) $$\dim(X^*) + \dim(V_i^*) \geq N-1 .$$

In any event, (2.8) holds. Indeed, the two M-contact loci X_M and $V_{i,M}$ lie in M, and they have only a finite number of points in common because S^p is finite. Hence

(2.9) $$\dim(X_M) + \dim(V_{i,M}) \leq N-1 .$$

Therefore, (2.8) holds by virtue of (1.2).

It is now evident that X and V_i will have a second contact with element (Q_i, L_i) where $Q_i \neq P_i$ but $L_i = H_i$ ($= M$) if and only if (i) equality holds in (2.9) and (ii) the reduced scheme underlying X_M is of degree ≥ 2 or that underlying $V_{i,M}$ is so; indeed, it may be assumed that X_M and $V_{i,M}$ are dimensionally transverse in M. Now, X_M and $V_{i,M}$ are linear if X and V_i are reflexive; hence, in this case, there is no such second contact.

(Example. In characteristic $q \geq 2$, the irreducible plane curve with affine equation $y = x^{mq+1}$, where m is prime to q, is such that every tangent line T is tangent at m distinct points; indeed, if T touches at (x,y) and if $z^m = 1$, then T also touches at (zx, zy). Let V be the cone in $\mathbb{P}^3$ over this curve, and let X vary in a non-trivial 1-parameter family of reflexive surfaces in $\mathbb{P}^3$ whose dual varieties are curves. Then the theorem implies that there is at least one X touching V, as $r_1(V) \neq 0$ by (1.8) and there is, obviously, at least one X* meeting a general line. Moreover, every X touching V will have at least m distinct contacts.)

The general analysis of the case in which X and V_i have a second contact with element (Q_i, L_i) distinct from (P_i, H_i) is carried out as follows. Let V_{p+1} be another copy of V_i. Then the corresponding set S^{p+1} contains the point

(2.10) $\qquad ((P_1, H_1), \cdots, (P_p, H_p), (Q_i, L_i), t)$

where t in T represents X. The point (2.10) lies in the intersection

(2.11) $\qquad S^{p+1} \cap (J \times T)$

where J is one of the 7 orbits of $\text{Aut}(\mathbb{P}^N)^{\times p}$ acting on $I^{\times(p+1)}$ diagonally on the i-th and (p+1)-th factor.

Of course, J is not the diagonal orbit, because (P_i, H_i) and (Q_i, L_i) are distinct. In fact, $P_i \neq Q_i$. Indeed, the two contacts are proper; so, if $P_i = Q_i$, then $H_i = L_i$. Thus, in particular, J is not a second of the 7 orbits. The dual case, in which $P_i \neq Q_i$ but $H_i = L_i$, has already been treated. Hence 4 orbits remain to be considered.

By the theorem of dimensional transversality, it may be assumed that the sets

(2.12) $\qquad (\text{pr}_1^{-1} CV_1) \cap \cdots \cap (\text{pr}_p^{-1} CV_p) \cap (\text{pr}_{p+1}^{-1} CV_{p+1})$ and F^{p+1}

meet $J \times T$ in subsets of complementary dimensions. Indeed, if the dimensions were less than complementary, then it could be assumed that (2.11) is empty. If they were more than complementary, then (2.11), being nonempty, would be of dimension ≥ 1, and so S^p would be infinite. In particular, therefore, by the less-than-complementary case, J is not the open orbit.

There are 2 similar orbits of codimension 1. If J is the one, then $P_i \in L_i$ but $Q_i \notin H_i$; if J is the other, then $P_i \notin L_i$ but $Q_i \in H_i$. Say that J is the first. By reason of dimension, the two sets of (2.12) each have an irreducible component containing (2.10) and contained in the closure of $J \times T$. Hence X, for example, has two irreducible components X' and X'' such that $P_i \in X'$ and $Q_i \in X''$ and such that every point

(2.13) $\qquad ((P,H), (Q,L)) \in (CX') \times (CX'')$

satisfies $P \in L$. Therefore, $X' \subset L_i$. Consequently, $T_{P_i} X \subset L_i$. Similarly, $T_{P_i} V_i \subset L_i$. Since the contact at P_i is proper, therefore $H_i = L_i$. Hence $Q_i \in H_i$, contrary to hypothesis. Thus, after all, J is not one, nor the other, of these 2 orbits.

Suppose, finally, that J is the orbit of codimension 2; in other words,

(2.14) $\qquad P_i \in L_i, Q_i \in H_i$ but $P_i \neq Q_i, H_i \neq L_i$.

Then, by reason of dimension, one of the two sets of (2.12), say the first, has an irreducible component containing (2.10) and contained in the closure of $J \times T$, and the other set, now F^{p+1}, meets the closure of $J \times T$ in a subset containing (2.10) and of codimension 1 in F^{p+1}. This case will now be analyzed until it appears that, if the characteristic is different from 2, then X is not reflexive. The case in which the roles of the two sets of (2.12) are interchanged is similar.

First, V_i has two irreducible components V' and V'' such that $P_i \in V'$ and $Q_i \in V''$ and such that every point

(2.15) $\qquad ((P,H), (Q,L)) \in (CV') \times (CV'')$

satisfies $P \in L$ and $Q \in H$. It follows that

(2.16) $\qquad V' \subset (T_{Q_i} V'')$ and $V'' \subset (T_{P_i} V')$.

Therefore, V' and V" are equal and linear.

Secondly, X has two irreducible components X' and X" such that the points

(2.17) $\qquad ((P,H), (Q,L)) \in (CX') \times (CX")$

satisfying $P \in L$ and $Q \in H$ form a subset of codimension at most 1; in fact, this subset has an irreducible component D of codimension at most 1 that contains the point $((P_i,H_i), (Q_i,L_i))$.

Consider the natural map, $f: D \to X' \times X"$. Its fiber over (P,Q) is contained in the set

(2.18) $\quad \{((P,H),(Q,L)) \in I^{\times 2} | H \supset (T_P X+Q), L \supset (T_Q X+P)\}$.

Now, in general, $Q \notin T_P X$. In fact, $X" \not\subset T_{P_i} X$. Indeed, otherwise $T_{Q_i} X \subset H_i$. However, $V" \subset H_i$. Hence, since the contact at Q_i is proper, $L_i = H_i$, contrary to hypothesis. Similarly, in general, $P \notin T_Q X$. Consequently, by reason of dimension, $f(D)$ is of codimension 1 in $X' \times X"$ and D is the full inverse image of $f(D)$ in $CX' \times CX"$. Hence

(2.19) $\qquad Q \in T_P X$ and $P \in T_Q X$ for all $(P,Q) \in f(D)$.

Moreover, $f(D)$ covers X'. Indeed, otherwise, $X" \subset T_{P_i} X$. Similarly, $f(D)$ covers X".

The dimensions of X' and X" are equal. Indeed, say that that of X' is less. Since $X" \cap T_{P_i} X$ has codimension 1 in X", therefore $T_{P_i} X \subset X"$. So $P_i \in X' \cap X"$. However, P_i is a simple point.

The line joining P and Q cannot lie in X for all (P,Q) in $f(D)$. Indeed, suppose the contrary. Then the fiber of $f(D)$ over P_i is a cone C. The tangent spaces $T_Q C$ are constant as Q runs along a generator through simple points of C. Since

(2.20) $\qquad\qquad T_Q C = (T_{P_i} X) \cap (T_Q X)$,

the corresponding T_QX rotate in a pencil. Now, either

(2.21) $\quad (V' \cap T_{P_i} X) = (V' \cap T_{Q_i} X) \quad \text{or} \quad (V' \cap T_{P_i} X) \subsetneq (V' \cap (T_{P_i} X + T_{Q_i} X))$.

In either case, it follows that V' touches X at every one of the infinitely many simple points R of X that lie on the line joining P_i and Q_i.

Finally, let E be a general linear space of codimension $\dim(X')-1$, and consider the curves $Y' = E \cap X'$ and $Y'' = E \cap Y''$. Let P be a general point of Y'. The fiber C of $f(D)$ over P is not a cone, but C is a hypersurface in $T_P X$. Hence the line $G = E \cap T_P X$ meets C in a point Q distinct from P. Obviously, G is equal to $T_P Y'$ and by (2.19) it is equal to $T_Q Y''$. Thus, if Y' and Y'' are distinct, then by symmetry they have the same set of tangent lines; if they coincide, then every tangent line makes two distinct contacts. In either case, they are not reflexive. It is not hard to prove (see Hefez-Kleiman [1983]) that, if a variety is reflexive, then so is a general hyperplane section, except in characteristic 2. Hence, if the characteristic is different from 2, then X is not reflexive.

To complete the proof of the theorem, there remains only one major task, to show that the number of points in the set S^p of (2.5), counted with natural multiplicity, is given by evaluating the product of the modules of the V_i. Since this number is equal to

(2.22) $\quad \int \text{pr}_1^*[CV_1] \cdots \text{pr}_2^*[CV_p] \cdot [F^p]$,

it suffices to establish the following lemma.

<u>Lemma.</u> For $i = 0, \cdots, N-1$ let A_i be an i-plane. Then, for any subvariety V of $\mathbb{P}^N$, the class $[CV]$ is given, modulo rational equivalence on I, by the formula

(2.23) $\quad [CV] = r_0(V)[CA_0] + \cdots + r_{N-1}(V)[CA_{n-1}]$.

Proof. The formula follows immediately from the definition (1.3) of the $r_i(V)$ and the following 2 facts: (a) for $i = 0,\cdots,N-1$ the elements

(2.24) $$w_i = pr_1{}^*c_1(\mathcal{O}(1))^i \, pr_2{}^*c_1(\mathcal{O}(1))^{N-1-i}$$

form a basis for the classes of N-cycles on I; (b) for $i = 0,\cdots,N-1$ the elements $[CA_i]$ form the dual basis. Now, (a) holds because $I/\mathbb{P}^N$ is a projective-space subbundle of the trivial bundle $\mathbb{P}^N \times \overset{\vee}{\mathbb{P}}{}^N$. Furthermore, (b) says, in other words, that

(2.25) $$r_i(A_j) = \delta_{ij} \qquad \text{(the Kronecker function)}.$$

Finally, (2.25) results directly from (1.5), (1.6) and (1.7).

3. Other Views and Approaches.

In any 1-parameter family, the number of varieties X touching a given variety V in general position in $\mathbb{P}^N$ is finite and is furnished by the module $r(V)$; this fact is the case $p = 1$ of Parts (a,i) and (a,ii) of the Theorem. Conversely, as will now be proved, $r(V)$ is furnished by N standard 1-parameter families, the linear pencils.

Fix a complete flag, in which B_j is a general j-plane,

(3.1) $$\emptyset = B_{-1} \subset B_0 \subset \cdots \subset B_N = \mathbb{P}^N.$$

For $i = 0,\cdots,N-1$ consider the linear pencil, the natural family parametrized by $T = \mathbb{P}^1$, of all (N-1-i)-planes X such that

(3.2) $$B_{N-i-2} \subset X \subset B_{N-i}.$$

The number of X touching a general j-plane A_j is δ_{ij}, by virtue of the projection formula applied to (2.22); indeed, the projection $pr_1 : I \times T \to I$ obviously carries F^1 onto the variety

(3.3) $$W_i = \{(P,H) \in I \mid P \in B_{N-i-2}, H \supset B_{N-i}\},$$

and the class $[W_i]$ obviously is the element w_i of (2.24), dual to $[CA_i]$. Since the number of X touching V is furnished by evaluating $r(V)$, it is just $r_i(V)$.

Segre ([1912] §37, p.924) gave another interesting description of the ranks; namely, $r_0(V)$ - the class of V - is the number of hyperplanes that contain a general (N-2)-plane and touch V, and $r_i(V)$ is the class of a general i-codimensional linear section of V. This description and the preceding one are equivalent; the equivalence is intuitively obvious geometrically and is easily proved formally.

The proof of the lemma may be described loosely as a proof by Schubert calculus, because it is based on the use of dual bases. In fact, Schubert introduced and developed the idea of dual bases; he presented it first in two research articles (1876,7) and then made it the grand theme of his book [1879]. Of course, Schubert's point of view is more down-to-earth. In essence, it is that a given enumerative condition may be expressed as a formal linear combination of standard or basic conditions, which he termed "characteristics" ([1879] p.282), and that the combining coefficients may be calculated by imposing the given condition in a dual series of problems; a condition may be expressed as a sum of two others when it is equivalent for the purposes of enumeration to (that is, it yields the same numbers as) the alternative condition of imposing one or the other.

From Schubert's point of view, the enumerative results above look as follows: the condition on a variety X varying in a 1-parameter family to touch a given variety V is equivalent to the alternative condition to touch one of $r_0(V)$ points, $r_1(V)$ lines, etc.; furthermore, $r_i(V)$ may be calculated as the number of (N-1-i)-planes X in a general linear pencil that touch V. Similarly, the Lemma and the bases in its proof look as follows: the condition on an element (P,H) to belong to a variety V is equivalent to the condition on it to belong to one of $r_0(V)$ points, $r_1(V)$ lines, etc.; and the conditions on (P,H) to belong to an i-plane for $i = 0, \cdots, N-1$ are dual to the conditions that P lies in a $(N-i-2)$-plane while H contains an $(N-i)$-plane.

Schubert ([1879] pp.50-1, 289-95) derives the expression for the number of varieties X in a 1-parameter family touching a given variety V in $\mathbb{P}^2$ and in $\mathbb{P}^3$, via a version of the Lemma, from the canonical dual bases on I, essentially as was done in Section 2. However, he obtains the bases differently. In fact, he proceeds in two similar but distinct ways, starting in both cases from the correspondence principle. One way the determination of a Künneth decomposition of the relative diagonal of $I/\check{\mathbb{P}}^N$, and the other involves that of the absolute diagonal of I. (See Grayson [1979] for a lovely up-to-date version of this work). Schubert's method was years ahead of its time and is still rather interesting.

Schubert [1879] goes on, at least in the specific cases he considers, to give the expression for the number of X in a p-parameter family touching given $V_1, \cdots, V_p$. He does not fuss, but implicitly he proceeds in a somewhat different way from that in Section 2. Essentially, he forms for each V the divisorial cycle on T,

(3.4) $$ZV = pr_{2*}((pr_1*[CV]) \cdot [F^1]);$$

its underlying set is the closure of the set of all points of T_0 that represent an X touching V. By additivity, the formula (2.23) for [CV] yields a similar formula for the class [ZV]. Now, in principle, the number of X touching $V_1, \cdots, V_p$ is equal to

(3.5) $$\int_T [ZV_1] \cdots [ZV_p] .$$

The desired expression follows immediately.

For the preceding argument to be rigorous and complete, the cycles $[ZA_j]$ must be locally principal, the cycles $[ZV_i]$ must intersect properly, and the intersection multiplicities must be investigated. Now, if T is replaced by a smooth, birationally equivalent variety, then any divisorial cycle on T will be locally principal. Moreover, suppose that there is an induced action of $\text{Aut}(\mathbb{P}^N)$ on T and that it has only finitely many orbits these conditions obtain in some important cases. Then, given several subvarieties of T such that each contains no orbit, their intersection will be proper and the intersection multiplicities may be qualified by applying the general transversality results, orbit by orbit, to the traces of the subvarieties.

Schubert (1874-9), inspired by Halphen (1873), developed the powerful idea embodied in (3.5); namely, the simultaneous imposition of several conditions is representable by a product that distributes over sums. Earlier, he and others used a cumbersome method, in which the several conditions are introduced successively. For example, to use this method to prove that the number of varieties X in a p-parameter family touching p given varieties $V_1, \cdots, V_p$ in general position is furnished by the product of their modules $r(V_0) \cdots r(V_p)$, assume the case p = 1, and consider the 1-parameter subfamily of those X touching $V_2, \cdots, V_p$. The number of X in the subfamily touching V_1 is, by the case p = 1, equal to

(3.6) $$r_0(V_1)\lambda_0 + \cdots + r_{n-1}(V_1)\lambda_{N-1},$$

where λ_i is the number of these X touching a general i-plane L_i. Now, fix i, and consider the (p-1)-parameter subfamily of those X touching L_i. The number of these X touching $V_2, \cdots, V_p$ is, on the one hand, λ_i, and on the other, by induction, furnished by the product of the modules $r(V_2) \cdots r(V_p)$. The desired result follows immediately. To be rigorous, one must define the subfamilies with enough care so that the multiplicities of appearance of the X can be controlled and compared.

De Jonquières (1861) gave a simple expression for the number of plane curves X in a 1-parameter family touching a given plane curve V, but it turned out to be valid in a restricted number of cases only.

Cremona ([1862] Nr. 111 bis a., Thm. I, p.170) refined de Jonquières' expression by involving the module $r(V)$. Chasles (1864, independently?) did something greater; he introduced the abstract notion of module as part of a general self-dual theory of enumeration. De Jonquières, Cremona and Chasles based their proofs on the correspondence principle. Zeuthen (1871) gave another proof. He used his own version of the generalized correspondence principle, which was discovered by Cayley (1866) and which was proved by Brill (1871) and himself independently. Coolidge ([1959] Thm. 14, p.436; Thm. 16, p.440) presents Zeuthen's proof, following Zeuthen's Lehrbuch. Then he treats the case of a several parameter family; he argues as above, but he unnecessarily reuses the correspondence principle at the induction step.

In addition to the correspondence principle, there is another major 19th century "principle of geometry", on which many enumerations were based. It was known successively as the principle of continuity (so named by Poincelet, 1822), the principle of special position (so named by Schubert, 1874) and the principle of conservation of number (so named by Schubert, 1876). The principle says, for example, that the number of X in a p-parameter family touching $V_1, \cdots, V_p$ remains the same when each V_i is varied continuously, even to the point of degeneration. Now, a contact between an X and a V_i corresponds to an intersection between CX and CV_i, and a motion of V_i induces a motion of CV_i. Hence, provided appropriate attention is paid to the way in which CV_i degenerates when V_i does (and more will be said below about this matter), then the present consequence of the principle is a consequence of the more abstract fact, that algebraic equivalence implies numerical equivalence, applied to (2.22).

One commonly used motion has been the flow provided by a certain family of degenerating automorphisms of $\mathbb{P}^N$, the homolography. The homolography is the family of degenerating homologies, whose centers are a fixed point P, whose axes are a fixed complementary hyperplane H and whose cross ratios are a variable number r; as $r \to 0$, the homologies degenerate into the projection from P to H. Schubert [1879] (Ex. 4, pp.13-14), for example, uses a homography and the principle of conservation of number to establish the expression for the number of plane curves in a 1-parameter family touching a given plane curve. It is possible, at least in characteristic 0, to follow this approach and obtain a second proof of the Lemma; the proof is presented next. On a technical level, this proof differs from the first one in an important way; it does not use the fact that the $[CA_j]$ generate the classes of $(N-1)$-cycles on I. However, its charm and an independent importance lie in its geometric content.

It can be proved using the graph construction of MacPherson ([1974] §5) that, in characteristic 0, if a subvariety V of $\mathbb{P}^N$ does not contain the center P of a homolography, then CV degenerates into a subscheme of I, whose reduced irreducible components are each of the form CW for an appropriate subvariety W of the axis H. If V is a hypersurface, then one of the W will be H itself. At any rate, proceed and take care of the W properly contained in H with a homolography within H. Repeating this procedure leads to the conclusion that on I the cycle $[CV]$ is rationally equivalent to a linear combination of the cycles $[CA_j]$, where A_j is a j-plane, for $j=0,\cdots,N-1$. The combining coefficients are the $r_j(V)$ by virtue of the characterization of them given at the beginning of this section, the principle of conservation of number and (2.25).

There is a third proof of the lemma. It is based on a formula of Schubert [1903] and Giambelli [1905], which was originally known of as a formula of special position. Today, it is called Porteous's formula, after Porteous [1971]. (See Kempf-Laksov [1974] for an algebro-geometric treatment of an even more general formula of wide applicability.)

The third proof runs as follows. First, it may be assumed that V is irreducible, say of dimension d. Next, consider the Nash modification, $f: V' \to V$. A point P of V' represents a d-plane T^-; if $f(P)$ is simple, then T is $T_{f(P)}V$; indeed, V' is just the closure of the graph of the Gauss map, which carries a simple point of V to its tangent d-plane. On V' there are two important exact sequences of locally free sheaves,

(3.7) $$0 \to K \to 0_{V'}{}^{N+1} \to B \to 0,$$

(3.8) $$0 \to A \to B \to f^*0_V(1) \to 0.$$

In fact, $\mathbb{P}(B)$ is the total space of the family of d-planes T; the inclusions $T \subset \mathbb{P}^N$ are specified by (3.7), and the inclusions $P \in T$ by (3.8).

Let I' denote the subvariety of $V' \times \check{\mathbb{P}}^N$ of (P,H) such that $f(P) \in H$. Consider its subvariety Z of those (P,H) such that $T \subset H$. It is easy to see that Z is the locus of zeros of a map

(3.9) $$u: p_2{}^*S \to p_1{}^*A \quad \text{where } S = 0_{\check{\mathbb{P}}^N}(-1)$$

and where p_1 and p_2 are the projections; indeed, the natural map from $p_2{}^*S$ to $p_1{}^*B$ lands in $p_1{}^*A$ because $P \in H$. Hence

(3.10) $$[Z] = c_d(p_1{}^*A - p_2{}^*S)$$

by Porteous's formula. So, by (3.9) there is a sheaf C on I such that

(3.11) $$[Z] = c_d(p_1{}^*B - p^*C)$$

where $p: I' \to I$ is the natural map. Obviously, p carries Z birationally onto CV. Hence, expanding (3.11) and using the projection formula yields

(3.12) $$[CV] = \Sigma_i \, p_*p_1{}^*c_{d-i}(B) \cdot s_i(C^v)$$

where $s_i(C^v)$ is the i-th Segre class, or inverse Chern class.

If V is a d-plane A_d, then B is trivial and (3.12) yields

(3.13) $$[CA_d] = pr_1^* c_1(\mathcal{O}_{\mathbb{P}^N}(1))^{N-d} \cdot s_d(C^\vee),$$

where $pr_1: I \to \mathbb{P}^N$ is the projection. Since $p_* p_1^* = pr_1^* f_*$, to establish the lemma, it remains to prove

(3.14) $$f_* c_{d-i}(B) = r_i(V) c_1(\mathcal{O}_{\mathbb{P}^N}(1))^{N-i}.$$

Now, by Bezout's theorem and the projection formula, (3.14) is equivalent to

(3.15) $$r_i(V) = \int c_{d-i}(B) f^* c_1(\mathcal{O}_V(1))^i [V'].$$

However, it is easy to see that $Z = \mathbb{P}(K^\vee)$. Hence, by (3.7)

(3.16) $$c_{d-i}(B) = s_{d-i}(K^\vee) = p_{1*} p_2^* c_1(\mathcal{O}_{\mathbb{P}^N}(1))^{N-1-i}.$$

Since $Z \to CV$ is birational, therefore (3.15) follows from the definition (1.3) of $r_i(V)$ and the projection formula.

The lemma may be generalized easily as follows. Fix $n \geq N-1-d$, where d is the minimum of the dimensions of the irreducible components of V. Let I_n denote the variety of pairs (P,L) where L is an n-plane in $\mathbb{P}^N$ and $P \in L$. Let $C_n V$ denote the closure in I_n of the set of pairs (P,L) such that P is a simple point of V and $L + T_P V \neq \mathbb{P}^N$. Then pulling (2.23) up to the (flag) manifold of triples (P,L,H) and pushing the result down to I_n yields the analogous formula,

(3.17) $$[C_n V] = r_{N-1-n}(V)[C_n A_{n-1-n}] + \cdots + r_{N-1}(V)[C_n A_{n-1}],$$

because the inverse image of CV projects birationally onto $C_n V$, while if W is a subvariety of $\mathbb{P}^N$ of dimension $< N-1-n$, then the inverse image of CW drops dimension under the projection to I_n.

Formula (3.17) is remarkable because, for one thing, it says that $[C_n V]$ is a somewhat special class on I_n. Indeed, $[C_n A_{n-1-n}], \ldots, [C_n A_{n-1}]$ do not generate the classes of codimension $n+1$ for $n < N-1$.

However, they are members of a rather lovely natural self-dual basis for all the classes. In fact, this basis on I_n is a special case; such a basis was found on any partial flag manifold by Ehresmann ([1934] §§19, 20), and more generally, on any G/P where G is a reductive algebraic group and P is a parabolic subgroup, by Chevalley (1953, unpublished but see Demazure [1974] Prop.1(a), Cor.(a), p.69). Using this basis on I_n, it is not hard to establish (3.17) directly. In fact, Zobel ([1958] §5), attributing the basis to Martinelli (1942), in this way obtained (3.17) in two cases, (i) n = 1 and (ii) d = N-1; correspondingly, he proved Parts (a,i) and (a,ii) of the Theorem for p = 1 in two cases, (i) either almost all X or V_1 is a curve and (ii) almost all X or V_1 is a hypersurface. (More will be said about Zobel's work in the next section.)

The other two proofs of the lemma also yield (3.17) directly, mutatis mutandis. (In the third proof, take S to be the universal subsheaf of the Grassmannian of n-planes, although, of course, not in the proof of (3.15), etc.) In fact, these two proofs work as well in the following even more general setting, where the proof using dual bases fails.

In the general setting, $\mathbb{P}^N$ is replaced by an arbitrary smooth, irreducible variety Y of dimension N. Correspondingly, I_n denotes the Grassmannian bundle $G_n(\Omega_Y^1)$, which parametrizes the pairs (P,L) where P ∈ Y and L is an n-dimensional vector subspace of the tangent space T_pY, and S denotes its universal subbundle; if $Y = \mathbb{P}^N$, then I_n is the same as before but S(1) is the earlier S. For an irreducible d-dimensional subvariety V of Y where $d \geq N-1-n$, let C_nV denote the closure in I_n of the set of pairs (P,L) such that P is a simple point of V and $L + t_pV \neq t_pV$. The Nash modification f: V' → V can be obtained by taking the closure of the graph of the "Gauss" map, which carries a simple point P of V to the pair (P, t_pV) of I_d. The sequences (3.7) and (3.8) are replased by a simple surjection, $f^*\Omega^1_Y \to \Omega$, where Ω is the locally free sheaf of rank d on V' induced by the universal quotient on I_d; if $Y = \mathbb{P}^N$, then $A = \Omega(1)$. The classes of dimension i on V,

(3.18) $$m_i(V) = f_*c_{d-i}(\Omega),$$

are called the <u>Mather-Chern classes</u> of V. The second and third proofs of the lemma generalize and yield Parts (a), (b) of the following proposition. Finally, the first assertion of Part (c) is a simple formal consequence of Part (b), because $m_d(V) = [V]$; the second assertion is trivial; and the third is a restatement of (3.7).

<u>Proposition</u>. In the general setting described above, the following statements hold.

(a) If V is degenerated to V_0 in a flat family of subschemes of Y, then, at least in characteristic zero, C_nV degenerates correspondingly to a subscheme of I_n, whose irreducible components are each of the form C_nW for an appropriate subvariety W of V_0. Moreover, if W is not a component of the scheme V_0, then W is contained in the singular locus of V_0.

(b) $[C_nV] = m_{N-1-n}(V)s_0(S^{\vee}) + \cdots + m_d(V)s_{d+1-N+n}(S^{\vee})$.

(c) As W runs through a set of subvarieties whose classes form a basis for the classes of dimension i on Y for $i = N-1-n,\cdots,d$, then $[C_nW]$ runs through a basis for the classes of codimension $n+1$ on I_n of the form

(3.19) $$m_{N-1-n}s_0(S^{\vee}) + \cdots + m_d s_{d+1-N+n}(S^{\vee})$$

where m_i is a class of dimension i on Y. In particular, $[C_nV]$ can be expressed as a linear combination of the $[C_nW]$. Moreover, if $Y = \mathbb{P}^N$ and W is an i-plane for $i = N-1-n,\cdots,d$, then the coefficient of $[C_nW]$ is $r_i(V)$.

4. Open problems.

There are several problems whose solution would mean a direct refinement of the Theorem. One problem concerns Part (c,iv), which presents a sufficient condition for there to be no X in the given family making two distinct proper contacts with one of the given varieties V_i. Now, the condition is convenient but not necessary. Indeed, the analysis in Section 2 yields more precise information. Namely, there is one narrow case in which such a bitangency actually does occur, and there is the possibility of a second case. The problem is to resolve this possibility.

In both cases of bitangency, the points of contact are distinct. In the first case, the hyperplanes of contact are equal; in the second case, they too are distinct. Furthermore, in the second case, one of the two bitangent varieties X or V_i has a component that is a linear space but not a hyperplane, and the other variety has the following property: there are two irreducible components, possibly equal but at any rate of the same dimension ≥ 2, and, in the product of these two components, there is a 1-codimensional locally closed subset that consists of all the pairs of distinct simple points such that the tangent spaces at the points are distinct, both spaces contain the line joining the points, and the line does not lie entirely in the variety.

In the second case, of course, the linear component contains both points of contact, and the other two components contain one each. Moreover, since both contacts are proper and since the hyperplanes of contact are distinct, it follows that the dimension of the linear component plus the common dimension of the other two is $\geq N$. Similarly, it follows that the common dimension of the two components is ≥ 2 and that their tangent spaces at the points of contact are distinct. (This assertion was made above but not proved in Section 2.) Conversely, it is easy to see that, whenever a variety that enjoys the above property, makes a proper contact with a linear space of complementary dimension, it must make a second proper contact, at a distinct point and with a distinct hyperplane of contact. The problem, then, is to find an example of a variety that enjoys the property or to prove that no such variety can exist.

Another problem is to strengthen Part (c,i) of the Theorem, which asserts only that each contact is proper. In some cases, more is known already. Katz [1973] proved essentially that in $\mathbb{P}^N$, in a 1-parameter family of hypersurfaces X , those touching a given variety V in general position intersect V in a variety X ∩ V with a nondegenerate quadratic singularity at the point of contact (and no other singularity) provided that (i) deg(X) = 1 and V is reflexive or (ii) deg(X) $\geq$ 2 and either the characteristic is not 2 or dim(V) is even. Furthermore, Ferrarese and Hefez (pvt. com.) proved that, if N = 2 and the characteristic is 0 , then the intersection multiplicities satisfy the relation.

(4.1) $\qquad i(P;X.V) = i((P,T_PV);CV.pr_{1*}[F^1]) - 1 ,$

whether or not V is in general position. The problem is to generalize these results.

There are several directions in which various special cases of the Theorem have been generalized, and there are several issues, one step away from the Theorem, on which work has been done. These directions and issues will now be discussed, and some open problems indicated. However, the relevant literature is extensive, and it would be impractical to cite more than some of the earliest and some of the most recent works. These work and their references should be consulted for more information.

There has been work in the theory of enumeration of contacts between the varieties of several families. Schubert ([1879] §14) proved, in an essentially rigorous way, the following result for plane curves, alongside a similar result for surfaces in $\mathbb{P}^3$: the points of contact between the curves in two 1-parameter families of plane curves with (1,0)-th and (0,1)-th characteristic numbers λ_0, λ_0' and λ_1, λ_1' trace a curve of degree (1st class)

(4.2) $$r_1 = \lambda_0 \lambda_0' + \lambda_1 \lambda_0' + \lambda_0 \lambda_0',$$

and their tangents envelop a curve of class (0th rank)

(4.3) $$r_0 = \lambda_0 \lambda_1' + \lambda_1 \lambda_0' + \lambda_1 \lambda_1'.$$

This work for curves and surfaces was generalized to hypersurfaces by Pieri (see Zeuthen-Pieri [1915] Fn. 121). In related work, Zobel [1958] found the number of contacts between varieties of any dimension in $\mathbb{P}^N$ varying in two families of appropriately related numbers of parameters. In fact, Zobel did something a little more general; he found the number of k-dimensional contacts. By definition, two varieties have a k-dimensional contact at a point P if they are smooth at P and their tangent spaces contain a common k-plane.

The preceding results have proofs similar to the proof of the Theorem. The original proofs themselves involve an intersection on I or I_k. It would seem better, however, to use a fibered product of varieties like F^P, if only to facilitate the passage to the case of three or more families. Perhaps, with fibered products, one could answer the question Zobel ([1958] §6) had in mind when he wrote this: "the problem of finding expressions for the numbers of varieties in a given system which have contacts of the kind studied here with varieties in two or more systems of suitable dimensions raises an important question in intersection theory which it will be more appropriate to discuss in a separate paper." Unfortunately, Zobel died young, and apparently that paper was not published (an obituary notice, including a list of publications, was given by Scott [1964]). Zobel (Fn(^{6})) did broach the matter of enumerative significance. He observed that the theory of general position and transversality available at the time yields the finiteness of the total number but not the multiplicities of appearance of the contacts, not even the expected values of 1 in characteristic 0. Of course, currently available theory will yield the 1's and more. It is an open problem to clean up and generalize all this work.

There has been a lot of work in the theory of enumeration of multiple and higher-order contacts. In connection with that theory, Zobel ([1958] §6) said this: "Arguments similar to those of §§2-4 can also be employed to find the number of [linear] spaces which are, at the same points, osculating to curves of two given systems, or osculating to curves of one given system and tangent to varieties of another given system...; these results may then be generalized by considering spaces which have more than ordinary contact with a variety other than a curve, for instance by containing the first two neighbourhoods of a point on the variety." Now, this approach will become more interesting after it has been used to recover and generalize some 19th century formulas.

In the plane, there is an explicit formula for the number of points where a curve in one general 1-parameter family osculates a curve in a second family; namely, if λ_0, λ_0' and λ_1, λ_1' denote the (1,0)-th and (0,1)-th characteristic numbers of the families, if k_1, k_1' denote the degrees of the loci of cusps, and if k_0, k_0' denote the classes of the loci enveloped by the inflectional tangents, then the formula is

(4.4) $\qquad \lambda_0 k_0' + \lambda_1 k_1' + k_0 \lambda_1' + k_1 \lambda_0' + 3\lambda_0 \lambda_0' + 3\lambda_1 \lambda_1'$.

Zeuthen (1879) found this formula and others of the same type for curves and surfaces via homology and the principle of conservation of number (see Zeuthen-Pieri [1915] p.274; note also the report on p.278 on the use of similar methods to enumerate the conics that cut or touch a given curve in $\mathbb{P}^N$ at several points.) It is an open problem to put this approach on a rigorous basis.

Schubert (1880) recovered (4.4) in particular as an application of his enumerative theory of triangles. Roberts and Speiser (see the announcement, Roberts-Speiser [1980])are currently working out Schubert's theory rigorously. However, it is an open problem to generalize the theory to cover coalescence among more than three points.

De Jonquières (1866) found a formula, which has become famous, for the number of plane curves of given degree that make a given number of contacts of given orders with a given curve and that pass through an appropriate number of points. A recent proof of the formula was given by Vainsencher [1981], who then went on using the same methods to obtain some related formulas for higher dimensional varieties. Moreover, Vainsencher studied the enumerative significance of the formulas. For higher dimensional varieties, there is much left open to do.

It was observed long ago that de Jonquières's formula yields via symbolic multiplication a formula in the case of several given curves. However, the variable curves that have an m-fold line as a component will be counted as making repeated m-fold contacts with each of the given curves. Thus, in some cases, the conditions may admit an infinite number of solutions, and the formula may lose its enumerative significance. Issues of this sort may be what Hilbert had in mind when he used the phrase, "limits of their validity", in the statement of his 15th problem. The problem is entitled, "Rigorous foundation of Schubert's enumerative calculus", and Hilbert wrote, "The problem consists in this: to establish rigorously and with an exact determination of the limits of their validity those geometrical numbers [formulas]..." (For some more information about Hilbert's 15th problem, see Kleiman [1976].) Yet rather than determining the limits of validity of the generalized de Jonquières formula, it would be better to refine the formula so that it is valid whenever the given curves are in general position. It is an open problem to find the right formula.

There can be an infinite number of irreducible curves each making a single contact of given order with a single given curve, although a finite number is expected and although the number of points of contact is finite. This possibility was pointed out by Hefez (pvt. com.), who provided the following example, which is valid in any characteristic:

in the 2-parameter family of conics that intersect the (smooth) curve $y = x+x^7 + y^8$ with multiplicity at least 3 at the origin, every conic in fact intersects with multiplicity at least 7. In a case like this, it would seem that de Jonquières's original formula holds no enumerative significance. Nevertheless, it turns out that, at least in characteristic 0, the formula yields the (weighted) number of points of contact. This fact can be proved by using a variant of the method used by Vainsencher, namely, the Wronskian method introduced by Halphen (1876) and developed in a general abstract form by Laksov [1981]; in fact, it suffice to apply Laksov's Thm 9. Moreover, the Wronskian method establishes the finiteness of the number of points. The case of several contacts with one or more given curves is open for investigation.

In positive characteristic, the issue is yet more involved. For instance, consider the smooth curve $y = x^{q+1} + y^q$, where q is a power of the characteristic. The tangent at a general point intersects the curve there with multiplicity q; the tangent at each of the q^3+1 points where $x^{q^2} = x$ intersects with multiplicity q+1. Now, de Jonquières's formula for the number of lines that make a contact of multiplicity $\geq$ 3 with a smooth plane curve of degree n is 3n(n-2), which is just the familiar Plücker formula. On the other hand, Thm. 9 of Laksov [1981] is valid in any characteristic, and in the present case it yields the formula n(n+(n-3)m), where m is the intersection multiplicity at a general point of the tangent with the curve. The two formulas coincide if and only if m = 2. Moreover, m = 2 if the curve is reflexive, and if not, then m is the inseparable degree of the map from the curve to its dual (possibly m = 2 in this case also). The case of variable conics has been analyzed by Hefez (pvt. com.) on the basis of Laksov's work, and this case is more complicated. Beyond that, the matter is open.

When the plane is embedded via the r-fold Veronese map, then the curves of degree r become the hyperplane sections; hence, as was observed long ago, de Jonquières's formula may be viewed as part of the theory of enumeration of exceptional secant linear spaces. For over 130 years, this theory has continually attracted a lot of attention. Success, however, has been somewhat limited. Aside from the isolated case of bitangent and tritangent planes to surfaces in $\mathbb{P}^3$ (see Vainsencher [1981] §8), success has been achieved only in two cases, (i) linear spaces of arbitrary dimension cutting curves and (ii) lines cutting varieties, especially hypersurfaces, special varieties of codimension 2, and surfaces. The methods employed in the two cases have been basically different. In case (i), they have, naturally enough, involved special properties of curves. In case (ii), on the other hand, they have been based on the ingenius idea of enumerating the coincidences of a given sort among a given number points on a variable line, an idea introduced and developed in this connection by Schubert (1877; [1879] §§33,4; [1886]). Some of the most recent work on both cases includes Arbarello-Cornalba-Griffiths- and Harris [1980], Colley [1983], Laksov [198], Ran [1982], and Vainsencher [1981]. The next cases open are these: lines cutting varieties of arbitrary dimension, and linear spaces of arbitrary dimension cutting surfaces.

Two special cases of the theorem have been generalized to families of transcendental varieties that are defined by algebraic differential equations. Such varieties were named <u>panalgebraic</u> by Loria [1901]. The first special case is the case of a 1-parameter (algebraic) family of plane curves X. The generalization is effected as follows. Identify the incidence correspondence I with the projectivization of the cotangent bundle of the plane. Then, in the notation of Section 2, the subset $\mathrm{pr}_1(F^1)$ of I is defined in inhomogenious coordinates by a polynomial equation

(4.5) $\qquad P(x,y,dy/dx) = 0$,

and this equation is satisfied by almost all X. Now, consider a plane curve V in general position, and consider the derivation of the expression for the number of X touching V. On applying the projection formula to (2.22), it becomes evident that what matters in the derivation is not that the X are algebraic but that $pr_1(F^1)$ is. Thus, the expression is valid as well for panalgebraic families - including various families of spirals, trigonometric curves, cycloids, evolutes, catenaries, etc. The second case which has been generalized is the case of a 1-parameter family of surfaces (see Zeuthen-Pieri [1915] n° 34). It is an open problem to treat several parameter families and higher dimensional varieties.

An algebraic partial differential equation accounts for the behavior of the conormal variety CV (= $C_{N-1}V$) of a subvariety V of an N-dimensional, smooth, irreducible variety Y. More precisely, the behavior is governed by the canonical contact structure on the projectivized cotangent bundle p: I → Y, the structure defined by the tautological "semi-1-form"

(4.6) $\omega: \mathcal{O}_I(-1) \to p^*\Omega_Y^1 \to \Omega_I^1$.

It is well-known (see Arnold [1978] Ex., p.355) and it easy to prove that CV satisfies the equation $\omega = 0$ in the sense that ω vanishes in Ω_{CV}^1. It is not hard to prove that, conversely, if D is a subvariety of I satisfying the equation $\omega = 0$ and if the restriction, D → pD, is smooth on a dense open subset of D (it is always so in characteristic 0), then D lies in CpD; whence, D = CpD if and only if D is of pure dimension N-1. Part (a) of the Proposition, which describes the behavior in characteristic 0 of $C_n V$ under a degeneration of V, follows now for n = N-1; the general case may be derived from this case, or it may be proved directly by replacing ω by the analogous semi-(N-n)-form. Similarly, the C. Segre-Wallace criterion for reflexiveness, recalled in Section 1, follows. Indeed, if $Y = \mathbb{P}^N$ and if ω^* is the "semi-1-form" on I viewed as the projectivization of the cotangent bundle of $\mathbb{P}^N_V$, then clearly $\omega+\omega^* = 0$; hence, CV satisfies $\omega^* = 0$. It is an open problem to algebrize and use more of the differential-ge .etric theory of contact manifolds. Perhaps in this way the theory of duality can be extended to nonembedded varieties.

Clebsch (1873) was the first to recognize a connection between differential equations and subvarieties of the incidence correspondence. However, Fouret (1874-8) independently founded the enumerative theory of panalgebraic curves and surfaces in a series of B.S.M.F. articles and Comptes Rendus notes. Zeuthen (1880) proposed a method based on this theory for recognizing whether or not certain differential equations are algebraically integrable. (See Zeuthen-Pieri [1915] n° 34.) Ferrarese is currently preparing an up-to-date expositon of the theory. The proof above of the C. Segre-Wallace criterion is essentially the originally one of Segre (1910), as presented in Kleiman [1980], but it is carried out in 3 steps of independent interest and usefulness. All of this work contributes toward a solution of a very important, little understood open problem, and that is to see what qualitative restrictions a differential equation puts on its solutions.

The behavior of $C_n V$ under a degeneration of V is a subject that appears to be related to parts of the complex-analytic theory summarized by Merle [1982], and it is an open problem to explore this issue. On a different tack, it is an open problem to understand the behavior of $C_n V$ in positive characteristic. A new approach will be required. Indeed, the equation $\omega = 0$ can have solutions in addition to those of the form CW. For example, Wallace ([1956] §7.3) showed how to construct infinitely many curves V such that the dual curve V* is equal to a given curve W and such that the separable and inseparable degrees of the map CV → W are equal to given integers s and q provided only q > 1; here, CV satisfies $\omega^* = 0$ but CV ≠ CW.

There is another open problem whose solution would mean a direct refinement of the theorem. It is to determine the relative positions of the p varieties $V_1, \cdots, V_p$ that are necessary and sufficient so that the varieties X in the given p-parameter family that touch them are counted with multiplicity 1 (or, more generally, mcq, see Part (d) of the Theorem and the remark below it). Possibly this (instead of what was suggested earlier) is what Hilbert had in mind when he asked in the statement of his 15th problem that numbers, like the total number of these X, be established "with an exact determination of the limits of their validity". At any rate, the multiplicities of appearance of the X's should be determined more generally when the relative positions of $V_1, \cdots, V_p$ are mildly degenerate.

For example, in $\mathbb{P}^2$ in any characteristic but 2, let X vary in the 5-parameter family of (reduced) conics, and let $V_1, \cdots, V_5$ be smooth conics. If no two V_i touch each other, if there is no pair of points such that each V_i passes through one of them or touches their join, and if there is no pair of lines such that each V_i touches one of them or passes through their intersection, then there are 3264 X's touching $V_1, \cdots, V_5$, each X is smooth and appears with multiplicity,

(4.7) $$\pi_{i=1}^{5} (4 - \text{card}(X \cap V_i)).$$

This result was proved in Fulton-MacPherson ([1977] §7). It was generalized to curves of arbitrary degree (without an explicit determination of the total number, of course) in Hefez-Sacchiero [1983]. However, there it is assumed that almost all X and all V_i are general curves of their degree. The case in which the V_i are given arbitrarily is probably the next case to consider.

Associated to an enumeration of algebro-geometric figures satisfying variable algebro-geometric conditions, there is a group, which reflects the complexity of the situation, and there is the problem of determining it. The group may be viewed as the Galois group of a field extension; namely, it is the extension of the field of definition of the generic conditions by the field of definition of the figures satisfying them. Alternatively, the group may be viewed as a monodromy group; namely, varying the parameters of the conditions along various closed paths yields a group of permutations of the figures satisfying the initial conditions. The equivalence of the two viewpoints was established by Harris [1979]. Harris worked exclusively over the complex numbers. It is an open problem to develop the theory in positive characteristic and in mixed characteristics.

Harris [1979] determined the group in these cases: the flexes and bitangents of a plane curve of degree d, the lines on a hypersurface of degree $2N-3$ in $\mathbb{P}^N$, and the 3264 conics touching 5 conics. In the cases of the flexes of a cubic, of the bitangents of a quartic, and of the lines on a cubic, the groups were found in the 19th century, and they are nontrivial. In each of the remaining cases, the group is the full symmetric group. Harris proved this result by showing that the group is doubly transitive and that it contains a transposition. A transposition exists because there is a stratum of parameter values where exactly two of the corresponding figures coalesce. In the case of the conics touching 5 conics, the existence of a transposition is therefore a consequence of the expression (4.7) of Fulton-MacPherson. Similarly, Hefez-Sacchiero [1983] proved that the group is the full symmetric group in the case of curves of degree $m \geq 2$ touching $p = \binom{N+m}{N} - 1$ general curves V_i of various degrees ≥ 2. Once again, the case in which the V_i are arbitrary (reduced) plane curves is probably the next case to consider.

Chasles felt that the characteristic numbers of a family suffice to describe it for enumerative purposes. He supported this contention with over 200 examples. Most were stated in affine and metric terms like focus, diameter, angle bisector, etc. However, they can be rephrased in terms of a line at infinity, a pair of circular points, etc. The effect is dramatic, but tangency is easily unmasked. On the other hand, Halphen

(1878) found some subtle conditions on conics. The number of smooth conics in a family that satisfy one of them is not simply equal to a linear combination of the characteristic numbers. Halphen rectified the situation by introducing additional characteristic numbers. Halphen's work can be justified by using Hironaka's theory of equivariant blowing-up (see Kleiman [1980] or by a direct analysis of the appropriate singular loci (Casas, pvt. com.). It remains to generalize the theory to p-parameter families of quadric n-folds in $\mathbb{P}^N$ in any characteristic. For varieties of higher degree, little is known. For example, it is an open problem to find a complete set of characteristic numbers for the (de Jonquières) condition on the plane curves in a family to make a given number of contacts of given order with a given curve V. Looked at from Schubert's point of view (see Section 3), the problem is to express the condition as a linear combination of certain standard conditions, independent of V, and to express the combining coefficients as the number of curves in certain standard families that satisfy the condition.

The Theorem, howsoever it is refined and generalized, can never be more than a first step in the rigorous verification of explicit numbers, like the three famous numbers mentioned at the very beginning. It is still necessary to find the characteristic numbers of the families. For the three numbers themselves, finding the corresponding characteristic numbers is by virtue of (1.12) also sufficient to complete the verification. For example, the characteristic numbers of the 5-parameter family of all conics in the plane are 1, 2, 4, 4, 2, 1; hence, the number of conics touching 5 others is

(4.8) $\qquad (2\lambda_0 + 2\lambda_1)^5 = 32\left(1 + 2\binom{5}{1} + 4\binom{5}{2} + 4\binom{5}{3} + 2\binom{5}{1} + 1\right) = 3264$.

Nevertheless, it is seldom easy to find the characteristic numbers.

For the family of all conics in the plane, some of the characteristic numbers were found by the ancient Greeks, and the rest, by Newton

(Principia, 1687). Zeuthen (1865), inspired by Cremona (1864), developed a method for finding the characteristic numbers of arbitrary families of conics. The key idea is to pass via basis-change relations between the basis of the two elementary conditions, to pass through a point and to touch a line, and the basis of the two degeneracy conditions, to be a line-pair and to be a double-line (see Kleiman [1980]). Suitably generalized, the method has been used in other cases as well. The characteristic numbers were found for the family of all conics in space by Chasles (1865), and for that of all quadrics in space by Zeuthen (1866). Schubert (1894) developed an algorithm for finding the numbers for the family of all quadric n-folds in $\mathbb{P}^N$. Recently, these numbers and the general algorithm were established rigorously in independent works, each with its own valuable point of view, by van der Waerden (pvt.ms., 1981), Vainsencher [1982], De Concini-Procesi [1982], and Laksov (pvt.ms., 1982). Thus, in particular, the rigorous verification of the numbers 3264 and 666, 841, 088 is complete.

Maillard (1871) and Zeuthen (1872) independently found the characteristic numbers of the 7-parameter family of cuspidal plane cubics, then those of the 8-parameter family of nodal cubics, then finally those of the 9-parameter family of all plane cubics. Schubert (1874, 5) refined this work and went on to find the characteristic numbers of the 12-parameter family of all twisted cubic space curves. Zeuthen (1873) found the characteristic numbers of the 14-parameter family of all plane quartics via a lengthy step-by-step determination of the numbers of various subfamilies of singular quartics. (For a discussion of all of this work, see Schubert [1879].) Recently, Sacchiero (pvt. com.), Speiser (pvt. com.), Sterz [1982], and Strømmer (pvt. com.) independently have begun the enumeration of plane cubics. Ellingsrud (pvt. com.), Harris (pvt. com.), Piene [1982], and Piene-Schlessinger [1982] have begun the enumeration of twisted cubics. Perhaps, before too long, the number 5, 819, 539, 783, 680 will be verified.

The 19th century work on finding the characteristic numbers of families of curves of higher degree is rich and lovely. Understanding it well enough to vindicate it and to continue it, is possibly the most important part of Hilbert's 15th problem remaining open. This part alone has the full stature of a Hilbert problem. It is, as Hilbert wrote, "so clear that you can explain it to the first man you meet on the street" and yet "from the discussion of [it] an advancement of science may be expected."

References

Arbarello, E.-Cornalba, M.-Griffiths, P.-Harris,J. [1980]: Topics in the theory of algebraic curves, to appear in the Princeton Math. Series.

Arnold, V. [1978]: Mathematical methods of classical mechanics, translated by K. Vogtmann and A. Weinstein, Graduate Texts in Math. 60, Springer, New York (1978).

Colley, S. [1983]: On the enumerative geometry of stationary multiple-points, thesis, Mass. Institute of Tech., Cambridge, Mass. (1983).

Coolidge, J. [1959]: A treatise on algebraic plane curves, Dover, New York (1959).

Cremona, L. [1862]: Introduzione ad una teoria geometrica delle curve piane, Bologna (1862) = Einleitung in einer Theorie der ebenen Curven, translated by M. Curtze, Greisswald (1865).

De Concini, C.-Procesi, C. [1982]: Complete symmetric varieties, preprint, Univ. di Roma (1982).

Demazure, M. [1974]: "Désingularisation des variétés de Schubert généralisées", Ann. Scient. Ec. Norm. Sup. (4) 7(1974), 53-88.

Ehresmann, C. [1934]: "Sur la topologie de certains espaces homogènes", Ann. of Math. 35 (1934), 396-433.

Fulton, W.-MacPherson, R. [1977]: "Defining algebraic intersections", Algebraic Geometry (Proc. Sympos., Univ. Tromsø, Tromsø Norway, 1977) Lecture Notes in Math., 687. Springer, Berlin (1978), 1-30.

Giambelli, G. [1905]: "La teoria delle formule d'incidence e di positione speciale e le forme binarie", Atti Acc. Scienze Torino XL (1904-5), 1041-62.

Grayson, D. [1979]: "Coincidence formulas in enumerative geometry", Com. in Algebra 7 (16) (1979), 1685-1711.

Harris, J. [1979]: "Galois groups of enumerative problems", Duke Math. J. 46 (1979), 685-724.

Hefez, A.-Kleiman, S. [1983]: "Notes on duality for projective varieties", to appear.

Hefez, A.-Sacchiero, G. [1983]: "The Galois group of the tangency problem for plane curves", to appear.

Hodge, W.-Pedoe, D. [1952]: Methods of algebraic geometry, vol. II, Cambridge Univ. Press. (1952), reprinted 1968.

Katz, N. [1973]: "Pinceaux de Lefschetz: théorème d'existence", SGA7II, Lecture Notes in Math. 340, Springer, Berlin (1973), 212-253.

Kempf, G.-Laksov, D. [1974]: "The determinantal formula of Schubert calculus", Acta Math. 132 (1974), 153-162.

Kleiman, S. [1974]: "The transversality of a generic translate", Compositio Math. 28 (1974), 287-297.

Kleiman, S. [1976]: "Problem 15. Rigorous foundation of Schubert's enumerative calculus", Mathematical Developments arising from Hilbert Problems, Proc. Sympos. in Pure Math., XXVIII, A.M.S., Providence (1976).

Kleiman, S. [1977]: "The enumerative theory of singularities of mappings", Real and complex singularities, Proc. Sympos. Oslo 1976, P. Holm editor, Sijthoff and Noordhoff (1977).

Kleiman, S. [1980]: "Chasles's enumerative theory of conics. A historical introduction", Studies in Algebraic Geometry, MAA Studies in Math. 20, A. Seidenberg editor (1980), 117-138.

Kleiman, S. [1981]: "Concerning the dual variety", Proc. 18th Scandinavian Congress of Math., Birkhaüser (1981).

Laksov, D. [1981]: "Wronskians and Plücker formulas for linear systems on curves", Institut Mittag-Leffler Report No. 11 (1981), to appear in Ann. Ecole Norm. Sup.

le Barz, P. [1982]: Quelques formules multi-sécantes pour les surfaces, preprint, Nice (1982).

Loria, G. [1901]: "Le curve panalgebriche", Memorie della R. Società delle Science di Praga (1901), ristampata con correzioni ed aggiunte nel Le Mat. pure ed applicate II, Num. 4-5 (1902), 1-24.

MacPherson, R. [1974]: "Chern classes for singular algebraic varieties", Ann. of Math. 100 (1974), 423-32.

Merle, M. [1982]: "Variétés polaires, stratifications de Whitney et classes de Chern des espaces analytiques complexes [d'après Lê-Teissier], Sém. Bourbaki, Nov. 1982, exp. 600.

Piene, R. [1978]: "Polar classes of singular varieties", Ann. Scient. Ec. Norm. Sup. 11 (1978), 247-76.

Piene, R. [1982]: "Degenerations of complete twisted cubics", Enumerative geometry and classical algebraic geometry. Proc. Sypos. Nice (1980), le Barz and Hervier editors, Birkhaüser, Boston (1982), 37-50.

Piene, R.-Schlessinger, M. [1983], On the Hilbert scheme compactification of the space of twisted cubics, preprint, Univ. of Oslo (1983).

Porteous, I. [1971]: "Simple singularities of maps", Liverpool singularity sympos. I, Lecture Notes in Math. 192, Springer, Berlin (1971), 286-307.

Ran, Z. [1982]: The class of a Hilbert scheme inside another, with applications to projective geometry and special divisors, announcement, Math. Dept. Brandeis Univ. (1982).

Roberts, J.-Speiser, R. [1980]: "Schubert's enumerative geometry of triangles from a modern viewpoint", Algebraic Geometry, Proc. of a Conference at Chicago Circle, Lecture Notes in Math. 862, Springer, Berlin (1981).

Schubert, H. [1879]: Kalkül der abzählenden Geometrie, Teubner, Leipzig (1879), reprinted with an introduction by S. Kleiman and a list of publications assembled by W. Burau, Springer, Berlin (1979).

Schubert, H. [1886]: "Die n-dimensional Verallgermeinerungen der fundamentalen Anzahlen unseres Raums", Math. Ann. 26 (1886), 26-51.

Schubert, H. [1903]: "Gleichungen zwischem Bedingungen bei special Lage linearer Räume", Mitt. math. Ges. Hamburg 4 (1903), 104.

Scott, D. [1964]: "Andrew Zobel", J. London Math. Soc. 39 (1964), 566-7.

Segre, C. [1892]: "Intorno alla storia del principio de correspondenza e dei sistemi di curve", Bibl. math., 6 (1892), 33-47 = Opere, vol. 1, 185-197.

Segre, C. [1912]: "Mehrdimensionale Raüme", Encyclopadie der Matematischen Wissenschaften, Teubner, Leipzig (1912-34) III, 2, 2, C7, 669-972.

Sterz, U. [1982]: "Beruhrungsvervollständigung für ebene Kurven dritter Ordnung", Beitr. Alg. Geom., in press.

Vainsencher, I. [1978]: "Conics in characteristic 2", Compositio Math. 36 (1978), 101-12.

Vainsencher, I. [1981]: "Counting divisors with prescribed singularities", Trans. A.M.S. 267 (1981), 399-422.

Vainsencher, I. [1982]: "Schubert calculus for complete quadrics", Enumerative geometry and classical algebraic geometry. Proc. Sypos. Nice (1980), le Barz and Hervier editors, Birkhaüser, Boston (1982), 199-235.

Urabe, T. [1981]: "Duality of numerical characters of polar loci", Publ. Res. Inst. Math. Sci. 17 (1981), 331-345.

Wallace, A. [1956]: "Tangency and duality over arbitrary fields", Proc. London. Math. Soc. 3 (1956), 321-342.

Zeuthen, H.-Pieri, M.: "Géométrie énumérative", Encyclopédie des science mathématique, Teubner, Leipzig (1915), III, 2, 260-331.

Zobel, A. [1958]: "On the contacts between the varieties of two systems", Rend. di Mat., 17 (1958), 415-422.

UN PROBLEME DU TYPE BRILL-NOETHER
POUR LES FIBRES VECTORIELS

Franco Ghione

Soit X une courbe complète, lisse, irréductible de genre g, définie sur un corps k algébriquement clos de caractérique zéro. Considérons un faisceau localement libre F sur X de degré d et de rang s. Dans cette note on veut étudier le sous-ensemble fermé de Pic_X^m:

$$W_m^r(F) = \{L \in \text{Pic}_X^m : \dim H^0(X, F \otimes L) \geq r+1\}$$

où r et m sont deux entiers fixés.

Dans le cas s = 1 on a le problème de Brill-Noether classique. En effet si N est un faisceau inversible de degré n, alors l'isomorphisme

$$\text{Pic}_X^m \xrightarrow{\sim} \text{Pic}_X^{m+n}$$

défini par multiplication avec N, donne pour tous $m \in \mathbb{Z}$, l'isomorphisme

$$W_m^r(F \otimes N) \xrightarrow{\sim} W_{m+n}^r(F)$$

et donc, si s = 1, on peut supposer $F = O_X$. En ce cas, alors, $W_m^r(O_X)$, que nous écrivons simplement avec W_m^r, est l'ensemble des séries linéaires sur X de degré m et de dimension au moins r. Pour W_m^r on sait que ([4], [3])

Recherche supportée par le "Ministero della Pubblica Istruzione".

(a) $$\dim W_m^r \geq \tau = g - (r+1)(r-m+g)$$

en plus, si dans la formule (a) il y a égalité, alors:

$$[W_m^r] \equiv \alpha \cdot \Theta^{g-\tau}$$

où $[W_m^r]$ dénote la classe fondamentale du sous ensemble férmé W_m^r dans l'anneau de Chow de Pic_X^m, Θ est la classe du diviseur teta, $\equiv$ signifie équivalence numérique des cycles, et α est un nombre entier donné par

$$\alpha = \prod_{i=0}^{r} \frac{i!}{(r-m+g+i)!}$$

(b) Si X est une courbe générique dans sa variété des modules, alors dans (a) on a égalité.

Dans le cas $s = 2$ on peut obtenir un résultat pareil à (a) en étudiant les séries linéaires des diviseurs de la surface réglée $\mathbb{P}(F)$, unisecants les génératrices [1].

Dans cette note on veut démontrer que, si F est un faisceau localement libre général, alors on a

(a) $$\dim W_m^r(F) \geq \tau(F) = g - (r+1)(r-\rho+g)$$

où $\rho = sm + d - (g-1)(s-1)$. En plus, si dans la formule (a) il-y-a égalité, alors

$$[W_m^r(F)] \equiv \alpha(F) s^{g-\tau(F)} \cdot \Theta^{r-\tau(F)}$$

où

$$\alpha(F) = \prod_{i=0}^{r} \frac{i!}{(r-\rho+g+i)!}$$

L'idée de la démonstration est celle de considérer le

schéma Quot_F^m qui paramétrise les quotients de F cohérents de rang $s-1$ et de degré $d+m$. Si G est un tel quotient on a la suite exacte:

$$0 \to I(G) \to F \to G \to 0$$

et $\underline{\text{Hom}}(I(G), O_X) = I(G)^{\vee}$ est un faisceau inversible de degré m. On a alors un morphisme fonctoriel

$$\Psi_m: \text{Quot}_F^m \to \text{Pic}_X^m$$

défini, avec abus de notations, par $\Psi_m(G) = I(G)^{\vee}$. Dans le cas $s = 1$ $\text{Quot}_{O_X}^m = \text{Hilb}_X^m = S^m(X)$ et le morphisme Ψ_m est celui considéré dans [4]. Dans le cas $s = 2$ on a la même situation considerée en [1].

En tous cas la fibre du morphisme Ψ_m est donnée par

$$\Psi_m^{-1}(L) = \mathbb{P}(H^0(X, F \otimes L)^{\vee})$$

et donc

$$W_m^r(F) = \{L \in \text{Pic}_X^m: \dim \Psi_m^{-1}(L) \geq r\}$$

Si on suppose alors que le schéma Quot_F^m soit lisse et irréductible, on peut appliquer la formule de Porteous pour évaluer la dimension de $W_m^r(F)$ et, dans le cas de bonne dimension, calculer la classe fondamentale du cycle $W_m^r(F)$ dans l'anneau de Chow de Pic_X^m. Maintenant l'hypothèse que Quot_F^m soit lisse et irréductible est vérifiée si F est général dans le sens de [2], et donc, si on connait les classes de Chern de Quot_F^m, on peut bien appliquer la formule de Porteous et démontrer le résultat annoncé. Donc le calcul des classes de Chern du schema Quot_F^m est la partie principale de cette note.

§1. UNE DESCRIPTION DU SCHEMA Quot_F^m

Dans ce paragraphe nous voulons donner une construction pour obtenir le schéma Quot_F^m comme une intersection complète de diviseurs dans le schéma $\text{Quot}_{O_X^s}^m$. La construction est donnée dans tous les détails dans [2].

Soit G un quotient cohérent de F de rang $s-1$ et de degré $d+m$. On a la suite canonique

$$0 \to I(G) \to F \to G \to 0$$

où $I(G)$ est un faisceau inversible de degré $-m$.
On a alors le morphisme fonctoriel, qui généralise le morphisme de Abel-Jacobi:

$$\Psi_m: \text{Quot}_F^m \longrightarrow \text{Pic}_X^m$$

qui transforme le quotient G dans la classe du faisceau inversible $I(G)^\vee$.

Considérons sur $X \times \text{Pic}_X^m$ le faisceau universel de Poincaré $\mathbf{L}$ et soient p_1, p_2 les projections canoniques

$$p_1: X \times \text{Pic}_X^m \longrightarrow X, \qquad p_2: X \times \text{Pic}_X^m \longrightarrow \text{Pic}_X^m.$$

Il existe un faisceau cohérent Q_F sur Pic_X^m ([2] 3.10) défini par

(1) $\qquad \underline{\text{Hom}}\,(Q_F, M) = p_{2*}(\mathbf{L} \otimes p_1^* F \otimes p_2^* M)$

tel que

$$\mathbb{P}(Q_F) = \text{Quot}_F^m$$

et le morphisme Ψ_m correspond à la projection canonique $\mathbb{P}(Q_F) \longrightarrow \text{Pic}_X^m$. On voit alors, immédiatement, que

(2) $$\Psi_m^{-1}(L) = \mathbb{P}(H^0(X, F \otimes L)^{\vee})$$

et $W_m^r(F)$ devient le sous-ensemble fermé de Pic_X^m défini par les $x \in \text{Pic}_X^m$ tels que $\dim \mathcal{Q}_F \otimes k(x) \geq r+1$. En plus on a l'isomorphisme fonctoriel

(3) $$W_m^r(F \otimes N) \simeq W_{m+n}^r(F)$$

pour chaque faisceau inversible N de degré n et pour chaque intier r et m.

On peut alors supposer $\deg \check{F} = -d \gg 0$ et on a alors ([2] 1.14) une suite exacte

(4) $$0 \to F \to \mathcal{O}_X^s \to \mathcal{O}_D \to 0$$

où $\mathcal{O}_D = \bigoplus_{i=1}^{-d} \mathcal{O}_{\{x_i\}}$, $x_i \in X$, $x_i \neq x_j$ pour $j \neq j$ et $\mathcal{O}_{\{x_i\}}$ est le faisceau concentré sur le point x_i, $\mathcal{O}_X/m_i = \mathcal{O}_{\{x_i\}}$. La suite (4) nous donne, pour chaque i, les projections $\mathcal{O}_X^s \to \mathcal{O}_{\{x_i\}}$ et donc les suites exactes

(5) $$0 \to F_i \to \mathcal{O}_X^s \to \mathcal{O}_{\{x_i\}} \to 0.$$

On a alors, pour chaque faisceau cohérent M sur Pic_X^m, la suite

$$0 \to p_{2*}(\mathbb{L} \otimes p_1^* F_i \otimes p_2^* M) \to p_{2*}(\mathbb{L} \otimes p_1^* \mathcal{O}_X^s \otimes p_2^* M) \to p_{2*}(\mathbb{L} \otimes p_1^* \mathcal{O}_{\{x_i\}} \otimes p_2^* M)$$

et, par la formule (1)

(6) $$\check{\mathbb{L}}_{x_i} \to \mathcal{Q}_{\mathcal{O}_X^s} \to \mathcal{Q}_{F_i} \to 0$$

où $\mathbb{L}_{x_i} = \mathbb{L}|_{x_i \times \text{Pic}_X^m}$, et cela défini une immersion

$$\alpha_i : \mathbb{P}(\mathcal{Q}_{F_i}) \to \mathbb{P}(\mathcal{Q}_{\mathcal{O}_X^s})$$

où le diviseur $\mathbb{P}(Q_{F_i}) =: Q_i$ est plongé dans $\mathbb{P}(Q_0 s) =: Y$ comme le schéma des zéros de la section de $O_{Q_0 s_X}(1) \otimes \psi_m^*(\mathbb{L}_{x_i})$ définie par le morphisme composé

$$O_Y \to \psi_m^*(Q_{0^s_X} \otimes \mathbb{L}_{x_i}) \to O_{Q_0 s_X}(1) \otimes \psi_m^*(\mathbb{L}_{x_i}).$$

En plus, par la formule (1), on déduit que

$$Q_{0^s_X} = Q_{0_X} \oplus \ldots \oplus Q_{0_X} \quad \text{(s-fois)}$$

et $\mathbb{P}(Q_{0_X}) = S^m(X)$. Donc, si $m \geq 2g-2$, $Q_{0^s_X}$ est localement libre de rang $s(m-g+1)$ et ses classes de Chern sont déduites des classes de Chern de Q_{0_X} qui sont bien connues.

Supposons maintenant que le faisceau de Poincaré $\mathbb{L}$ soit normalisé dans le point $x_0 \in X$:

$$\mathbb{L}_{x_0} = O_{Pic^m_X}$$

et soit Q un diviseur de Y dans la classe définie par $O_{Q_0 s_X}(1)$. On a alors l'équivalence algébrique

$$Q_i \equiv Q.$$

Les suites (4) et (5) nous donnent l'isomorphisme ([2] 1.2)

$$\mathbb{P}(Q_F) = \overset{-d}{\underset{i=1}{X}} \mathbb{P}(Q_{F_i}),$$

et, si F est général dans sa varieté des modules, l'intersection des diviseurs $\mathbb{P}(Q_{F_i}) = Q_i$ est définie et transversale [2] et $\mathbb{P}(Q_F)$ est irréductible (si de dimension positive) [2]. On a alors la

Proposition 1

Soit $Q = Q_{0_X}$ le faisceau sur Pic_X^m tel que $\mathbb{P}(Q) = S^m(X)$, soit $Y = \mathbb{P}(Q \oplus \ldots \oplus Q)$ (s-fois) et Q un diviseur de Y dans la classe définie par $\mathcal{O}_Y(1)$. Supposons que $\deg \check{F} = -d \gg 0$ sur $s \geq 2$, $m \geq 2g-2$, il existe alors une immersion fermée

$$\alpha: \text{Quot}_F^m \to Y$$

telle que

$$\text{Quot}_F^m = \bigcap_{i=1}^{-d} {}_Y Q_i$$

où le Q_i sont des diviseurs lisses de Y algébriquement équivalents à Q. En plus l'intersection des diviseurs Q_i est définie et transversale dans tous les points et Quot_F^m est irréductible si sa dimension est positive.

Si $s = 1$ la proposition peut s'enoncer également en regardant $d = 0$, $Y = \mathbb{P}(Q)$ et $\text{Quot}_F^m = Y$.

§2. LES CLASSES DE SEGRE DU SCHEMA Quot_F^m

Soit V une varieté lisse et irréductible définie sur le corps k et soit F un faisceau localement libre de rang r sur V. Nous dénotons par $C(t,F)$ le polynôme de Chern:

$$C(t,F) = 1 + c_1(F)t + c_2(F)t^2 + \ldots + c_r(F)t^r;$$

le polinôme de Segre est défini par

$$S(t,F) = 1 + s_1(F)t + s_2(F)t^2 + \ldots + s_d(F)t^d$$

où $d = \dim V$ et

$$C(t,\check{F}) \cdot S(t,F) = 1.$$

Pour les classes de Segre on a la formule

$$\pi_*(\xi^{r-1+i}) = s_i(F)$$

où $\pi: \mathbb{P}(F) \to V$ est la projection canonique, et $\xi = c_1(\mathcal{O}_V(1))$ est la classe de Grothendieck de F. Enfin, si Ω_V est le faisceau des differentiels de V sur k, alors les classes caractéristiques de V sont données par

$$C(t,V) =: C(t,\check{\Omega}_V) = 1 + c_1(V)t + \ldots + c_d(V)t^d$$

et donc

$$C(t,V) = S(t,\Omega_V)^{-1}.$$

On veut maintenant calculer $S(t,\Omega_V)$ pour la variété $V = \text{Quot}_F^m$. Avec les mêmes notations que dans le paragraphe précédent, nous avons le diagramme suivant

$$\text{Quot}_F^m = \mathbb{P}(Q_F) \xhookrightarrow{\alpha} Y = \mathbb{P}(Q \oplus \ldots \oplus Q)$$

$$\Psi_m \qquad\qquad \pi$$

$$\text{Pic}_X^m$$

En utilisant les suites exactes canoniques

$$0 \to \pi^* \Omega_{\text{Pic}_X^m} \to \Omega_Y \to \Omega_{Y|\text{Pic}_X^m} \to 0$$

$$0 \to \Omega_{Y|\text{Pic}_X^m} \to \pi^*(Q)^s \otimes \mathcal{O}_Y(-1) \to \mathcal{O}_Y \to 0$$

on obtient

$$C(t,\Omega_Y) = (1-\xi t)^{s(m-g+1)} \cdot C\left(\frac{t}{1-\xi t}, \pi^*(Q^s)\right)$$

et donc

$$S(t,\Omega_Y) = \left[S(\frac{t}{1+\xi t}, \pi^*Q)/(1+\xi t)^{m-g+1}\right]^s$$

Pour calculer les classes de Segre de Quot_F on utilise le fait que, si F est général, $\mathbb{P}(Q_F)$ est l'intersection complète des diviseurs Q_i et donc on peut calculer le fibré normal de $\mathbb{P}(Q_F)$ dans Y. On trouve la

Proposition 2

On a

$$S(t,\Omega_{\text{Quot}_F^m}) = \prod_{i=1}^{-d}(1+\zeta_i t)\left[S(\frac{t}{1+\zeta t}, \psi_m^*(Q))/(1+\zeta t)^{m-g+1}\right]^s$$

où $\zeta = \alpha^*\xi$ et $\zeta_i = \alpha^*([Q_i])$. En particulier on trouve, dans l'anneau d'équivalence numérique

$$S(t,\Omega_{\text{Quot}_F^m}) \equiv S(\frac{t}{1+\zeta t}, \psi_m^*(Q))^s/(1+\zeta t)^{\rho-g+1}$$

où

$$\rho = s(m-g+1) + g - 1 + d = \dim \text{Quot}_F^m$$

et nous supposons toujours $\rho \geq 0$.

§3. APPLICATIONS DE LA FORMULE DE PORTEOUS

Pour chaque série de Laurent

$$S(t) = \ldots + s_{-2}t^{-2} + s_{-1}t^{-1} + s_0 + s_1 t + s_2 t^2 + \ldots$$

à coefficients dans un anneau gradué, nous posons

$$\Delta_{a,r}(S(t)) = \begin{vmatrix} s_a & s_{a+1} & \cdots & s_{a+r-1} \\ s_{a-1} & s_a & \cdots & s_{a+r-2} \\ \cdots & \cdots & \cdots & \cdots \\ s_{a-r+1} & s_{a-r+2} & \cdots & s_a \end{vmatrix}$$

Soit, maintenant, $f: X \to Y$ un morphisme de variétés lisses, projectives à fibres (dans le sens de schémas) lisses.

Soit $\rho = \dim X$ et $g = \dim Y$. Considérons le sous-ensemble fermé $G^r \subset X$,

$$G^r = \{x \in X | \dim f^{-1}f(x) \geq r\} \quad \rho-g \leq r \leq \rho.$$

Nous avons les résultats suivants (Parteous):

i) Si $G^r \neq \emptyset$ alors $\operatorname{codim}_X G^r \leq r(g-\rho+r)$

ii) Si $G^r \neq \emptyset$ et chaque composant irréductible de G^r est de codimension $r(g-\rho+r)$ alors

$$[G^r] = \Delta_{g-\rho+r,r}(C(t,f^*\Omega_Y^\vee) \cdot S(t,\Omega_X))$$

iii) Si $G^r = \emptyset$ alors

$$\Delta_{g-\rho+r,r}(C(t,f^*\Omega_Y^\vee) \cdot S(t,\Omega_X)) = 0.$$

Nous voulons appliquer ces résultats au morphisme

$$\Psi_m: \operatorname{Quot}_F^m \to \operatorname{Pic}_X^m.$$

Etant donné que $\Omega_{Pic_X^m}$ est trivial, il suffit de calculer le déterminant $r \times r$

$$\Delta_{g-\rho+r,r}(S(t,\Omega_{Quot_F^m})) \quad \text{pour} \quad \rho-g \leq r \leq \rho.$$

En utilisant la proposition 2 on a

$$\Delta_{g-\rho+r,r}(S(t,\Omega_{Quot_F^m})) \equiv$$

$$\equiv \Delta_{g-\rho+r,r}((1+\zeta t)^{g-1-\rho} \cdot S(\frac{t}{1+\zeta t}, \Psi_m^*(Q))^s) =$$

$$= \Delta_{g-\rho+r,r}((1+\zeta t)^{g-\rho}(1-\zeta\frac{t}{1+\zeta t}) \cdot S(\frac{t}{1+\zeta t}, \Psi^*(Q))^s) = [\,4 \quad \text{lemma 13}]$$

$$= \Delta_{g-\rho+r,r}((1-\zeta t) \cdot S(t, \Psi_m^*(Q))^s) \equiv$$

$$\equiv \Delta_{g-\rho+r,r}((1-\zeta t)\exp(\Psi_m^*(\theta) \cdot st)) =$$

$$= \alpha(F) \cdot \sum_{i=0}^{r} \frac{(g-\rho+r-1+i)!}{i!(r-i)!} \zeta^i s^{\rho-r-\tau-i} \Psi_m^*(\theta^{\rho-r-\tau-i})$$

où $\tau = g-(r+1)(r-\rho+g)$ et

$$\alpha(F) = \prod_{i=0}^{r} \frac{i!}{(g-\rho+r+i)!}.$$

En particulier

$$(\Psi_m)_*(\zeta^r \cdot \Delta_{g-\rho+r,r}(S(t,\Omega_{Quot_F^m}))) \equiv \alpha(F) s^{g-\tau} \cdot \theta^{g-\tau}$$

et donc, si $\tau \geq 0$, $\Delta_{g-\rho+r,r}(S(t,\Omega_{Quot_F^m})) \neq 0$.
Nous pourrions alors conclure avec la

Proposition 3

Soit F un faisceau général de rang s et de dégré d. Soient m et r deux entiers tels que $\rho = sm+d-(s-1)(g-1) \geq 0$ et

$\rho-g \leq r \leq \rho$. Alors on a que

$$\dim W_m^r(F) \geq \tau = g-(r+1)(r-\rho+g).$$

En particulier si $\tau \geq 0$ alors $W_m^r(F) \neq \emptyset$. En plus si $\dim W_m^r(F) = \tau \geq 0$ alors

$$[W_m^r(F)] \equiv \alpha(F) \cdot s^{g-\tau} \cdot \Theta^{g-\tau}.$$

La proposition (3) dans le cas classique ($s = 1$, $d = 0$) donne le "théorème d'existence" de Kleiman-Laksov qui correspond à la partie (a) du problème de Brill-Noether. Il nous semble alors naturel de poser la question (b) même dans le cas de fibrées de rang plus grand que un. Précisement on peut conjecturer que:

Conjecture

Soit X un courbe à modules généraux et F un faisceau localement libre sur X de dégre d, de rang s et général. Alors

$$\dim W_m^r(F) = g-(r+1)(r-sm-d+(s-1)(g-1)+g).$$

REFERENCES

[1] Ghione F.: *La conjecture de Brill-Noether pour les surfa̱ ces réglées*. Proc. of the Week of Algebraic Geometry, Bucharest 1980, Teubner-Text Zur Math. band 40, 63-79 (1981).

[2] Ghione F.: *Quot scheme over a smoot curve*. (Preprint Univ. di Napoli) serie III, n. 33, (1982).

[3] Griffiths P., Harris J.: *On the variety of special linear systems on a general algebraic curve*. Duke Math. Journal 47, 233-272 (1980).

[4] Kleiman S., Laksov D.: *Another proof of the existence of special divisors*. Acta Math. 132, 163-176 (1974).

Franco Ghione
Dipartimento di Matematica
Università di Roma II
Roma - Italy

On the Construction of Rational Surfaces with Assigned Singularities

Silvio Greco and Angelo Vistoli

Introduction

In this paper we deal with the following problem, which was very popular early in this century: given a singular point x_o of the complex analytic surface X, does there exist a rational algebraic surface Y with a singular point y_o such that the germs (X,x_o) and (Y,y_o) are isomorphic?
This problem is related with the attempt of generalizing to surfaces the parametrization techniques which are well known for curves.
Much work in this direction was done by several classical authors, such as Del Pezzo, Enriques, C. Segre, Hensel, Jung and others (see $/\bar{E}/$, book 4, ch. 4, section 39 for a historical account).
Later Franchetta came back to the problem and provided a number of significant positive answers, by using double coverings of the plane (see $/\bar{F1}/$, $/\bar{F2}/$).
In this paper we try a different approach and we get some further positive examples.
The general idea, suggested by R. Hartshorne, is the following: let $X \to X_o$ be a resolution of the normal singular point x_o, with exceptional curve E. Embed E in a rational surface Y in such a way that the two embeddings are equivalent (i.e. E has biholomorphic neighbourhood in either embedding). Then $E \subset Y$ can be collapsed to a singularity equivalent to x_o in a surface Y_o.
Two difficulties arise: how to find $E \subset Y$, and when is Y_o an algebraic surface.
In section 2 we show that if E is irreducible and nonsingular of genus g, then two embeddings of E are equivalent provided that the two normal bundles are isomorphic and their degree is less than $4 - 4g$. This implies a positive answer to the embedding question in a number of cases, listed in 2.5.
In section 3 we give a contractability criterion, which allows, in connection with the previous results, to show that if E is a nonsingular elliptic curve, then x_o is equivalent to a singularity of some rational algebraic surface.

I: Preliminaries

1.1 Let $\bar{X}$, $\bar{Y}$ be two complex analytic surfaces, and let $x_o \in \bar{X}$, $y_o \in \bar{Y}$ be two normal singular points.

We say that the singularities $(\bar{X}, x_o)$, $(\bar{Y}, y_o)$ are <u>equivalent</u> if the corresponding germs of analytic spaces are isomorphic, i.e. if $O_{\bar{X},x_o}$ is isomorphic to $O_{\bar{Y},y_o}$ as a $\mathbb{C}$-algebra.

This notion can be studies by means of the equivalence of embeddings.

1.2 Let E be an analytic curve, and let $i: E \to X$, $j: E \to Y$ be closed embeddings in two analytic surfaces X, Y. We say that i and j are <u>n-equivalent</u> if there is an isomorphism of nonreduced analytic spaces $(E, \underline{O}_X/I^n) \simeq (E, \underline{O}_Y/J^n)$, where I, J are the ideals of the embeddings.

We say that i and j are <u>equivalent</u>, and we write $i \sim j$, if thay are n-equivalent for all $n > 0$.

1.3 <u>Theorem</u>: Two normal singularities $(\bar{X}, x_o)$, $(\bar{Y}, y_o)$ are equivalent if and only if there are resolutions of x_o and y_o such that the exceptional curves are isomorphic (as reduced analytic spaces) and their embeddings are equivalent.

See $/\!$ L $\!/$ for a proof.

1.4 We say that a singularity $(\bar{X}, x_o)$ <u>belongs to a rational surface</u> (resp. to a rational algebraic surface) if there exists an equivalent singularity $(\bar{Y}, y_o)$ with $\bar{Y}$ compact and rational (resp. projective and rational).

II: Equivalence of embeddings

From now on C is an irreducible smooth projective curve of genus g. We give criteria for the equivalence of certain embeddings of C, with some application to our problem.

If $C \to X$ is an embedding in a nonsingular surface, we denote by $N_{X/C}$ the normal bundle, which is an invertible sheaf on C. Recall that the degree of $N_{X/C}$ is C^2, the self-intersection of C in X.

2.1 **Proposition**: Let $i: C \to X$, $j: C \to Y$ be embeddings of C into smooth analytic surfaces. Then:

(i) if $i \sim j$, then $N_{X/C} \simeq N_{Y/C}$ (as vector bundles over C).

(ii) the converse is true if $\deg N_{X/C} < 4-4g$.

Proof: (i) $N_{X/C} = (I/I^2)^v$ ("v" means "dual"), where I is the ideal of the embedding.

(ii) By $/\bar{L}/$, th. 6.8 (with $A_i = A_i = C$ and $s_o = 2$), it is sufficient to show that i and j are 2-equivalent. Now any 2-thickening C' of C is an extension

$$0 \to L \to \underline{O}_{C'} \to \underline{O}_C \to 0$$

where $L = I/I^2 = N_{X/C}^v = N_{Y/C}^v$.

Thus it suffices to show that all such extensions are isomorphic.

Since C is smooth the isomorphism classes of such extensions are classified by $H^1(C, \omega_C^v \otimes L)$ (see $/\bar{H}/$, III, ex. 4.10).

Now by Serre duality and the degree condition we have $h^1(\omega_C \otimes \omega_C \otimes L^v) = 0$; hence there is only one extensions (up to isomorphism).

2.2 **Proposition**: Let $i: C \to X$ be an embedding, where X is a smooth projective surface, and let $P \in C$. Let $X' \xrightarrow{p} X$ be the blow up with center P, and let $j: C \to X'$ be the embedding of C as proper trasform. Then

$$N_{X'/C} \simeq N_{X/C} \otimes_{\underline{O}_C} \underline{O}_C(-P).$$

Proof: Let H be a very ample divisor on X. We may assume that $\underline{O}_X(C+H)$ is generated by global sections ($/\bar{H}/$, 5.17), so that there is an effective divisor D, linearly equivalent to C+H, and such that $P \notin \text{supp}(D)$. Since C is linearly equivalent to D-H, we have:

$N_{X/C} \simeq \underline{O}_C(C \cdot (D-H))$

Moreover if $E = p^{-1}(P)$ we have $j(C) \sim p^*(D-H)-E$, whence

$N_{X'/C} = \underline{O}_C(C \cdot (p^*(D-H)-E))$
$= \underline{O}_C(C \cdot (p^*(D-H))) \otimes \underline{O}_C(-C \cdot E)$
$= N_{X/C} \otimes \underline{O}_C(-P)$

which is our claim.

2.3 **Proposition:** Let $i: C \to X$, $j: C \to Y$ be two embeddings of C into smooth surfaces, with Y projective, and assume $\deg N_{X/C} < 4-4g$, $\deg N_{Y/C} \geq 3-3g$. Then i is equivalent to an embedding $j': C \to Y'$ where Y' is obtained from Y by blowing up a finite number of points.

Proof: Put $N_{X/C} = \underline{O}_C(D)$, $N_{Y/C} = \underline{O}_C(B)$ for suitable divisors D and B. We have
$h^o(B-D) \geq \deg(B-D) - g + 1 > 3-3g-4+4g-g+1 = 0$.
Thus there is a divisor $A \geq 0$ such that $A \sim B-D$, i.e. $B \sim A+D$.
Let $Y' \to Y$ be the blow up of all the points of A. By 2.2 we have $N_{Y'/C} = N_{Y/C} \otimes \underline{O}_C(-A) = N_{X/C}$, and the conclusion follows by 2.1.

2.4 **Theorem:** Let $(\bar{X}, x_o)$ be a two-dimensional normal singularity and assume that there is a resolution $X \to \bar{X}$ with exceptional curve C, and $C^2 < 4-4g$. Assume further that C can be embedded into a projective rational surface with self-intersection $\geq 3-3g$. Then $(\bar{X}, x_o)$ belongs to a rational surface.

Proof: By 2.4 the embedding $C \to X$ is equivalent to an embedding $C \to Y$, with Y rational projective. Since $N_{Y/C}$ is negative there exists a contraction of C in Y (see $/\!\_L\_/$, 4.9). The conclusion follows from 1.3.

Now we list some cases in which 2.4 can be applied.

2.5 **Corollary:** The conclusion of 2.4 is true if $C^2 < 4-4g$ and, moreover, one of the following conditions holds:
(i) C is a plane curve;
(ii) C is hyperelliptic;
(iii) $g \leq 7$.

Proof: (i) is obvious.
(ii) if C is hyperelliptic, then it is birationally equivalent to a plane curve C' of degree $g+2$ with an ordinary g-uple point P. By blowing up P we see that C can be embedded in a rational surface with self-intersection $(g+2)^2 - g^2 > 0$, and we can apply 2.4.

(ii) by a theorem of Halphen (see $/\bar{H}\bar{/}$, IV, 6.1) C can be embedded in P^3 with degree g+3. Let C' be a generic projection in P^2. C' has $n = \frac{1}{2}(g+2)(g+1)-g$ nodes.

By blowing up these nodes we can embed C in a rational surface with self-intersection $(g+3)^2-4n = (g+3)^2-2(g+2)(g+1)+4g = -g^2+4g+5$, and it is easy to see that if $g \leq 7$ we can apply 2.4.

2.6 <u>Remarks</u>: By the above results we can see that:

(i) Two singularities having a resolution with exceptional curve isomorphic to P^1 are equivalent if and only if the self-intersections are the same. Hence any such singularity is equivalent to the singularity at the vertex of the affine cone corresponding to a rational normal curve in P^n. In particular it belongs to a rational algebraic surface.

(ii) Embeddings of non singular elliptic curves are determined by the normal bundle. Hence any singularity which can be resolved with a non singular elliptic curve as exceptional curve belongs to a rational <u>analytic</u> surface. In the next section we shall see that it belongs to a rational <u>algebraic</u> surface.

III: Examples of singularities which belong to rational algebraic surfaces

We begin with the following contractability criterion:

3.1 <u>Proposition</u>: Let X be a smooth irreducibile algebraic surface, and let $E \subset X$ be a reduced curve with components $E_1, \ldots, E_n$. Assume that there exists a divisor H such that

(a) $H \cdot E = 0$

(b) H-D is ample, for some suitable $D = \Sigma r_i E_i$, $r_i > 0$.

Then E is algebraically contractible.

<u>Proof</u>: We use the same technique as in $/\bar{A}\bar{/}$, 2.3.

We may assume that H-D is very ample and that $H^1(X, \underline{O}_X(H-D)) = 0$.

By (a) we have $\underline{O}_D(H) = \underline{O}_D$, and hence we get the exact sequence

$$0 \to \underline{O}_X(H-D) \to \underline{O}_X(H) \to \underline{O}_D \to 0$$

which shows that $\Gamma(\underline{O}_X(H)) \to \Gamma(\underline{O}_D)$ is surjective. This implies that $|H|$ has no base point on D. Moreover H-D is very ample and hence $|H|$ defines a morphism $f: X \to P^N$ for some N, such that $f(E)$ is a point, and which is an isomorphism outside E. Let Y be the normalization of $f(X)$. Then f factors through $X \to Y$, and it is easy to see that this gives the contraction of E.

3.2. **Corollary**: Let $C \subset P^2$ be an irreducible curve of degree c, and let $P_1, \ldots, P_n \in C$ with multiplicities $e_{P_i}(C) = a_i$. Let $H \subset P^2$ be a curve not containing C and of degree h, with $b_i = e_{P_i}(H)$. Assume that the following are satisfied:

(i) $\Sigma_i a_i b_i = hc$

(ii) $b_i > a_i$ for all $i = 1,\ldots,n$

(iii) For any irreducible curve $D \subset P^2$ of degree d with $e_{P_i}(D) = r_i$ we have
$$(h-c)d > \Sigma_i (b_i - a_i) r_i$$

(iv) $h^2 - \Sigma b_i^2 + c^2 - \Sigma a_i^2 > 0$.

Let X be the surface obtained by blowing up $P_1,\ldots,P_n$, and let C' be the proper transform of C. Then C' is algebraically contractible in X.

Proof: Let H' be the proper transform H in X, and let us show that H' verifies the assumption of 3.1, with $D = C'$.

By (i) we have $H' \cap C' = \emptyset$; we have to prove that $H'-C'$ is ample. For this we use the Nakai criterion (see $/\overline{H}/$, V, 1.10).

By (iv) we have
$(H'-C')^2 = H'^2 + C'^2 = h^2 - \Sigma b_i^2 + c^2 - \Sigma a_i^2 > 0$.

It remains to show that $H'-C'$ intersects every irreducible curve D' of X. Let $E_1,\ldots,E_n$ be the exceptional divisors of the blow up. If $D' = E_i$, then $(H'-C') \cdot E_i = b_i - a_i > 0$ by (ii). If $D' \neq E_i$ for all i's, then D' is the stict transform of an irreducible curve D of P^2. Put $d = \deg D$, $r_i = e_{P_i}(D)$.

Then $(H'-C') \cdot D' = H' \cdot D' - C' \cdot D' = H \cdot D - \Sigma b_i r_i - C \cdot D + \Sigma a_i r_i = (h-c)d - \Sigma(b_i - a_i) r_i > 0$ by (iii).

3.3 **Theorem**: Let $(\overline{X}, x_o)$ be a normal two-dimensional singularity. Assume that there exists a resolution $X \to \overline{X}$ whose exceptional curve C is elliptic and nonsingular.

Then $(\bar{X}, x_o)$ belongs to a rational algebraic surface.

Proof: We embed C in P^2 in such a way that we can apply at the same time 3.2. and 2.1.

Let $s = -C^2$ and let $N_{X/C} = \underline{O}_C(-D)$. We may assume that D is in the form $Q_1 + \ldots + Q_s$, where the Q_i's are all distinct points: if $s = 1$, then D is just a point, and if $s > 1$, $|D|$ has no base points, and we can apply Bertini's theorem.

Now we distinguish three cases, according to the residues of s mod. 3.

Case 1: $s = 3r$. By a classical theorem (see $/\bar{C}\_7$, pp. 132-133) $|D|$ has a divisor of the form $3rP$, where P is a point.

Embed C in P^2 by $|3P|$. Then $|D| = |3rP|$ consists of all the divisors cut out by the curves of degree r. In particular $D = C.C^r$, for a suitable curve C^r.

Let C^3 be a cubic curve which cuts C in 9 distinct points $P_1, \ldots, P_9$ different from the Q_i's, and let $H = 3C^3 + 3C^r$.

Let Y be obtained by blowing up $Q_1, \ldots, Q_s, P_1, \ldots, P_9$. An easy computation proves that the assumption of 3.2 are satisfied.

Thus the proper transform of C in Y is algebraically contractible.

By 2.2 we have

$$N_{Y/C} = N_{P^2/C}(-\Sigma Q_i - \Sigma P_i) = \underline{O}_C(\Sigma P_i - \Sigma Q_i - \Sigma P_i) = N_{X/C}.$$

Now the conclusion follows by 2.1 and 3.2.

Case 2: $s = 3r-1$

As above there is a point P such that $3rP \sim 2Q_1 + \ldots + Q_s$.

Embed C in P^2 by $|3P|$, so that $2Q_1 + Q_2 + \ldots + Q_s$ is cut out by a curve of degree r. Consider the linear system of all the curves of degree $r+3$ having a triple point at Q_1, a double point at Q_2 and passing through $Q_3, \ldots, Q_s$. Its dimension is at least $\frac{1}{2}r(r+3)+3$, while the dimension of the system of the curves of degree r having a double point at Q_1 and passing through Q_2 is $\frac{1}{2}r(r+3)-4$, if $r > 1$. Hence we see that, if $r > 1$, these curves cut out on C, outside $3Q_1 + 2Q_2 + Q_3 + \ldots + Q_s$, a g_7^6; and if $r = 1$, it is easy to see that this linear series has just one base point, aligned with Q_1 and Q_2, and the remaining points form a g_6^5. Let then $P_1 + \ldots + P_7$ be a divisor in this series, cut out by C^{r+3}, such that all the P_1's are distinct. Let C^3 be any cubic curve cutting out distinct points $P_1, \ldots, P_7$, P_8, P_9 on C.

Since $2Q_1+Q_2+\ldots+Q_s$ is cut out by a curve of degree r, we have that $P_8+P_9 \sim Q_1+Q_2$. Hence there exists a curve C^r of degree r, not containing C, such that $C.C^r =$
$= Q_1+Q_3+\ldots+Q_s+P_8+P_9$. Let $H = C^{r+3}+C^r+2C^3$. Then deg $H = 2r+9$, $e_{Q_1}(H) = 4$, $e_{Q_i}(H) = 2$ for $i = 2,\ldots,s$, $e_{P_i}(H) = 3$, and a straightforward calculation shows that H verifies the conditions of 3.2. The conclusion follows as in the previous case.

Case 3: s = 3r-2

The argument is similar to the previous one. Start with $3Q_1+Q_2+\ldots+Q_s \sim 3rP$, and proceed as in case 2, with the curves of degree r+2 having a double point at Q_1 and passing through $Q_2,\ldots,Q_s$.

References

/A/ Artin, M.: Some numerical criteria for contractability of curves on algebraic surfaces, Am. J. Math. 64, 1962, 485-496

/C/ Coolidge J.: A treatise on algebraic plane curves, Dover books in Adv. Math, New York, 1959

/E/ Enriques, F. and Chisini, O.: Teoria geometrica delle equazioni e delle funzioni algebriche, Zanichelli, Bologna, 1934

/F1/ Franchetta, A: Sui punti doppi isolati delle superfici algebriche, Note I e II, Rend. Acc. dei Lincei, 1946

/F2/ Franchetta, A.: Osservazioni sui punti doppi isolati delle superfici algebriche, Rend. Mat. e Appl., 1946

/H/ Hartshorne, R.: Algebraic Geometry, Springer-Verlag, New York, 1977

/L/ Laufer, H.: Normal two-dimensional singularities, Ann. Math. Stud., Princeton University Press, Princeton, 1971

Silvio Greco
Dipartimento di Matematica
Politecnico di Torino
Corso Duca degli Abruzzi 24
10129 Torino
Italy

Angelo Vistoli
Istituto di Geometria
Università di Bologna
P.zza di Porta San Donato 5
40127 Bologna
Italy

This paper was written with the financial support of M.P.I. (the Italian Ministry of Education).

POSTULATION DES COURBES GAUCHES

par

Laurent GRUSON et Christian PESKINE

Pour classifier les courbes (lisses et connexes) de P_3 [1], on veut caractériser les composantes irréductibles du schéma de Hilbert H_g^d (courbes de degré d et genre g), donner les dimension de ces composantes et enfin construire explicitement une courbe assez générale de chaque composante. C'est le programme que Halphen pense réaliser dans son célèbre mémoire sur les courbes gauches. Il nous parait maintenant certain que les méthodes qu'il développe sont insuffisantes. Rappelons par exemple, ([4], app.2), qu'une des deux composantes de H_{18}^{13} échappe à sa classification: en effet si la composante (de dimension 52) dont la courbe générale n'est pas tracée sur une quartique apparait clairement, on ne trouve nulle trace de l'autre composante (de dimension 52) pour laquelle la courbe générale est tracée sur une surface cubique reglée et rencontre 4 fois une génératrice. L'étude des cinq composantes irréductibles de H_{30}^{16} ([4], app.2) démontre mieux encore la complexité du problème décrit plus haut. Si dans ces deux cas la classification reste claire, c'est parcequ'il existe une classification complète des surfaces cubiques (une courbe tracée sur une surface quartique n'est pas, dans ces cas, générale dans sa composante irréductible). Il semble improbable qu'une pareille chance se renouvelle en plus grand degré; plus précisément, nous pensons qu'il existe des composantes du schéma de Hilbert dont la courbe générale est tracée sur la surface générale d'une famille particulière, et non classée, de surfaces. Ces précautions étant prises, elles justifient,

---

[1] Espace projectif de dimension 3 sur un corps algébriquement clos de caractéristique nulle.

en partie, notre interêt pour le problème qui suit:

"Déterminer le plus petit entier n, dépendant de d et g, tel que toute courbe de H_g^d est contenue dans une surface de degré $\leq n$".

Notons

$G(d,s) = \sup\{g(C),$ pour $C \subset P_3$ courbe lisse connexe de degré d et non contenue dans une surface de degré $< s\}$.

Théorème A: 1) Si $s(s-1) < d$, on a

$G(d,s) = 1 + [d(d+s^2-4s) - r(s-r)(s-1)]/2s$, où $d+r \equiv 0$ (s) et $0 \leq r < s$; de plus $g(C) = G(d,s)$ si et seulement si C est liée à une courbe plane de degré r par des surfaces de degrés s et $(d+r)/s$.

2) Si $s^2 - 2s + 3 \leq d \leq s(s-1)$, on a

$G(d,s) = s^3 - 5s^2 + 9s - 6 + \nu(\nu+2s-3)/2$, où $\nu = d - (s^2-2s+3)$; de plus si $g(C) = G(d,s)$, la courbe C est arithmétiquement normale tracée sur une surface de degré s.

3) Si $s^2 - 2s + 2 = 2$, on a $G(d,s) = 1 + (s-3)d = s^3 - 5s^2 + 8s - 5$, et $g(C) = G(d,s)$ si et seulement si C est la courbe des zéros d'une section de $E(s-1)$ où E est un fibré de corrélation nulle (i.e. $C_1(E) = 0$, $C_2(E) = 1$).

4) Si $[(s+2)^2 + 2]/3 \leq d < s^2 - 2s + 2$, alors $G(d,s)$ est inconnu.

5) Si $d < [(s+2)^2 + 2]/3$, on a $G(d,s) \leq 1 + (s-1)d - \binom{s+2}{3}$.

Nous voulons faire ici une analyse rapide des méthodes utilisées pour établir ces résultats et attirer l'attention du lecteur sur l'absence de solution au problème dans la 4eme région. Signalons que l'inégalité du 5 (sur laquelle nous ne reviendrons plus) se déduit directement du théorème de Clifford qui démontre que dans cette région le faisceau inversible $O_C(s-1)$ est non spécial ([7],3.3); cette inégalité est elle la meilleure possible? C'est le problème de l'existence de courbes de postulation générale étudié par Hirschowitz (courbes rationelles) et Ballico-Ellia ($g \leq d-3$), mais non résolu dans cette généralité.

Nous appelons indice de spécialité d'une courbe C de P_3, et notons e(C) (où e lorsqu'il n'y a pas de confusion possible), le plus grand entier n tel que $h^1(O_C(n)) \neq 0$. Il est clair qu'on a $ed \leq 2g-2$.

Posons

$E(d,s) = \sup\{e(C)$, pour $C \subset P_3$ courbe lisse connexe de degré d et non contenue dans une surface de degré $< s\}$.

Ce nombre est implicitement étudié par Halphen.

Théorème B: 1) Si $s(s-1) < d$, on a $E(d,s) = s + [d/s] - 4$

2,3) Si $s^2 - 2s + 2 \leq 2 \leq s(s-1)$, on a $E(d,s) = 2s - 6$.

4) Si $[(s+2)^2 + 2]/3 \leq d < s^2 - 2s + 2$, on a
$$E(d,s) \leq [s - 7 + \sqrt{12d - 3s^2 - 6s + 1}]/2$$

On voit que pour les 3 premières régions la valeur de $E(d,s)$ est celle suggérée par le théorème A. Nous avons ici une majoration de $E(d,s)$ dans la 4eme région dont on espère que c'est la meilleure possible (on vérifie facilement que l'inégalité fournie pour cette région est aussi valable et est la meilleure possible pour les 2eme et 3eme régions). Décrivons maintenant les trois méthodes utilisées pour établir ces résultats.

La première méthode d'Halphen.

Soient $p: \tilde{P} \to P_3$ l'éclatement d'un point général de P_3 et $q: \tilde{P} \to P_2$ le morphisme de projection. Halphen propose d'étudier les fibrés $q_*(p^*J(n))$ (où J est l'idéal d'une courbe de P_3) sur P_2, en particulier de minorer leur deuxieme classe de Chern, donc de majorer le genre de la courbe.

La méthode de Castelnuovo.

Si H est un plan, et si J_H est l'idéal dans H du groupe de points $H \cap C$, on vérifie facilement $g(C) \leq \sum_1^{e+1} h^1(J_H(n))$. Castelnuovo propose donc d'étudier la postulation d'une section plane générale de C.

La deuxième méthode d'Halphen.

Par définition de e, il existe une extension

$$0 \to O_{P_3}(-e-4) \to M^V \to O_{P_3} \to O_C \to 0,$$

où M est un O_{P_3}-module réflexif de rang 2 de classes de Chern $c_1 = e+4$, $c_2 = d$ et $c_3 = 2g-2-ed$. Il est alors naturel de chercher quelles conditions sur les classes de Chern d'un O_{P_3}-module réflexif de rang 2 entraîne l'existence d'une section non nulle de ce module.

Comme nous le verrons, cette dernière méthode est particulièrement adéquate pour démontrer le théorème B; les deux premières qui toutes deux permettent de prouver A sont illustrées au mieux par les démonstrations de la majoration du genre des courbes gauches ($s=2$ avec nos notations) fournies par Halphen et Castelnuovo.

Majoration du genre d'une courbe gauche suivant Halphen.

Une projection générale n'ayant que des points doubles, l'application naturelle $q_*(p^*O_{P_3}(1)) \to q_*(p^*O_C(1))$ est surjective. Notons $M = q_*(p^*J(1))$ et considérons la suite exacte

$$0 \to O_{P_2}(-2) \oplus O_{P_2}(-3) \to M^V(-2) \to q_*(p^*\omega_C) \to 0. \text{ Comme } M^V = M(d-3)$$

(car M est réflexif de rang 2 et $c_1(M) = 1-d$), on en déduit $h^o(M(d-4-e)) = 0$, $h^o(M(d-3-e)) \neq 0$ et $h^o(M(d-2-e)) \geq 4$. Ceci montre que si M n'est pas semi-stable, on a $M(d-3-e) = O_{P_2} \oplus O_{P_2}(-1)$, avec $e = (d-4)/2$; dans ce cas C est intersection complète d'une quadrique et d'une surface de degré $d/2$ et on a $g(C) = [(d-2)/2]^2$. Si M est semi-stable, on a $c_1(M(d-3-e)) \geq 0$, donc $d - 5 - 2e \geq 0$, ce qui entraîne $H^o(M(e)) = 0$; on en déduit $h^o(q_*(p^*O_{P_3}(1))(e)) \leq h^o(O_C(e+1))$, soit $g \leq (e+1)(d-e-3)$ (et comme $e \leq (d-5)/2$ ceci implique $g \leq [(d-2)/2]^2$). Halphen remarque de plus que pour $e \geq 1$ l'égalité $g = (e+1)(d-3-e)$ a lieu seulement si C est tracée sur une quadrique (l'assertion inverse étant évidente). En effet, si cette égalité a lieu on a $H^1(M(e)) = 0$, donc par dualité $H^1(M(d-4-e)) = 0$; remarquant alors que le module gradué $\oplus_{n \geq o} H^o(O_C(n))$ est engendré, sur l'anneau $\oplus_{n \geq o} H^o(O_{P_2}(n))$ par ses éléments de degré $\leq e+2$, on en déduit $H^1(M(d-5)) = 0$, soit par dualité $H^1(M(1)) = 0$, c'est à dire une application surjective $H^o(q_*(p^*O_{P_3}(1))(1)) \rightarrow H^o(O_C(2)) \rightarrow 0$ qui montre bien que C est tracée sur une quadrique.

Majoration du genre d'une courbe gauche suivant Castelnuovo.

Pour un groupe de points $\Gamma \subset P_2$, Castelnuovo introduit l'indice de séparation de Γ, le nombre $\chi_\Gamma = \max\{n, h^1(J_\Gamma(n)) \neq 0\}$, où J_Γ est l'idéal de Γ. Il remarque que si Γ ne contient pas 3 points alignés on a $\chi_\Gamma \leq [d^o\Gamma/2] - 1$ et montre dans ce cas, par une récurrence simple sur $d^o\Gamma$, que $\sum_1^{r+1} h^1(J_\Gamma(n)) \leq (r+1)(d^o\Gamma-r-3)$ pour $r+1 \leq \chi_\Gamma$. Appliquant ce résultat a une section plane générale d'une courbe gauche, on trouve $g \leq (e+1)(d-e-3)$ en vérifiant immédiatement $e+1 \leq \chi_{C \cap H}$ lorsque $C \cap H$ est le groupe de points section de C par le plan H. La méthode de Castelnuovo ne permet pas de montrer que pour $e \geq 1$ l'égalité n'a lieu que pour les courbes tracées sur une quadrique; par contre elle démontre l'inégalité pour le genre arithmétique d'une courbe intègre.

Voyons les difficultés rencontrées pour étendre ces démonstrations aux cas $s > 2$. Dans la preuve d'Halphen, après l'utilisation du lemme des trisécantes, le point crucial est la semi-stabilité de $q_*(p^*J(1))$

(sauf l'exception interpretée). Il faudrait donc interpreter l'instabilité de $q_*(p^*J(n))$ pour $n < s$; si ceci est faisable pour $s = 3$ (on démontre alors $g \leq (e+1)(2d-3e-6)/2$ et $e \leq (d-3)/3$, sauf pour les intersections complètes d'une cubique et d'une surface de degré $d/3$ pour lesquelles on a bien $e = (d-3)/3$ mais $g = 1+(e+1)(2d-3e-6)/2)$, la classification devient trop complexe lorsque s grandit. On peut néanmoins montrer que la restriction d'un fibré $q_*(p^*J(n))$ (avec $n < s$) à une droite générale L de P_2 est de la forme $\overset{n}{\underset{0}{\oplus}} O_L(r_i)$ où $\{r_i, 0 \leq i \leq n\}$ est un ensemble connexe d'entiers. Remarquons alors que $\overset{n}{\underset{0}{\Sigma}} r_i = c_1(q_*(p^*J(n))) = (n+1)n/2 - d$ et que $c_2(q_*(p^*J(n))) \geq \underset{i < j}{\Sigma} r_i r_j$ ([2], l'égalité ayant lieu si et seulement si $q_*(p^*J(n)) = \overset{n}{\underset{0}{\oplus}} O_{P_2}(r_i))$. Cette dernière inégalité démontre A.1 pour tout s (en l'appliquant à $n = s-1$) en remarquant que le genre maximum est donné par la suite connexe de r_i "d'amplitude maximum" vérifiant $\overset{s-1}{\underset{0}{\Sigma}} r_i = s(s-1)/2 - d$. Il est clair que cette inégalité n'est la meilleure possible que pour $d > s(s-1)$, car pour $s(s-1) \geq d$ un r_i positif apparait dans la suite d'entiers d'amplitude maximum.

Dans cette description, guère plaisante, on n'a rien dit sur la seule difficulté, la connexité de l'ensemble d'entiers $\{r_i\}$; bien que $q_*(p^*J(n))$ ne soit pas toujours semi-stable le lecteur devine bien qu'il s'agira d'une nouvelle variante du théorème de Grauert-Mulich. Cette variante est de même nature que celle que nous utiliserons dans l'extension de la méthode de la section plane, mais nous décrirons plutot cette dernière qui nous parait moins rebutante. Rappelons que le point crucial de la preuve de Castelnuovo est le lemme des trisécantes; d'eventuels raffinements de ce lemme sont surtout pénibles lorsqu'on est amené à étudier des sous groupes de points d'une section plane générale. Nous voulons tourner, en partie, cette difficulté par un artifice en introduisant le caractère d'un groupe de points du plan. Si Γ est un tel groupe de points, contenu dans une courbe Σ de degré σ

et non dans une courbe de degré moindre, soit $q : \Sigma \to P_2$ une projection de Σ de sommet hors de Σ sur une droite à l'infini. Si J_Σ est l'idéal de Γ dans Σ, nous appellerons caractère de Γ la suite d'entier $n_0 \geq n_1 \geq \ldots \geq n_{\sigma-1} \geq \sigma$ vérifiant $q_*(J_\Sigma) = \bigoplus_0^{\sigma-1} O_{P_1}(-n_i)$. On vérifie facilement que ce caractère, de longueur σ, est indépendant de Σ et du centre de projection. Remarquons qu'on a $d^o\Gamma = \sum_0^{\sigma-1}(n_i - i)$ et $\sum_{n \geq 1} h^1(J_\Gamma(n)) = 1 + \sum_0^{\sigma-1}(n_i-i)(n_i+i-3)$;

Proposition: Soit $\Gamma_S \subset S \times P_2$ une famille intègre, plate au dessus de S, de groupes de points plans. Le caractère d'une fibre générale est une suite connexe d'entiers.

C'est bien sûr celui de la fibre générique (le caractère est défini même si le corps de base n'est pas algébriquement clos); la connexité de celui-ci se déduit du fait que son cône projetant dans $P_2(k(\eta))$ (où η est le point générique de S) est intègre. ([3],3.2).

Remarque 1: La section plane générique d'une courbe intègre étant intègre, une section plane générale a un caractère connexe. Si la courbe est gauche, cette section n'est évidemment pas alignée, i.e. son caractère est de longueur au moins 2. L'inégalité
$\sum_1^{e+1} h^1(J_{H \cap C}(n)) \leq (e+1)(d-3-e)$, où $H \cap C$ est la section plane considérée, s'en déduit immédiatement.

Remarque 2: Si C est une courbe gauche lisse connexe, on sait (lemme des trisécantes) que la congruence de ses bisécantes est intègre; donc le groupe des points doubles D d'une projection générale a un caractère connexe. D'autre part, par définition de e une adjointe de degré minimum a degré $d-3-e$; le caractère de D est alors de la forme $d-2 = n_0 \geq n_1 \geq \ldots \geq n_{d-4-e} \geq d-3-e$ (avec $n_0 = d-2$ car les points, séparés par les courbes de degré $d-3$, ne le sont pas par les courbes de degré $d-4$ pour une courbe projetée). Il est clair que $d^o D \geq \sum^{d-4-e}(m_i - i)$, avec $m_i = \sup(d-2-i, d-3-e)$, on en déduit directement $g \leq (e+1)(d-e-3)$, l'egalité ayant lieu si et seulement si $n_i = m_i$ pour $0 \leq i \leq d-4-e$; il n'est pas difficile de verifier que pour $e \geq 1$ ceci entraîne que la courbe est tracée sur une quadrique.

Après cette digression, concluons en signalant que le théorème A 1,2, se déduit de la variante qui suit du théorème de Grauert-Mulich exactement comme nous avons déduit, dans la remarque 1, la majoration du genre des courbes gauches du fait qu'une section plane générale n'était pas alignée.

Lemme (Laudal,[5]): Si C est une courbe intègre de P_3 non contenu dans une surface de degré t, et si toute section plane est contenue dans une courbe de degré t on a $t^2 + 1 \geq d^oC$.

Rappelons que la variété d'incidence (plans, points) est $\text{Proj}(\text{Sym}.\Omega^V(-1))$, où Ω est le module des differentielles sur P_3. L'hypothese exprime alors que pour $\alpha \gg 0$ il existe une section non nulle de $J(t) \otimes \text{Sym}_\alpha \Omega^V(-1)$, où J est l'idéal de C. On démontre alors que pour t puis α minimums et pour un plan général H, l'application composée induite.

$$O_H \oplus \Omega_H(1) \longrightarrow \text{Sym}_\alpha(O_H \oplus \Omega_H(1)) \longrightarrow J(t) \otimes O_H$$

a un conoyau à support fini; la majoration annoncée s'en déduit facilement. On peut de plus prouver que si $t^2 + 1 = d$ la courbe C est variété des zéros d'une section de $E(t)$ où E est un fibré de corrélation nulle, ce qui démontre A.3.

Il nous reste à expliquer comment la deuxième méthode proposée par Halphen permet de démontrer le théorème B. On applique cette fois ci directement le théorème de restriction semi-stable de Grauert-Mulich-Spindler: Si E est un module réflexif semi-stable sur P_3, la restriction de E à une droite générale L est de la forme

$\bigoplus_1^{rg(E)} O_L(r_i)$ où $\{r_i\}$ est un ensemble connexe d'entiers. Comme corollaire immédiat de cet énoncé on vérifie:

(*) $4c_2(E) - c_1(E)^2 \geq 0$ si $rg(E) = 2$, inégalité stricte si E n'est pas décomposable,

(**) $3c_2(E) - c_1(E)^2 \geq 0$ si $rg(E) = 3$.

Si on considére l'extension décrite plus haut

$$0 \to O_{P_3}(-e-4) \to M^V \to O_{P_3} \to O_C \to 0,$$

le théorème B 1 se déduit immédiatement de (*).

Montrons maintenant que l'inégalité énoncée en B.4 est toujours valable (elle démontre essentiellement aussi B.2,3). Il faut d'abord vérifier que cette inégalité est équivalente à l'inégalité $c_1(M^V(s-1))^2 + 2c_1(M^V(s-1)) + 3 - 3c_2(M^V(s-1)) \leq 0$. Il reste à démontrer le résultat suivant:

Théorème (Hartshorne[8] 0.1): Soit N un module réflexif de rang 2 sur P_3 de classes de Chern c_i. Si $c_1^2 + 2c_1 + 3 - 3c_2 > 0$ et $c_1 \geq -2$, on a $H^0(N) \neq 0$.

Rappelons que $\chi(N) = (c_1+4)(c_1^2+2c_1+3-3c_2)/6 + c_3/2$ (où c_3 est toujours positif), et démontrons le théorème par récurrence sur c_1. Remarquons d'abord qu'il est élémentaire pour $c_1 \leq 0$; en effet, dans ces cas l'inégalité entraîne $c_2 \leq 0$, donc l'instabilité de N ou la décomposition $N = O_{P_2}^2(-1)$ (d'après (*)) mais cette décomposition contredit aussi l'hypothèse, donc N est instable et $H^0(N) \neq 0$. Supposons donc $c_1 > 0$. Comme $\chi(N) > 0$, on peut supposer $h^2(N) \neq 0$, ce qui entraîne l'existence d'une extension non scindée $0 \to O_{P_3}(-4) \to E \to N \to 0$, où E est un module réflexif de rang 3, de classes de Chern c_i' avec $c_1' = c_1 - 4$ et $c_2' = c_2 - 4c_1$. On vérifie immédiatement que l'inegalité de l'énoncé implique $c_1'^2 - 3c_2' > 0$, donc E est instable. S'il existe une application non nulle $O_{P_3}(n) \to E$ avec $n > (c_1-4)/3 \geq -1$, elle induit évidemment une section non nulle de N. Sinon il existe une application non nulle $E \to O_{P_3}(n)$ avec $n < (c_1-4)/3$. Si l'application induite $O_{P_3}(-4) \to O_{P_3}(n)$ est nulle, il existe une application non nulle $N \to O_{P_3}(n)$ qui démontre l'énoncé. Sinon, considérons l'homomorphisme injectif $N' \to N$ où N' est le noyau de $E \to O_{P_3}(n)$. Comme on a pris soin de prendre n minimum, on vérifie $c_2(N') \leq c_2 + (n+4)(n-c_1)$. Comme l'extension n'est pas scindée, on a $c_1(N') = c_1 - 4 - n < c_1$; d'autre part $n < (c_1-4)/3$ entraîne $c_1(N') \geq -2$. On prouve enfin facilement $c_1(N')^2 + 2c_1(N') + 3 - 3c_2(N') > (n+4)(c_1-2n+2) \geq 0$, ce qui démontre $H^0(N') \neq 0$, par récurrence, donc $H^0(N) \neq 0$.

Bibliographie

[1] Castelnuovo, G. : Sui multipli di una serie lineare ... Rend.circ.Mat. Palermo, t. VII, 1893.

[2] Elencwajg, G.
Forster, O. : Bounding cohomology groups of vector bundles on P_n. Math.Ann. 246, 1980.

[3] Gruson, L.
Peskine, C. : Genre des courbes de l'espace projectif. Algebraic Geometry. Lecture Notes in Math. n° 687, Springer.

[4] - " - : Genre des courbes de l'espace projectif (II). Ann.Scient.Ec.Norm.Sup., 4e série, t.15, 1982.

[5] - " - : Section plane d'une courbe gauche: Postulation. Enumerative Geometry ... Progress in Math. Vol. 24. Birkhäuser.

[6] Halphen, G. : Mémoire sur la classification des courbes gauches algébriques. Œuvres complètes t.III.

[7] Hartshorne, R. : On the classification of algebraic space curves. Vector Bundles ... Progress in Math. Vol. 7. Birkhäuser.

[8]] - " - : Stables reflexive sheaves II. Invent. math. 66, 165-190 (1982).

Projective Geometry of Elliptic Curves

Klaus Hulek
Department of Mathematics
Brown University
Providence, RI 02912
USA

Table of Contents

0. Introduction
I. The elliptic normal curve $C_n \subseteq \mathbb{P}_{n-1}$
II. An abstract configuration
III. Examples
IV. Normal bundles of elliptic space curves
V. Open problems

0. Introduction

In this paper I want to discuss some geometric aspects of elliptic curves. Roughly speaking, the paper is divided into two parts. In the first three chapters we shall discuss certain properties of elliptic normal curves. In chapter IV we shall then apply these results to a certain problem concerning the normal bundle of elliptic space curves of degree 5.

The contents in more detail: Chapter I is concerned with the explicit construction of functions embedding a given elliptic curve C as a linearly normal curve of degree n in $\mathbb{P}_{n-1}$. These functions will essentially be n-fold products of translates of the Weierstrass σ-function. They will be chosen in such a way that the symmetries of the embedded curve $C_n \subseteq \mathbb{P}_{n-1}$ take on a particularly simple form. In particular, we shall see that the curve C_n is invariant under the Heisenberg group H_n and that H_n operates on C_n by translation with n-torsion points. In chapter II we shall then construct an abstract configuration which is associated to the Heisenberg group in prime dimension p. It is a generalization of the well-known configuration associated to the 9 points of inflection of a plane cubic. To illustrate this we shall discuss the cases $n = 3,4$ and 5 in some detail.

Chapter IV is concerned with the normal bundle of elliptic space curves of degree 5. About two years ago Ellingsrud and Laksov classified the normal bundles of such curves. In their result a central role is played by a certain one-dimensional family of quintics in $\mathbb{P}_4$. We want to apply the results of chapters I to III to get some more information about these hypersurfaces. Our main result is that these quintics form a linear family of hypersurfaces whose equations are invariant under the Heisenberg group H_5. Therefore, they are closely related to the configuration described in chapter II. Due to limited space, we shall have to restrict ourselves to sketches of proofs rather than give full proofs in chapter IV.

Finally, we shall discuss some open problems in chapter V.

I. <u>The elliptic normal curve</u> $C_n \subseteq \mathbb{P}_{n-1}$

In this chapter we want to collect some material concerning the symmetries of elliptic normal curves $C_n \subseteq \mathbb{P}_{n-1}$. Practically all of this was classically known. An excellent reference is an article by Bianchi [1] which was published in Mathematische Annalen in 1880. There he mainly treats the case of a plane elliptic cubic and of an elliptic quintic in $\mathbb{P}_4$, but he also looks at the general case of an elliptic normal curve of odd degree. The even degree case was treated by A. Hurwitz in [6].

(I.1) Let C be an elliptic curve with fixed origin $\mathcal{O}$. Moreover, let

$$\Gamma = \{n_1\omega_1 + n_2\omega_2 : n_1, n_2 \in \mathbb{Z}\}$$

be a lattice such that $C = \mathbb{C}/\Gamma$. The n-torsion points of C are then given by

$$P_{pq} = \frac{p\omega_1 + q\omega_2}{n} \ ; \ p,q \in \mathbb{Z}.$$

By abuse of notation we can write $p,q \in \mathbb{Z}_n$. The n-torsion points form a subgroup $G_n \subseteq \mathbb{C}$ and by identifying $\frac{\omega_1}{n}$ with $(1,0)$ and $-\frac{\omega_2}{n}$ with $(0,1)$ we fix an isomorphism

$$G_n \cong \mathbb{Z}_n \times \mathbb{Z}_n.$$

(n = 3)

(I.2) Next we want to describe explicitly a set of functions embedding C as a linearly normal curve of given degree. These functions are chosen in such a way that the symmetries of the embedded curve take on a particularly simple form.

First recall that the Weierstrass σ-function is defined by

$$\sigma(z) := z \prod_{\omega \in \Gamma - \{0\}} (1 - \frac{z}{\omega}) e^{(\frac{z}{\omega} + \frac{z^2}{2\omega^2})}$$

It has the property that it has simple zeroes exactly at the points of the lattice. With respect to translation by ω_1 and ω_2 the following fundamental formulas hold:

(1) $$\sigma(z + \omega_1) = -e^{\eta_1(z + \frac{\omega_1}{2})} \sigma(z)$$

(2) $$\sigma(z + \omega_2) = -e^{\eta_2(z + \frac{\omega_2}{2})} \sigma(z).$$

Here η_1 and η_2 are the period constants of the Weierstrass ζ-function. On the curve C itself σ defines a section $\sigma \in \Gamma(O_C(0))$.

For what follows next we shall have to distinguish between the case of odd and even degree. So let us first fix an <u>odd</u> integer $n \geq 3$. For $p,q \in \mathbb{Z}$ we set

$$\sigma_{pq}(z) := \sigma(z - \frac{p\omega_1 + q\omega_2}{n}).$$

Moreover we define the following constants

$$\omega := -e^{-\frac{n-1}{2} \frac{\eta_2 \omega_1}{n}}, \quad \Theta := e^{-\frac{\eta_1 \omega_1}{2n}}.$$

Finally we get functions x_m, $m \in \mathbb{Z}$ as follows

$$x_m(z) := \omega^m \Theta^{m^2} e^{m\eta_1 z} \sigma_{m,0}(z) \cdot \ldots \cdot \sigma_{m,n-1}(z) .$$

Next let $n \geq 4$ be an <u>even</u> integer. Then we write

$$\tilde{\sigma}_{pq}(z) := \sigma(z - \frac{p\omega_1 + q\omega_2}{n} - \frac{1}{2}(\omega_1 + \frac{\omega_2}{n})) .$$

Similarly as above we next define constants

$$\tilde{\omega} := e^{-\frac{1}{2}(\eta_1 \omega_1 + \eta_2 \omega_1)}, \quad \tilde{\Theta} := \Theta = e^{-\frac{\eta_1 \omega_1}{2n}}$$

which give rise to the functions

$$x_m(z) := \tilde{\omega}^m \tilde{\Theta}^{m^2} e^{m\eta_1 z} \tilde{\sigma}_{m,0}(z) \cdot \ldots \cdot \tilde{\sigma}_{m,n-1}(z) .$$

Using the fundamental formulas (1) and (2) it is now easy to check that

(3) $$x_{n+m}(z) = x_m(z)$$

for all integers m. Hence we have in both cases defined a set of n functions $\{x_m ; m \in \mathbb{Z}_n\}$ which are a product of suitably adjusted σ-functions. The choice of these functions is justified by the following theorem.

(I.3) <u>Theorem</u>: <u>The functions</u> x_i <u>define</u> n <u>linearly independent sections</u> $x_i \in \Gamma(\mathcal{O}_C(n\mathcal{O}))$ <u>and the map</u>

$$z \longmapsto (x_0(z) : \ldots : x_{n-1}(z))$$

embeds C <u>as a linearly normal curve</u> $C_n \subseteq \mathbb{P}_{n-1}$ <u>of degree</u> n. If $\varepsilon = e^{\frac{2\pi i}{n}}$ then the following formulas hold:

(i) $x_i(-z) \sim (-1)^n x_{-i}(z)$

(ii) $x_i(z - \frac{\omega_1}{n}) \sim x_{i+1}(z)$

(iii) $x_i(z + \frac{\omega_2}{n}) \sim \varepsilon^i x_i(z)$.

<u>Here</u> ~ <u>means that equality holds up to a common nowhere vanishing function independent of</u> i.

<u>Remarks</u>: (i) It is sufficient to know the fundamental formulas (1) and (2) to prove this result.

(ii) At z = 0 formula (i) holds absolutely, i.e. ~ can be replaced by equality.

(I.4) We next want to rephrase the above result in a slightly different terminology. To do this we consider the vector space

$$V = \mathbb{C}^n$$

and denote its standard basis by $\{e_m\}_{m \in \mathbb{Z}_n}$. We define elements $\sigma, \tau \in GL(V)$ by

$$\sigma(e_i) := e_{i-1}$$

$$\tau(e_i) := \varepsilon^i e_i.$$

The automorphisms σ and τ do not commute but one finds

$$[\sigma,\tau] = \varepsilon \cdot id_V$$

<u>Definition</u>: The subgroup $H_n \subseteq GL(V)$ generated by σ and τ is called the <u>Heisenberg group of dimension</u> n. The representation of H_n defined by the inclusion is called the <u>Schrödinger representation</u> of the Heisenberg group.

<u>Remarks</u>: (i) For a more general definition of the Heisenberg group and its Schrödinger representation see Igusa's book [7, p. 10]. Instead of an arbitrary locally compact group we have just considered $\mathbb{Z}_n$ here.

(ii) The centre of the Heisenberg group H_n equals

$$\mu_n = \{\varepsilon^m \cdot id_V;\ m \in \mathbb{Z}\}$$

and the group H_n is a central extension

$$1 \longrightarrow \mu_n \longrightarrow H_n \longrightarrow \mathbb{Z}_n \times \mathbb{Z}_n \longrightarrow 1 .$$

where σ and τ are mapped to $(1,0)$ and $(0,1)$ respectively. The order of H_n is n^3. In fact if $n = p$ is a prime number then H_p is the unique group of order p^3 with exponent p.

To rephrase our result we finally consider the <u>involution</u>

$$\iota : \mathbb{C}^n \longrightarrow \mathbb{C}^n$$

$$e_m \longmapsto e_{-m} .$$

Then theorem (I.3) can be expressed as follows.

(I.5) **Theorem:** (i) The involution ι leaves the elliptic normal curve $C_n \subseteq \mathbb{P}_{n-1}$ invariant (as a curve) and operates on it as the involution with respect to the origin $\mathcal{O}$.

(ii) Similarly the Heisenberg group H_n leaves the curve $C_n \subseteq \mathbb{P}_{n-1}$ invariant and operates on it by translation with n-torsion points.

Remark: We can look at the situation from an even more abstract point of view. The group $G_n \cong \mathbb{Z}_n \times \mathbb{Z}_n$ of n-torsion points operates on C by translation. This operation can be extended to an operation of G_n on $\mathbb{P}_{n-1} = \mathbb{P}\,(\Gamma(\mathcal{O}_C(n\mathcal{O})))$ in the following way

$$\sum_{i=1}^{s} P_i \longmapsto \sum_{i=1}^{s} (P_i + P)$$

where + denotes the addition on the elliptic curve. We thus get an irreducible projective representation

$$\rho : G_n \longrightarrow PGL(n,\mathbb{C}) \ .$$

On the other hand, the Heisenberg gtoup H_n is a representation group of the group $G_n \cong \mathbb{Z}_n \times \mathbb{Z}_n$. i.e. each irreducible projective representation of $\mathbb{Z}_n \times \mathbb{Z}_n$ can be lifted to a linear representation of H_n and vice versa. Theorem (I.3) then tells us that the above projective representation of $\mathbb{Z}_n \times \mathbb{Z}_n$ lifts to the Schrödinger representation of H_n.

II. An abstract configuration

In this section we shall describe an abstract configuration which can be associated to the Heisenberg groups H_p where $p \geq 3$ is a prime number.

(II.1) First note that there are exactly $p+1$ subgroups $\mathbb{Z}_p \subseteq \mathbb{Z}_p \times \mathbb{Z}_p$. They are generated by $(0,1)$ and $(1,\ell)$, $\ell \in \mathbb{Z}_p$ respectively. We shall first determine all hyperplanes $H \subseteq \mathbf{P}_{p-1}$ which are invariant under one of these subgroups.

Clearly

$$\tau(H) = H$$

if and only if H is one of the following hyperplanes

$$H_k = \{x_{-k} = 0\} .$$

Note that

$$H_k = \sigma^k(H_0) .$$

Next we shall determine all hyperplanes H such that

$$\tau^\ell \sigma(H) = H .$$

We first remark that, because of σ, the equation of any such H must be of the form

$$x_0 + \sum_{m=1}^{p-1} \lambda_m x_m = 0 .$$

It is easy to check that the invariance under $\tau^\ell \sigma$ is equivalent to

$$\lambda_1^p = 1$$
$$\lambda_m = \lambda_1^m \cdot \varepsilon^{\frac{1}{2}m(m-\ell)} \qquad \text{for} \quad m = 2,\ldots,p-1 \;.$$

Hence we can set

$$\lambda_1 = \varepsilon^{-\frac{1}{2}(p-1)\ell - k}$$

for some $k \in \mathbb{Z}_p$ and the other λ_m's then become

$$\lambda_m = \varepsilon^{\frac{m}{2}(m-p)\ell - mk}$$

Then the p hyperplanes

$$H_{k\ell} = \{ \sum_{m=0}^{p-1} \varepsilon^{\frac{m}{2}(m-p)\ell - mk} x_m = 0 \} \; ; \; k = 0,\ldots,p-1$$

are the hyperplanes invariant under $\tau^\ell \sigma$. Note that

$$H_{k\ell} = \tau^k(H_{0\ell}) \;.$$

We can sum up our results as follows:

<u>Proposition:</u> <u>For each of the $p+1$ subgroups $\mathbb{Z}_p \subseteq \mathbb{Z}_p \times \mathbb{Z}_p$ there are exactly p hyperplanes which are invariant under this subgroup.</u>

(II.2) Next we shall again consider the involution

$$\iota : \mathbb{C}^p \longrightarrow \mathbb{C}^p$$

$$e_m \longmapsto e_{-m} .$$

This involution defines a decomposition of $\mathbb{C}^p$ into eigenspaces

$$\mathbb{C}^p = E^+ \oplus E^-$$

where

$$E^+ = \langle e_0, e_1 + e_{p-1}, \ldots, e_{\frac{p-1}{2}} + e_{\frac{p+1}{2}} \rangle$$

$$E^- = \langle e_1 - e_{p-1}, \ldots, e_{\frac{p-1}{2}} - e_{\frac{p+1}{2}} \rangle .$$

Clearly $\dim E^+ = \frac{1}{2}(p+1)$ and $\dim E^- = \frac{1}{2}(p-1)$.

Lemma: $\quad E^- = H_0 \cap H_{00} \cap \ldots \cap H_{0,p-1}$

Proof: (1) We shall first prove that E^- is contained in this intersection. Clearly $E^- \subseteq H_0$. Furthermore recall that $H_{0\ell}$ is given by

$$\sum_{m=0}^{p-1} \lambda_m^\ell x_m = 0$$

where

$$\lambda_m^\ell = \varepsilon^{\frac{1}{2}m(m-p)\ell}$$

Our assertion now follows immediately since

$$\lambda_{p-m}^{\ell} = \varepsilon^{\frac{1}{2}(p-m)(-m)\ell} = \varepsilon^{\frac{1}{2}m(m-p)\ell} = \lambda_m^{\ell}.$$

(ii) To finish the proof we shall show that $\frac{1}{2}(p+1)$ of the hypersurfaces H_0 and $H_{0\ell}$ are independent. To do this we have to examine the matrix

$$\begin{pmatrix} 1 & 1 & \lambda_0 & \lambda_0^2 & \cdots & \lambda_0^{p-1} \\ 0 & 1 & \lambda_1 & \lambda_1^2 & \cdots & \lambda_0^{p-1} \\ \vdots & \vdots & \vdots & & & \vdots \\ 0 & 1 & \lambda_{p-1} & \lambda_{p-1}^2 & \cdots & \lambda_{p-1}^{p-1} \end{pmatrix}$$

Using the well known formula for the Vandemonde determinant it will be sufficient to see that $\frac{1}{2}(p+1)$ of the λ_m's are different. Therefore we look at

$$\lambda_{2k} = \varepsilon^{k(2k-p)} = \varepsilon^{2k^2}.$$

It suffices to see that

$$2k^2 \not\equiv 2\ell^2 \mod p$$

if $k,\ell \in \{0,\ldots,\frac{p-1}{2}\}$ are different. But this is clearly so since otherwise

$$p \mid 2(k-\ell)(k+\ell)$$

which is impossible. This finishes the proof.

(II.3) Our next step is to define for all $k, \ell \in \mathbb{Z}_p$ the subspaces

$$E_{k\ell} := \tau^k \sigma^\ell (E^-).$$

Lemma: $E_{k\ell} \cap E_{k'\ell'} = 0$ __if__ $(k,\ell) \neq (k',\ell')$.

__Proof.__ It will be enough to show that

$$E_{00} \cap E_{-k,-\ell} = 0$$

if $(k,\ell) \neq (0,0)$. To see this assume that

$$x = \sum_{m=0}^{p-1} x_m e_m \in E_{00} \cap E_{-k,-\ell}.$$

Since $x \in E_{00}$ it follows that

(1) $$x_m = -x_{-m}.$$

On the other hand, since $x \in E_{-k,-\ell}$ it follows that $\tau^k \sigma^\ell(x) \in E_{00}$. This is equivalent to

(2) $$x_{m+\ell} \varepsilon^{2mk} = -x_{-m+\ell}$$

If $\ell = 0$ and $k \neq 0$ it follows immediately from (1) and (2) that $x = 0$. Hence assume $\ell \neq 0$. By (1) it follows that $x_0 = 0$. Setting $m = -\ell$ in (2) one gets $x_{2\ell} = 0$ which because of (1) implies $x_{-2\ell} = 0$. Using (2) again, this time for $m = -3\ell$ we find $x_{4\ell} = 0$. Proceeding in this way one finds $x = 0$.

(II.4) We can now sum up the situation as follows: We have found $p(p+1)$ hyperplanes which we have denoted by H_k and $H_{k\ell}$ respectively. Moreover we have constructed p^2 subspaces $E_{k\ell}$ of dimension $\frac{1}{2}(p-1)$. Now each of the spaces $E_{k\ell}$ is contained in exactly $p+1$ of the hyperplanes and is in fact their common intersection. On the other hand, each of the hyperplanes H_k and $H_{k\ell}$ contains exactly p of the subspaces $E_{k\ell}$ and is indeed spanned by any two of them. In particular we cay say:

Proposition: The $p(p+1)$ hyperplanes H_k and $H_{k\ell}$ together with the p^2 subspaces $E_{k\ell}$ form a configuration of type $(p^2_{p+1}, p(p+1)_p)$.

(II.5) So far we have said nothing about the relation of this configuration to the elliptic normal curve C_p. Because of

$$x_m(0) = -x_{-m}(0)$$

it follows that the (projective) space $E_{00} = E^-$ contains the origin $\mathcal{O}$. Hence each of the spaces $E_{k\ell}$ goes through exactly one of the p-torsion points of C_p.

Since the hyperplanes H_k and $H_{k\ell}$ are invariant under some subgroup $\mathbb{Z}_p \subseteq G_p$ it follows that they each contain exactly p of the p-torsion points. On the other hand the hyperplanes are determined by these points. The exact relation is the following:

$$H_k \ni \{\frac{k\omega_1 - m\omega_2}{p} ; m \in \mathbb{Z}_p\}$$

$$H_{k\ell} \ni \{\frac{m(\omega_1 - \ell\omega_2) + k\omega_2}{p} ; m \in \mathbb{Z}_p\}$$

We can summarize this as follows.

Proposition: *Each of the hyperplanes* H_k *and* $H_{k\ell}$ *intersects* C_p *in exactly* p *of the* p-*torsion points. The union of all* p *hyperplanes belonging to a fixed subgroup* $\mathbb{Z}_p \subseteq G_p$ *contains all* p^2 *hyperosculating points of* C_p.

III. Examples

In this section we want to illustrate the results of the preceding two sections in the case of elliptic normal curves of low degree.

(III.1) <u>n = 3</u>. In this case $C_3 \subseteq \mathbb{P}^2$ is a plane cubic curve. Its equation must be - at least up to a scalar - invariant under both the Heisenberg group H_3 and the involution ι. It then needs only elementary considerations to see that C_3 must be given by a cubic equation of type

$$x_0^3 + x_1^3 + x_2^2 + ax_0 x_1 x_2 = 0 .$$

i.e. in Hesse normal form.

Next we want to describe the configuration determined by H_3 and ι. Each of the 4 subgroups $\mathbb{Z}_3 \subseteq G_3$ gives rise to 3 invariant lines, i.e., to a triangle. Each of these triangels intersects C in 3 points of inflection and each of the triangles contains all 9 points of inflection. The 9 subspaces $E_{k\ell}$ have (affine) dimension 1 hence coincide with the 9 points of inflection. Hence we get the classically well known "Wendepunktskonfiguration" associated to an elliptic cubic. It is of type $(9_4, 12_3)$.

In the following picture we want to describe how the invariant lines are related to the 3-torsion points.

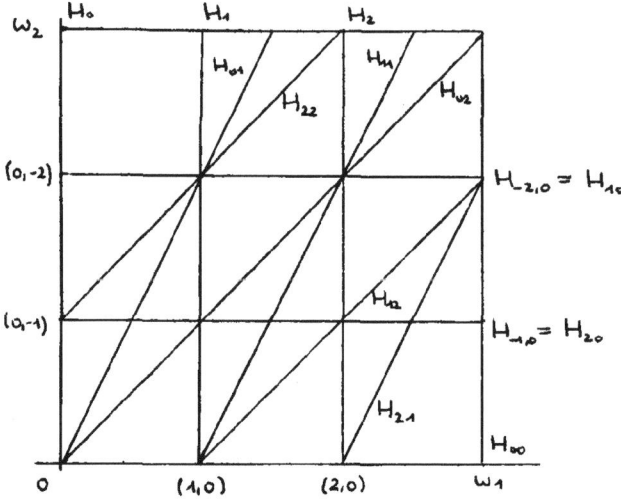

(III.2) <u>n = 4</u>. It is well known that the elliptic quartic curve $C_4 \subseteq \mathbb{P}_3$ is cut out by a pencil of quadric surfaces. To determine this pencil one has to look at the H_4-module

$$S^2 V = H^0(O_{\mathbb{P}^3}(2)) \ .$$

It splits up into a sum of eigenspaces

$$S^2 V = \bigoplus_{i=1}^{5} V_i$$

where

$$V_1 = \langle x_0^2 + x_2^2, x_1^2 + x_3^2 \rangle$$
$$V_2 = \langle x_0^2 - x_2^2, x_1^2 - x_3^2 \rangle$$
$$V_3 = \langle x_1 x_3, x_0 x_2 \rangle$$
$$V_4 = \langle x_0 x_1 + x_2 x_3, x_1 x_2 + x_0 x_3 \rangle$$
$$V_5 = \langle x_0 x_1 - x_2 x_3, x_1 x_3 - x_0 x_3 \rangle$$

Now $V_1 \cong V_3$ as H_4-modules whereas no other two of the direct summands are isomorphic. Clearly the pencil of quadrics which cuts out C_4 must be invariant under H_4. It can be neither V_1 nor V_4 or V_5. It cannot be V_1 since C_4 does not lie in a plane. To exclude V_4 and V_5 note that $x_i(\frac{1}{2}(\omega_1+\frac{\omega_2}{4})) = 0$ if and only if $i = 0$. One then concludes easily that C_4 is the intersection of the quadrics

$$Q_0 = x_0^2 + x_2^2 + 2a\, x_1 x_3$$

$$Q_1 = x_1^2 + x_3^2 + 2a\, x_0 x_2$$

where

$$a = -\frac{x_2^2(z_0)}{2x_1(z_0)x_3(z_0)} \quad \text{with} \quad z_0 = \frac{\omega_1}{2}+\frac{\omega_2}{8}.$$

The pencil of quadrics

$$Q = \lambda Q_0 + Q_1$$

contains four singular quadrics which are given by

$$\lambda = \pm\frac{1}{a},\ \pm a.$$

The vertices of these quadric cones can be easily computed to be

$$S_1 = (0:1:0:-1)$$
$$S_2 = (0:1:0:1)$$
$$S_3 = (1:0:-1:0)$$
$$S_4 = (1:0:1:0)$$

Next note that the involution $\iota : e_m \mapsto e_{-m}$ defines a decomposition

$$V = E^- \oplus E^+$$

where

$$E^- = \langle e_1 - e_3 \rangle$$
$$E^+ = \langle e_0, e_2, e_1 + e_3 \rangle .$$

It follows that S_1 is just the point defined by E^- whereas the other vertices lie in the plane determined by E^+. This has the following consequence. The projection from S_1 defined a 2:1 map

$$\pi : C_4 \longrightarrow C \cong \mathbb{P}_1$$

where C is a plane conic. The projection π induces an isomorphism

$$\pi^* : \Gamma(\mathcal{O}_{\mathbb{P}_1}(2)) \cong E^+ .$$

From this one concludes that the branch points of π are the 4 points of 2-torsion on C. In other words, the vertex S_1 lies on the tangents through the origin and the points related to $\mathcal{O}$ by half-periods. In this way the 16 tangents to the 4-torsion points can be grouped into 4 sets of 4 tangents which all go through one of the vertices.

(III.3) $\underline{n = 5}$. In this case we have an elliptic quintic $C_5 \subseteq \mathbb{P}_4$. Again any such curve is cut out by quadric hypersurfaces. Indeed look at the exact sequence

$$0 \longrightarrow I_C(2) \longrightarrow O_{\mathbb{P}_4}(2) \longrightarrow O_C(2) \longrightarrow 0 .$$

One checks easily that the map

$$\Gamma(O_{\mathbb{P}_4}(2)) \longrightarrow \Gamma(O_C(2))$$

is surjective and concludes that

$$h^0(J_C(2)) = 5 .$$

Hence there are 5 quadric hypersurfaces through C. To determine these we have to look at the H_5-module

$$S^2 V = \Gamma(O_{\mathbb{P}_4}(2)) .$$

One finds easily that

$$S^2 V = \overline{V}^{\oplus 3}$$

where $\overline{V}$ is the 5-dimensional representation of H_5 given by

$$\overline{\rho} : H_5 \longrightarrow GL(V)$$

$$\overline{\rho}(\sigma) = \sigma$$

$$\overline{\rho}(\tau) = \tau^2 .$$

It follows that the quadrics containing C must be of the form

$$Q_0 = x_0^2 + ax_2x_3 + bx_1x_4$$

$$Q_i = \sigma^i(Q_0) ; \qquad i = 1,\ldots,4 .$$

One then finds that

$$a = - \frac{x_0^2\left(\frac{\omega_1}{5}\right)}{x_2\left(\frac{\omega_1}{5}\right)x_3\left(\frac{\omega_1}{5}\right)} = - \frac{x_4^2(0)}{x_1(0)x_2(0)} = - \frac{x_1(0)}{x_2(0)}$$

Similarly one sees that

$$b = \frac{x_2(0)}{x_1(0)} = -\frac{1}{a} .$$

Hence C is contained - and in fact equals - the intersection of the quadrics

$$Q_0 = x_0^2 + ax_2x_3 - \frac{1}{a} x_1x_4$$

$$Q_1 = x_1^2 + ax_3x_4 - \frac{1}{a} x_0x_2$$

$$Q_2 = x_2^2 + ax_0x_4 - \frac{1}{a} x_1x_3$$

$$Q_3 = x_3^2 + ax_0x_1 - \frac{1}{a} x_2x_4$$

$$Q_4 = x_4^2 + ax_1x_2 - \frac{1}{a} x_0x_3$$

Next we want to discuss the <u>configuration</u> associated to the Heisenberg group H_5 and the involution ι. We get 6 sets each consisting of 5 hypersurfaces. These form what was classically called the 6 "fundamental pentahedra". The subspaces $E_{k\ell}$ define 25 skew lines $L_{k\ell}$ whose equations are

$$x_{-k} = x_{1-k} + \varepsilon^{2\ell}x_{4-k} = \varepsilon x_{2-k} + x_{3-k} = 0 .$$

Altogether we get a configuration of type $(25_6, 30_5)$. It is also worthwhile to note the following aspect of this configuration. The 6 fundamental pentahedra determine 6 quintic forms, namely

$$Q_0 = \prod_{k=0}^{4} x_k$$

$$Q_i = \prod_{k=0}^{4} (\sum_{m=0}^{4} \varepsilon^{\frac{m}{2}(m-p)i-mk} x_m) ; \quad i = 1,\ldots,5.$$

One checks easily that the quintic forms $Q_0,\ldots,Q_5$ are invariant under the Heisenberg group H_5. Although perhaps a bit more tedious one can also check that these quintics are linearly independent. On the other hand, it is straightforward group theory (see [6]) that the quintic forms invariant under H_5 form an affine 6-dimensional space $\Gamma_H(O_{\mathbb{P}_4}(5)) \subseteq \Gamma(O_{\mathbb{P}_4}(5))$. The intersection of these quintics can easily be shown to consist of 25 skew lines.

Hence we get

Proposition: The six fundamental pentahedra determine a basis of the space $\Gamma_H(O_{\mathbb{P}_4}(5))$ of invariant quintic forms and the 25 skew lines $L_{k\ell}$ are the common intersection of these quintics.

I want to conclude this section with the following remark. If one projects C_5 from the origin one gets an elliptic normal curve $C_4 \subseteq \mathbb{P}_3$. This projection is compatible with the involution. It maps $L_{00} = \mathbb{P}(E^-)$ to a point S_1 which is given by the 1-dimensional eigenspace of ι which belongs to the eigenvalue -1. Hence S_1 is the vertex of a quadratic cone through C_4. But this means that L_{00} is the singular line of a rank 3 quadric in $\mathbb{P}_4$ which goes through C_5. Corresponding to the four quadric cones through C_4 there are 4 such lines. But L_{00} is distinguished by the fact that it is contained in the osculating plane to C_5 at $\mathcal{O}$. An analogous interpretation can, of course, be given for the other lines $L_{k\ell}$ too.

IV. Normal bundles of elliptic space curves

Here I want to discuss a relation between the material presented in sections I to III and a problem concerning the normal bundle of elliptic space curves $C' \subseteq \mathbb{P}_3$ of degree 5. In their paper [4] Ellingsrud and Laksov classified normal bundles of elliptic quintics (for a precise statement see (IV.2)). Their result depends on a certain 1-dimensional family of quintic hypersurfaces Y_M. Ellingsrud and Laksov themselves pointed out that it would be desirable to have a good understanding of these quintics. It is the purpose of this section to help towards this goal. In fact it was by working on this problem that I was led to study the symmetries of elliptic normal curves.

In order to keep this paper to a reasonable length I do not want to give all the details of all the proofs in this section. I shall, however, try and outline how to prove the stated results and I trust that this will enable the interested reader to fill in the necessary details himself.

(IV.1) We start by remarking that every elliptic quintic $C' \subseteq \mathbb{P}_3$ is the projection of some elliptic normal curve $C_5 \subseteq \mathbb{P}_4$. It is the starting point of Ellingsrud and Laksov's paper to fix some such normal curve $C = C_5$ and to classify the normal bundle of a projection according to the centre of projection. If $P \in \mathbb{P}_4$ then we shall denote the projection of C from P by C_P. If $P \in \mathbb{P}_4 - \operatorname{Tan} C$ then the normal bundle of C_P is defined as

$$N_P := \pi_P^* T_{\mathbb{P}_3} / T_C .$$

Here $\pi_P : C \longrightarrow \mathbb{P}_3$ is the projection map.

We are now ready to formulate the result of Ellingsrud and Laksov.

(IV.2) **Theorem (Ellingsrud/Laksov):** To each line bundle $M \in \text{Pic}^0(C)$ of degree 0 one can associate a quintic hypersurface $Y_M \subsetneq \mathbf{P}_4$ with the following properties:

(i) $Y_0 = \text{Sec } C$

(ii) $Y_M = Y_{M^{-1}}$ and $Y_M \neq Y_{M'}$ otherwise

(iii) Each point $P \in \mathbf{P}_4 - \text{Sec } C$ belongs to a unique hypersurface Y_M.

(iv) If $P \in Y_M - \text{Sec } C$ then either

    (a) $M^2 \neq 0_C$ and $N_P(-2) = M \oplus M^{-1}$ or if

    (b) $M^2 = 0_C$ then there is an open, non-empty set of points $P \in Y_M - \text{Sec } C$ such that $N_P(-2)$ is indecomposable, isomorphic to the non-trivial extension $\text{Ext}^1_{0_C}(M,M)$.

Remarks: (i) For a possible definition of the quintics Y_M see (IV.4).

(ii) I have been told that Ellingsrud has recently shown the existence of curves with $N_P(-2) = M \oplus M$ where $M^2 = 0_C$ but $M \neq 0_C$.

As I have said before it is the purpose of this section to gather some more information concerning the quintic hypersurfaces Y_M. We can sum up our results as follows.

(IV. 3) **Theorem:** (i) The quintic hypersurfaces Y_M form a linear family, i.e. the map $M \longmapsto Y_M$ consists of the covering induced by the involution on $\text{Pic}^0(C)$ followed by a linear embedding of $\mathbf{P}_1$.

(ii) The equations defining the Y_M are invariant under the Heisenberg group H_5, i.e. the Y_M are linear combinations of the fundamental pentahedra.

(iii) The intersection of the quintics Y_M consists of two components

$$\bigcap Y_M = \text{Tan } C \cup F$$

where deg Tan C = 10 and F is another ruled surface of degree 15 containing the 25 skew lines $L_{k\ell}$.

We shall devote the following three paragraphs to a discussion of these statements.

(IV. 4) We shall first recall the definition of the quintics Y_M. To do this set

$$L := \mathcal{O}_C(H) = \mathcal{O}_C(5\sigma)$$

and

$$V := H^0(L).$$

Moreover, if $P \in \mathbf{P}_4$, let $V_P \subseteq V$ be the corresponding hyperplane. For any $P \in \mathbf{P}_4 - \text{Tan } C$ one has the following commutative and exact diagram over C.

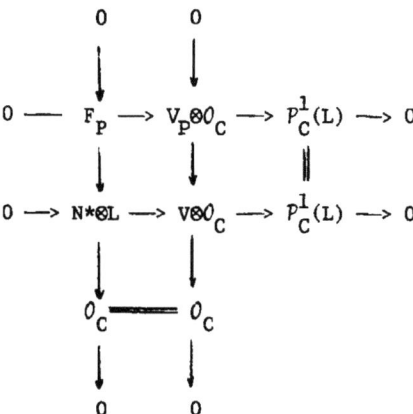

Here $N^* = N^*_{C/\mathbf{P}_4}$ is the conormal bundle of C in $\mathbf{P}_4$ and $P^1_C(L)$ is the bundle of first principal parts of L. Note that

$$F_P = N_P^* \otimes L.$$

In order to vary the point P one considers the product

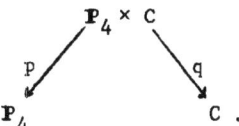

This gives rise to a diagram

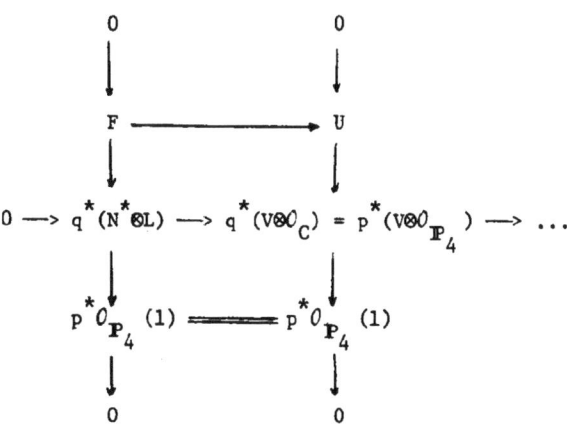

where the left hand vertical row is exact over $(\mathbb{P}_4 - \text{Tan } C) \times C$. Tensoring this row with $q^*(L \otimes M)$ where $M \in \text{Pic}^0 C$, one gets a sequence

(1_M) $0 \longrightarrow F \otimes q^*(L \otimes M) \longrightarrow q^*(N^* \otimes L^2 \otimes M) \longrightarrow p^* O_{\mathbb{P}_4}(1) \otimes q^*(L \otimes M) \longrightarrow 0$.

Applying p_* leads to a morphism

$$\phi_M : H^0(N^* \otimes L^2 \otimes M) \longrightarrow \mathcal{O}_{\mathbb{P}_4}(1) \otimes H^0(L \otimes M) \ .$$

Because of

$$h^0(N^* \otimes L^2 \otimes M) = h^0(L \otimes M) = 5$$

this map can be viewed as a 5×5 matrix with entries linear forms. Then Ellingsrud and Laksov define their quintics as

$$Y_M := \{\det \phi_M = 0\} \ .$$

The hypersurfaces Y_M then have the property that

$$Y_M = \{P; \ h^0(F_P \otimes L \otimes M) \neq 0\} \ .$$

We claim that the Y_M form a <u>linear family</u>. Now if one wanted to use assertion (iii) of the theorem of Ellingsrud and Laksov one can deduce this fact - at least in characteristic 0 - easily from the fact that a general point lies on a unique member of this family. However, we want to choose a different approach which in fact provides more information. For this purpose we consider the product $\mathbb{P}_4 \times C \times \text{Pic}^0 C$ and denote by E the Poincaré bundle over $C \times \text{Pic}^0 C$. Then we have a sequence

$$0 \longrightarrow F \boxtimes L \boxtimes E \longrightarrow N^* \boxtimes L^2 \boxtimes E \longrightarrow \mathcal{O}_{\mathbb{P}_4}(1) \boxtimes L \boxtimes E \longrightarrow 0 \ .$$

Here $\boxtimes$ denotes the tensor product of the pullbacks to $\mathbb{P}_4 \times C \times \text{Pic}^0 C$. The restriction of this sequence to $\mathbb{P}_4 \times C \times \{M\}$ is just (1_M). Next let

$$f : \mathbb{P}_4 \times C \times \mathrm{Pic}^0 C \longrightarrow \mathbb{P}_4 \times \mathrm{Pic}^0 C$$

be the projection. Then $f_*(N^* \boxtimes L^2 \boxtimes E)$ and $f_*(O_{\mathbb{P}_4}(1) \boxtimes L \boxtimes E)$ are locally free of rank 5 and by Grothendieck-Riemann-Roch one finds

$$\Lambda^5 f_*(N^* \boxtimes L^2 \boxtimes E) = q^* O_C(2\sigma)$$

$$\Lambda^5 f_*(p^* O_{\mathbb{P}_4}(1) \boxtimes L \boxtimes E) = p^* O_{\mathbb{P}_4}(1) \otimes q^* O_C(4\sigma).$$

From this one concludes readily that the map

$$\phi : \mathrm{Pic}^0 C \longrightarrow \mathbb{P}_N := \mathbb{P}(\Gamma(O_{\mathbb{P}_4}(5)))$$

$$M \longmapsto Y_M$$

admits a factorization

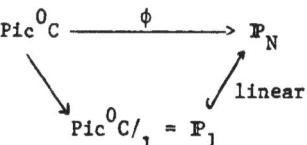

Note that this also gives another proof of part (ii) of Ellingsrud and Laksov's theorem.

(IV. 5) We next want to outline how to prove the <u>invariance</u> of the equations of the Y_M under the Heisenberg group H_5. To do this we first note that the Y_M are invariant as hypersurfaces. This follows from their geometric meaning since C is invariant. Hence it will be sufficient to show invariance for one of the equations, eg. for Y_0. Remember that

$$Y_0 = \{\det\phi_0 = 0\}$$

where the map

$$\phi_0 : H^0(N^* \otimes L^2) \longrightarrow 0_{\mathbb{P}_4}(1) \otimes H^0(L)$$

was described in the preceding paragraph. On the other hand the exact sequence

$$0 \longrightarrow I_C^2(2) \longrightarrow I_C(2) \longrightarrow N^*(2) \longrightarrow 0$$

gives rise to an isomorphism

$$H^0(I_C(2)) \cong H^0(N^*(2)) .$$

In (III. 3) we explicitly found a basis $Q_0,\ldots,Q_4$ of $H^0(I_C(2))$. Using this basis we find that

$$\det\phi_0 = \det\left(\frac{\partial Q_i}{\partial x_j}\right)$$

is an equation for Y_0. Making use of the special form of the quadrics Q_i it is then not difficult to prove invariance.

(IV. 6) Finally, I want to discuss the <u>common intersection</u> of the hypersurfaces Y_M. The fact that the tangent surface Tan C is contained in this intersection was already noticed by Ellingsrud and Laksov [4, §10]. The ruled surface F can be described as follows: We have seen before that there is a (projectively) 4-dimensional family of quadrics through C. Among these there is a 1-dimensional family of quadrics whose rank is 3. Then F is nothing but the union of the singular lines of these quadrics, i.e.

$$F = \bigcup_{\substack{Q \supseteq C \\ \text{rank } Q = 3}} \text{sing}(Q).$$

Indeed if $P \in F$ is a general point then $C_P \subseteq \mathbb{P}_3$ is a quintic curve lying on a quadric curve Q'. Moreover C_P has a node and passes through the vertex P_0 of Q' where it is smooth. One finds

$$N_P(-2) = \mathcal{O}_C(P_0) \oplus \mathcal{O}_C(-P_0)$$

which implies that

$$h^0(F_P \otimes L \otimes M) \neq 0$$

for all $M \in \text{Pic}^0 C$. Hence P is in the intersection of the Y_M. Conversely, if $P \in \text{Sec } C - (F \cup \text{Tan} C)$ then C_P again is a quintic curve with a node this time lying on a smooth quadric surface. Hence there is an exact sequence

$$0 \longrightarrow \mathcal{O}_C \longrightarrow N_P(-2) \longrightarrow \mathcal{O}_C \longrightarrow 0$$

from which one concludes that

$$h^0(F_P \otimes L \otimes M) = 0$$

if $M \neq \mathcal{O}_C$. It should be remarked that the description of F as a union of singular lines of quadrics is practically contained in [4]. It was also known to G. Sacchiero.

We now want to give a second description of F. To do this we consider the second symmetric product S^2C of the curve C. Note that by means of the map

$$S^2C \longrightarrow C$$

$$\overline{(P,Q) \longmapsto P+Q}$$

the surface S^2C becomes a $\mathbb{P}_1$-bundle over C. Indeed it is the unique $\mathbb{P}_1$-bundle $\mathbb{P}(E)$ where E is an indecomposable rank 2 bundle of odd degree over C. This is equivalent to saying that S^2C is the unique $\mathbb{P}_1$-bundle over C where the minimal self-intersection number of a section C_0 is 1. We shall next construct a map

$$\phi : S^2C \longrightarrow \mathbb{P}_4$$

whose construction is very much inspired by [4]. To do this we employ the following convention: For any point $P \in C$ we denote by t_P the (up to a scalar) unique section in $H^0(\mathcal{O}_C(P))$. Now assume that $2P_1 + 3P_2 \not\sim H \not\sim 3P_1 + 2P_2$. Then we define $Q, R \in C$ by

$$Q \sim H - 2(P_1 + P_2)$$

$$R \sim 3(P_1 + P_2) - H$$

and set

$$\phi(P_1, P_2) := H^0(L(-P_1-P_2))t_{P_1}t_{P_2} \oplus \mathbb{C}t_R^2 t_Q^3 \subseteq V.$$

On the other hand, if $2P_1 + 3P_2 \sim H$ or $3P_1 + 2P_2 \sim H$ we set

$$\phi(P_1,P_2) := H^0(L(-P_2))t_{P_2}$$

or

$$\phi(P_1,P_2) := H^0(L(-P_1))t_{P_1}$$

respectively. One then has the following

Proposition. The map

$$\phi : S^2C \longrightarrow \mathbf{P}_4$$

$$\overline{(P_1,P_2)} \longmapsto \phi(P_1,P_2)$$

is a birational morphism of S^2C onto the ruled surface F. It is 1:1 outside the section

$$D := \phi^{-1}(C) = \{\overline{(P_1,P_2)};\ 2P_1 + 3P_2 \sim H\}$$

and the map

$$\phi_D : D \longrightarrow C$$

is a 4:1 covering.

Sketch of Proof: In order to prove this proposition we first have to see that ϕ maps S^2C onto F. To see this let $P_1, P_2 \in C$ be two general points. Then we define sections in $V = H^0(L)$ by

$$y_0 := t_R^2 t_Q^3$$

$$y_1 := t_Q t_{P_1}^2 t_{P_2}^2$$

$$y_2 := t_Q^2 t_R t_{P_1} t_{P_2} .$$

We can then define a quadric surface

$$y_0 y_1 - y_2^2 = 0 .$$

Clearly this surface has rank 3 and contains C. Its singular line is the image under ϕ of the fibre of $S^2 C$ through the point (P,P). Hence ϕ maps $S^2 C$ to F. Using techniques very much like those in [4, prop. 2] we can, in fact, show that each rank 3 quadric through C arises in this way and hence that ϕ is onto. Moreover, one can see in this way that there is exactly one such quadric for each fibre of $S^2 C$, i.e. that the rank 3 quadrics through C are parametrized by C itself.

To see that ϕ is birational we want to show that ϕ is 1:1 outside $D = \phi^{-1}(C)$. But this is so because each point $P \notin C$ lies on the singular line of at most one rank 3 quadric through C. Otherwise projection from P would give a degree 5 curve C_P contained in the intersection of two quadric surfaces in $\mathbb{P}_3$. Finally look at the map

$$\phi_D : D \longrightarrow C .$$

The image of a point $\overline{(P_1,P_2)} \in D$ is given by

$$\phi(P_1,P_2) = P_2 .$$

On the other hand, for a fixed point $P_2 \in C$ the equation

$$2P_1 \sim H - 3P_2$$

has four solutions. This shows that ϕ_D is a 4:1 map. Another way of seeing this goes as follows. The assertion says that there are exactly four lines of F through each point $P \in C$. But this is equivalent to saying that the quartic elliptic curve $C_P \subseteq \mathbb{P}_3$ lies on precisely four different quadric cones. The latter we have seen in (III.2).

We finally want to describe how the 25 <u>skew lines</u> $L_{k\ell}$ can be found in this picture. To do this let $\overline{\Delta}$ be the image of the diagonal $\Delta \subseteq C \times C$ under the canonical projection onto S^2C. Then

$$\overline{\Delta} \cdot D = 50$$

and set-theoretically

$$\overline{\Delta} \cap D = \{\overline{(P,P)}; \ 5P \sim H\}$$

where each of these points has to be counted with multiplicity 2. Note that if $5P \sim H$ then

$$\phi(P,P) = P \in C.$$

Proposition: <u>The image of the 25 fibres through the points</u>

$$\bar{\Delta} \cap D = \{\overline{(P,P)};\ 5P \sim H\}$$

<u>under the map</u> ϕ <u>are the skew lines</u> $L_{k\ell}$.

Proof: Let $P \in C$ be a five torsion point and let L_P be the image under ϕ of the fibre of S^2C through $\overline{(P,P)}$. In order to describe L_P we choose a point $S \in C$ such that $S \neq P$ but $2S \sim 2P$. By the construction of ϕ the line L_P is given by

$$L_P = \{w_0 = w_1 = w_2 = 0\}$$

where

$$w_0 = t_P^5$$
$$w_1 = t_P^3 t_S^2$$
$$w_2 = t_P t_S^4.$$

To see that L_P is one of the lines $L_{k\ell}$ we have to show that L_P lies in all those hyperplanes H_k and $H_{k\ell}$ of our configuration which go through P.

But any such hypersurface is of the form

$$H = \{w = 0\}$$

where

$$w = t_P t_{P+P'} \cdots t_{P+4P'}$$

for some five torsion point $P' \neq \mathcal{O}$. So we have to see that w is in the span of the sections w_i. To see this note that

$$\lambda w_0 + \mu w_1 = t_P^3 (\lambda t_P^2 + \mu t_S^2).$$

Since $P \neq S$ the sections $t_P^2, t_S^2 \in H^0(\mathcal{O}_C(2P))$ form a basis. Hence we can choose λ and μ such that

$$\lambda t_P^2 + \mu t_S^2 = t_{P+P'} t_{P+4P'}.$$

It follows that

$$w_3 := t_P^3 t_{P+P'} t_{P+4P'} \in \text{span}(w_0, w_1)$$

Similarly we see that

$$w_4 := t_P t_S^2 t_{P+P'} t_{P+4P'} \in \text{span}(w_1, w_2)$$

Finally applying this argument once more to w_3 and w_4 we see that

$$w \in \text{span}(w_3, w_4) .$$

This proves the proposition.

V. Open problems

To conclude this paper I should like to sketch three problems which arise naturally out of the material presented above.

<u>Problem 1</u>: We have just seen that the hypersurfaces Y_M are defined by quintic forms which are invariant under the Heisenberg group H_5; i.e., they determine a 2-dimensional subspace in $\Gamma_H(O_{\mathbb{P}_4}(5))$ which has dimension 6. The question arises how to describe this subspace. One condition is that the quintics must contain the curve C. This is not the case for a general invariant quintic as one can see by looking at the fundamental pentahedra. But it is not true that this is a sufficient condition for an invariant quintic to be one of the Y_M's. I hope to return to this problem at some later date.

<u>Problem 2</u>: It would be desirable to have a more geometric way to understand the behaviour of the normal bundle of elliptic quintic curves in $\mathbb{P}_3$. In their papers [2,3] Eisenbud and Van de Ven describe how the splitting of the normal bundle of a rational curve can be realized by looking at certain ruled surfaces and cones over plane curves. It would be very nice to have an analogous picture for those elliptic quintics whose normal bundle decomposes. Moreover, this should lead to an understanding why the normal bundles of certain "special" elliptic quintics are indecomposable.

Problem 3: This concerns a possible relation between the Heisenberg group in prime dimension p and vector bundles. If $p = 3$ then H_3 is the symmetry group of a plane cubic and hence is trivially related to the line bundle $\mathcal{O}_{\mathbb{P}_2}(3)$. On the other hand, if $p = 5$ then the Heisenberg group H_5 plays a central role in the construction of the Horrocks-Mumford bundle F on $\mathbb{P}_4$ (see [5]). Moreover if $C \subseteq \mathbb{P}_4$ is the elliptic normal curve and if Tan C is its tangent surface then there is a section $s \in \Gamma(F)$ whose zero-set equals Tan C. This means that one can recover F from Tan C via the Serre construction. The question now is whether one can associate to any Heisenberg group H_p some vector bundle on $\mathbb{P}_{p-1}$. The first step would be to associate a rank 3 bundle on $\mathbb{P}_6$ to H_7.

Postscript: Until very recently I did not know whether the configuration which I have described in chapter II and which was first mentioned to me by W. Barth was classically known or not. At least it seemed to have been forgotten. It was only by chance that I found a paper by C. Segre published in 1886 in Mathematische Annalen in which he describes this configuration. I have now included C. Segre's paper as number [8] among the list of references.

Second postscript (December 1982): In the meantime I have been able to determine the affine 2-dimensional subspace $U \subseteq \Gamma_H(I_C(5))$ which belongs to the linear family of quintics Y_M. It can be written as the intersection of two 3-dimensional spaces, namely

$$U = \Gamma(I_{\text{Tan } C}(5)) \cap \Gamma_H(I_C^2(5)).$$

All quintic hypersurfaces through the tangent surface Tan C have equations which are H_5-invariant. These quintics are closely related to the Horrocks-Mumford bundle F.

References

[1] Bianchi, L.: Ueber die Normalformen dritter und fünfter Stufe des elliptischen Integrals erster Gattung. Math. Ann. 17, 234-262 (1880).

[2] Eisenbud, D., Van de Ven, A.: On the normal bundle of smooth rational space curves. Math. Ann. 256, 453-463 (1981).

[3] Eisenbud, D., Van de Ven, A.: On the variety of smooth rational space curves with given degree and normal bundle, Inv. Math. 67 (1982).

[4] Ellingsrud, G., Laksov, D.: The normal bundle of elliptic space curves of degree 5 . 18th Scand. Congress of Math. Proc. 1980. Ed. E. Balslev, pp. 258-287, Birkhäuser 1981.

[5] Horrocks, G., Mumford, D.: A rank 2 vector bundle on $\mathbb{P}_4$ with 15,000 symmetries. Topology 12, (1973).

[6] Hurwitz, A.: Ueber endliche Gruppen linearer Substitutionen, welche in der Theorie der elliptischen Transcendenten auftreten. Math. Ann. 27, 183-233 (1886).

[7] Igusa, J.: Theta Functions. Berlin, Heidelberg, New York. Springer-Verlag 1982.

[8] Segre, C.: Remarques sur les transformations uniformes des courbes elliptiques en elles-mêmes. Math. Ann. 27, 296-314 (1886).

Klaus Hulek
Department of Mathematics
Brown University
Providence, RI 02912
USA

Linkage of General Curves of Large Degree

by

Robert Lazarsfeld and Prabhakar Rao

Introduction.

Our purpose is to describe the liaison class of a general curve in $\mathbb{P}^3$ of degree much larger than its genus. In particular, we prove a conjecture of Joe Harris ([H], p.80) to the effect that such a curve can be linked only to curves of larger degree and genus.

Recall that two curves $X, Y \subseteq \mathbb{P}^3$ are <u>directly linked</u> if X is residual to Y in the complete intersection of two surfaces; they are <u>linked</u> if the one can be obtained from the other by a succession of direct linkages. Classically, linkage was seen as a method for producing interesting examples of space curves starting from simpler ones. Later work on linkage-or liaison, as it is also called-has largely focused on the equivalence relation it generates. Apéry [A] and Gaeta [G] proved that a curve $X \subseteq \mathbb{P}^3$ is linked to a complete intersection if and only if it is projectively Cohen-Macaulay; the analogous statement in higher dimensions was proved by Peskine and Szpiro [P-S]. The theorem of Apéry and Gaeta was generalized by the second author, who studied the <u>deficiency module</u>

$$M(X) = \bigoplus_{n \in \mathbb{Z}} H^1(\mathbb{P}^3, \mathcal{I}_X(n))$$

of $X \subseteq \mathbb{P}^3$, a finite module over the homogeneous coordinate ring $S = k[T_0, T_1, T_2, T_3]$. Specifically, it was shown in [R] that two curves $X, Y \subseteq \mathbb{P}^3$ are linked if and only if the module $M(X)$ of X coincides up to grading with either the module $M(Y)$ of Y or its dual $M(Y)^{\vee}$. Moreover, any finite S-module M arises as the deficiency module of some curve in $\mathbb{P}^3$. Thus one has an essentially complete picture, from a cohomological point of view, of the various liaison equivalence classes that can occur for curves in $\mathbb{P}^3$.

It is natural to ask, however, for a clearer geometric understanding of the curves that exist within a given liaison class. In the present paper, we consider the linkage class of a general smooth irreducible curve $X \subseteq \mathbb{P}^3$ of sufficiently large degree. Our main result (§3) states that if Y is any curve linked to X, other than X itself, then $\deg(Y) > \deg(X)$ and $p_a(Y) > p_a(X)$. Somewhat more precisely, we distinguish between even linkage-i.e., liaison involving an even number of direct linkages-and odd linkage, defined similarly. We show that if Y is evenly linked to X, then it is a deformation of the curve obtained by taking the union of X and certain complete intersection curves. If Y is oddly linked to X, then it arises in an analogous manner from the curve Z directly linked to X by irreducible surfaces of lowest possible degree.

The questions we consider here were first raised by J. Harris (cf [H]). A priori, one could hope-as some of the classical geometers apparently did-that techniques of liaison could be used to study space curves inductively, by linking a given curve to a (possibly very special) curve of lower degree or genus. Believing that at least for general curves such an approach is fundamentally flawed, Harris suggested that a general curve should in various senses be minimal in its liaison class. Our results may be seen, then, as giving additional support (if any is needed) to the philosophy that there is no easy way to get one's hands on a "general" curve. Some suggestive results in the direction of Harris's conjectures were obtained for lines and rational curves by Migliore [M]; at least indirectly these have contributed substantially to the present paper, as has work of Schwartau [S].

Most of our results are stated for an arbitrary curve $X \subseteq \mathbb{P}^3$ subject only to the condition that it not lie on any surfaces of degree e+4 or less, e being the largest integer such that $h^1(X, O_X(e)) \neq 0$. The generality assumption is used only in §3 to keep the curves in question off surfaces of low degree. It seems likely that similar results hold for curves $X \subseteq \mathbb{P}^3$ general in the sense of Brill-Noether theory. What is missing is even a weak approximation to the maximal rank conjecture (cf [H], p.79).

We are grateful to L. Ein, J. Harris, J. Migliore, P. Schwartau and M. Stillman for suggestions and encouragement. We also wish to thank C. Ciliberto for allowing this paper to appear in these proceedings even though we did not participate in the conference itself.

§0. Notation and Conventions.

(0.1). We work over an algebraically closed field k of arbitrary characteristic. A <u>curve</u> $X \subseteq \mathbb{P}^3$ is a subscheme of pure dimension one, without embedded points. Thus X is (locally) Cohen-Macaulay. I_X is the ideal sheaf of X, and I(X) its homogeneous ideal.

(0.2). If F is a coherent sheaf on $\mathbb{P}^3$, we let

$$H^i_*(\mathbb{P}^3, F) = \bigoplus_{n \in \mathbb{Z}} H^i(\mathbb{P}^3, F(n)),$$

so that $H^i_*(\mathbb{P}^3, F)$ is a graded module over the homogeneous coordinate ring S. We write simply $\mathcal{O}$ for the structure sheaf $\mathcal{O}_{\mathbb{P}^3}$.

(0.3). Given a curve $X \subseteq \mathbb{P}^3$, we say that X lies on a surface F if $F \in I(X)$. If $F, G \in I(X)$ meet properly, then they link X to a curve Y whose scheme structure is determined as in [P-S]. If $0 \to P \to N \to 0 \to \mathcal{O}_X \to 0$ is a locally free resolution of $\mathcal{O}_X$, with $H^1_*(\mathbb{P}^3, P) = 0$, then $\mathcal{O}_Y$ has a resolution

(0.4) $\quad 0 \longrightarrow N^v(-f-g) \longrightarrow \mathcal{O}(-f) \oplus \mathcal{O}(-g) \oplus P^v(-f-g) \longrightarrow \mathcal{O} \longrightarrow \mathcal{O}_Y \longrightarrow 0$

where f and g denote respectively the degrees of F and G ([P-S], Propn. 2.5).

§1. Curves minimal in their even liaison class.

Given a curve $X \subseteq \mathbb{P}^3$, set

$$e(X) = \max\{n \mid H^1(X, \mathcal{O}_X(n)) \neq 0\}.$$

Our goal in this section and the next is to show that if X does not lie on any surfaces of degree $e(X) + 4$ or less, then X is in various senses minimal in its even liaison class. For example, we will see that X has smaller degree and arithmetic genus than any other curve to which it is evenly linked (Corollary 1.5). The basic idea is that given any two evenly linked curves $X, Y \subseteq \mathbb{P}^3$, there exist vector bundle maps

$$\bigoplus_{i=1}^{r} \mathcal{O}(-a_i) \xrightarrow{u} E \quad \text{and} \quad \bigoplus_{i=1}^{r} \mathcal{O}(-b_i) \xrightarrow{v} E$$

which drop rank respectively on X and Y. The crucial fact is that if X lies on no surfaces of degree $\leq e(X) + 4$, then $b_i \geq a_i$, and at least one inequality is strict if $X \neq Y$ (Lemma 1.2). This allows us to compare the numerical invariants of X to those of Y, and to describe geometrically how Y is obtained from X. Under mild additional hypotheses, analogous statements can be made for odd linkage (Proposition 1.6). We remark that the importance of the integer $e(X)$ in questions of liaison was known already to Gaeta.

We start by recalling a useful representation of a given curve as a determinantal locus.

Lemma 1.1. Let $X \subseteq \mathbb{P}^3$ be a curve. Then there is an exact sequence

$$0 \longrightarrow P \xrightarrow{u} N \longrightarrow I_X \longrightarrow 0,$$

where N is a vector bundle, with $H^2_*(\mathbb{P}^3, N) = 0$, and P is a direct sum of line bundles of degrees $\geq -e(X) - 4$.

Proof. We use a construction similar to one used by Sernese [Se] (cf also [GLP] §2). Consider the graded S-algebra $R = H^0_*(\mathbb{P}^3, \mathcal{O}_X)$. Since $R_n = 0$ for $n \ll 0$, R necessarily has a minimal generator in degree zero, which we may take to be the identity. The sheafification of a minimal free S-resolution of R therefore takes the form

$$0 \longrightarrow P_2 \xrightarrow{p_2} P_1 \xrightarrow{\binom{p_1}{\phi}} P_0 \oplus \mathcal{O} \xrightarrow{(p_0, \varepsilon)} \mathcal{O}_X \longrightarrow 0,$$

where each P_i is a direct sum of line bundles, and $\varepsilon : 0 \longrightarrow 0_X$ is the natural map. One then obtains, using the snake lemma, the following commutative diagram of exact sequences of sheaves:

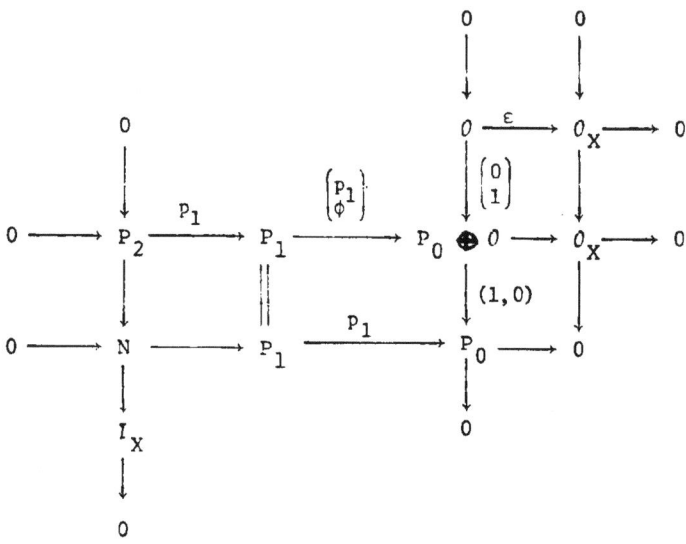

Here N is of course defined as the kernel of p_1; the vanishing of $H^2_*(\mathbb{P}^3, N)$ follows from the vanishings of $H^1_*(\mathbb{P}^3, P_0)$ and $H^2_*(\mathbb{P}^3, P_1)$. By duality one has $\text{Ext}^2_S(R, S(-4)) = H^1_*(\mathbb{P}^3, 0_X)^\vee$, and hence all the summands of P_2 have degrees $\geq -e(X) - 4$. Taking $P = P_2$, the lemma follows.

Remark. Keeping the notation of the previous proof, observe that $M(X) = H^1_*(\mathbb{P}^3, N)$. Hence the map induced by p_1 on global sections gives a presentation of the deficiency module of X:

$$H^0_*(\mathbb{P}^3, P_1) \longrightarrow H^0_*(\mathbb{P}^3, P_0) \longrightarrow M(X) \longrightarrow 0$$

Moreover, if X lies on no surface of degree $e(X) + 3$, so that $H^0(\mathbb{P}^3, N(t)) = H^0(\mathbb{P}^3, P_2(t))$ for $t \leq e(X) + 3$, then this is actually a minimal presentation of M(X). In practice, this yields a convenient method for computing a presentation of M(X)-or at least determining the number of generators and relations in each degree-in concrete examples.

For instance, suppose that X is the disjoint union of d lines $L_i = \{A_i = B_i = 0\}$, where A_i and B_i are linear forms. Then $H^0_*(\mathbb{P}^3, \mathcal{O}_X)$ is resolved by taking the direct sum of the Koszul complexes formed from A_i and B_i. Hence M(X) can be described as the S-module given by generators $e_1, \ldots, e_d$ in degree 0, subject to the relations

$$\sum_{i=1}^{d} e_i = 0$$

$$A_i \cdot e_i = 0, \quad B_i \cdot e_i = 0 \quad (1 \leq i \leq d).$$

Using the first relation to eliminate one of the generators, one obtains a minimal presentation having the form $S(-1)^{2d} \longrightarrow S^{d-1} \longrightarrow M(X) \longrightarrow 0$. The disjoint union of d complete intersections is treated almost identically. Similarly, the deficiency module of a smooth rational curve $X \subseteq \mathbb{P}^3$ of degree $d \geq 4$ and not on a quadric surface has a minimal presentation of the form $S(-2)^{2d-3} \longrightarrow S^{d-3}(-1) \longrightarrow M(X) \longrightarrow 0$. We refer to Migliore [M] for a more geometric discussion of the linkage properties of lines and rational curves.

Now suppose that $Y \subseteq \mathbb{P}^3$ is evenly linked to X. Then we may repeatedly apply (0.4) to the exact sequence of (1.1) to obtain an exact sequence

(*) $\quad\quad 0 \longrightarrow B \longrightarrow N \oplus F \longrightarrow \mathcal{I}_Y(\delta) \longrightarrow 0$

for some $\delta \in \mathbb{Z}$, where B and F are direct sums of line bundles. On the other hand, if $A = P \oplus F$, then of course we have

(**) $\quad 0 \longrightarrow A \xrightarrow{u \oplus 1} N \oplus F \longrightarrow I_X \longrightarrow 0.$

Our main technical lemma allows us to compare the degrees of the summands of A and B provided that X lies on no surfaces of degree $e(X) + 3$ or less. Specifically, in the next section we shall prove

Lemma 1.2. In the notation just introduced, write

$$A = \bigoplus_{i=1}^{r} O(-a_i), \quad \text{with} \quad a_1 \leq a_2 \leq \cdots \leq a_r$$

and

$$B = \bigoplus_{i=1}^{r} O(-b_i), \quad \text{with} \quad b_1 \leq b_2 \leq \cdots \leq b_r.$$

If X lies on no surfaces of degree $e(X) + 3$, then

$$b_i \geq a_i \quad \text{for all} \quad 1 \leq i \leq r.$$

If moreover X lies on no surfaces of degree $e(X) + 4$, and if $Y \neq X$, then $b_i > a_i$ for at least one index i.

Note that the integer δ in (*) is just the sum $\sum_{i=1}^{r} (b_i - a_i)$. Observe also that at least when $F = 0$ so that $A = P$, the lemma is highly plausible: for then the hypothesis on X implies that the free submodule $H^0_*(\mathbb{P}^3, P) \subseteq H^0_*(\mathbb{P}^3, N)$ consists of the lowest degree generators of $H^0_*(\mathbb{P}^3, N)$.

Before proceeding, we note the amusing

Corollary 1.3. Let $X \subseteq \mathbb{P}^3$ be a curve not lying on any surface of degree $e(X) + 4$ or less. Then X is the only curve whose deficiency module is isomorphic to $M(X)$ (with the given grading), and for $n > 0, M(X)(n)$ cannot be realized as the deficiency module of any curve. In particular, X is determined by its module.

Proof. If $Y \subseteq \mathbb{P}^3$ is a curve whose module coincides with $M(X)$ up to grading, then Y is evenly linked to X (by [R]), and we have

$$M(Y) = H^1_*(\mathbb{P}^3, N)(-\delta) = M(X)(-\delta),$$

δ being the integer introduced above. But under the hypothesis on X, Lemma 1.2 asserts that $\delta > 0$ unless $Y = X$.

Remarks.

(1) By contrast, using a construction which we shall review below, Schwartau [S] has shown that given any curve $X \subseteq \mathbb{P}^3$, and any $n < 0$, there exist infinitely many curves $Y \subseteq \mathbb{P}^3$ with $M(Y) = M(X)(n)$.

(2) We shall check in §3 that the hypothesis of the corollary is satisfied when X is a general curve of sufficiently large degree. In this form, the result had been conjectured by J. Harris ([H], p.80). The last statement of the corollary was established by Migliore [M] when X is a union of lines.

(3) At least in a special case, there is a simple geometric argument showing that the hypothesis on X is necessary for the validity of the result. Specifically, suppose that X is reduced, and lies on a smooth surface $S \subseteq \mathbb{P}^3$ of degree $f + 4 \leq e(X) + 4$. Then one has an exact sequence

$$0 \longrightarrow \mathcal{O}_S \xrightarrow{\cdot X} \mathcal{O}_S(X) \longrightarrow \omega_X(-f \cdot H) \longrightarrow 0,$$

where H denotes the hyperplane divisor class on S. Since
$f \leq e(X)$, duality shows that $H^0(X,\omega_X(-f \cdot H)) \neq 0$. In view of the
vanishing of $H^1(S,0_S)$, it follows that X moves in a non-trivial
linear system on S. But any two curves in such a linear system
are evenly linked, and their deficiency modules are isomorphic
(with the same grading).

Our next object is to interpret geometrically the conclusion
of Lemma 1.2. To this end, we need first to describe certain
"basic double linkages."

Given a curve $X \subseteq \mathbb{P}^3$, let F be a surface of degree f
containing X, and choose any surface H of degree $h \geq 1$ meeting F
properly. If G is a general surface through X of sufficiently
large degree, we may use F and G to link X to a curve X^*, and then
use F and G·H to link X^* to a curve Y. Y does not depend on the
surface G, and we will say that it is obtained from X by a <u>basic
double linkage</u> using F and H. This is a special case of the construction of liaison addition introduced by Schwartau [S]. Set
theoretically, Y is the union of X and the complete intersection
of F and H. Evidently $\deg(Y) = \deg(X) + f \cdot h$, and it follows from
([S], p.91) that $p_a(Y) - p_a(X) = hf \cdot (h+f-4)/2 + h \cdot \deg(X)$. Observe
that this difference is always non-negative, and is strictly positive
unless X is a line, and $h = f = 1$.

The geometric meaning of Lemma 1.2 is summarized in

<u>Proposition 1.4.</u> <u>Let $X \subseteq \mathbb{P}^3$ be a curve not contained in any
surface of degree $\leq e(X) + 3$, and let $Y \subseteq \mathbb{P}^3$ be any curve evenly
linked to X. Then there exists a sequence of curves</u>

$$X = X_1, X_2, \ldots, X_{m-1}, X_m$$

such that X_{i+1} is obtained from X_i by a basic double linkage using suitable surfaces $F_i \in I(X_i)$ and H_i, and such that Y is a deformation of X_m through curves having a fixed deficiency module. Moreover if X lies on no surface of degree $e(X) + 4$, and if $Y \neq X$, then at least one non-trivial basic double linkage must occur.

The statement of the Proposition is illustrated in Figure 1. Observe that, conversely, any curve Y obtained from X as indicated is evenly linked to X (by virtue of [R]).

Proof. Under the hypothesis on X, the assertion of Lemma 1.2 is that X and Y arise via exact sequences

$$0 \longrightarrow \bigoplus_{i=1}^{r} \mathcal{O}(-a_i) \xrightarrow{u} E \longrightarrow I_X \longrightarrow 0$$

$$0 \longrightarrow \bigoplus_{i=1}^{r} \mathcal{O}(-b_i) \xrightarrow{v} E \longrightarrow I_Y(\delta) \longrightarrow 0,$$

where E is a vector bundle of rank r+1, with $H^2_*(\mathbb{P}^3, E) = 0$, and

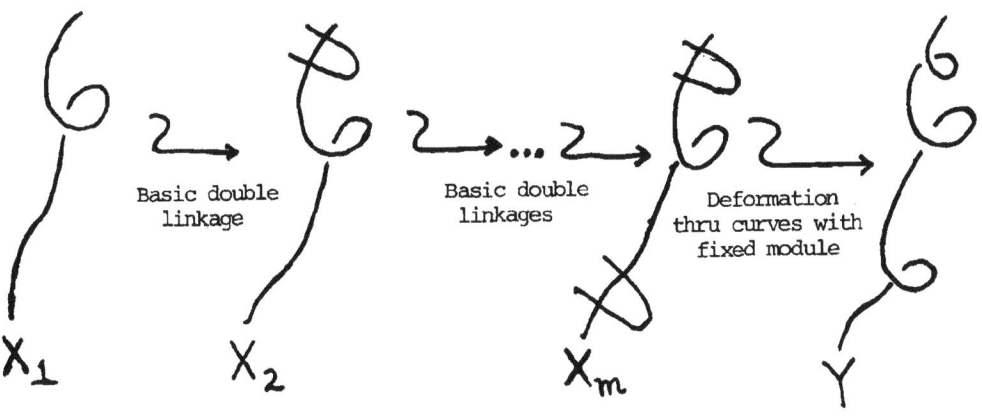

Figure 1

$$\delta_i = b_i - a_i \geq 0 \text{ for all } 1 \leq i \leq r.$$

We argue by descending induction on $\delta = \sum \delta_i$ that this set-up implies the assertion of the Proposition.

Suppose first that $\delta = 0$, so that $a_i = b_i$ for $1 \leq i \leq r$. Given $t \in k$, let

$$w_t = tu + (1-t)v \in \mathrm{Hom}(\bigoplus \mathcal{O}(-a_i), E).$$

Then for general $t \in k$, the vector bundle map w_t drops rank along a curve X_t, and it is elementary that the curves X_t fit together to form a flat family of subschemes of $\mathbb{P}^3$, parametrized by a Zariski open set $U \subseteq \mathbb{A}^1$ containing 0 and 1. For $t \in U$ one has

$$M(X_t) = H_*^1(\mathbb{P}^3, E),$$

and so Y is a deformation of X through curves with fixed deficiency module.

Assuming then that $\delta > 0$, let u_i and v_i be the ith components of u and v respectively, and denote by s_i the image of v_i in $\mathrm{Hom}(\mathcal{O}(-b_i), I_X)$:

$$\begin{array}{ccccccccc}
 & & & & & & \mathcal{O}(-b_i) & & \\
 & & & & & & \downarrow v_i \searrow^{s_i} & & \\
0 & \longrightarrow & \bigoplus_{i=1}^{r} \mathcal{O}(-a_i) & \xrightarrow{u=(u_1,\ldots,u_r)} & E & \longrightarrow & I_X & \longrightarrow & 0.
\end{array}$$

As before we suppose that the integers $\{a_i\}$ and $\{b_i\}$ are non-decreasing in i.

Let $\ell \in [1,r]$ be the largest integer such that $\delta_\ell > 0$. Re-indexing the $\{b_i\}$ if necessary, we may assume that either $\ell = r$ or $b_{\ell+1} > b_\ell$. We assert that for some index $j \le \ell$, the section $s_j \in H^0(\mathbb{P}^3, I_X(b_j))$ is non-zero. In fact, one has $a_i = b_i > b_\ell$ for $i > \ell$, and if $s_j = 0$ for all $j \le \ell$, then the first ℓ components of v would factor through those of u:

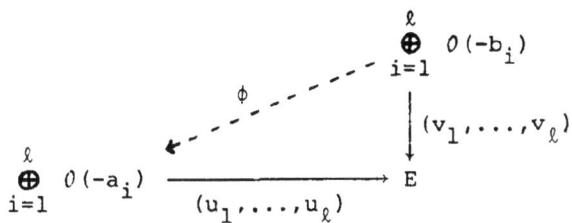

But $\sum_{i=1}^{\ell} a_i < \sum_{i=1}^{\ell} b_i$ since $\delta > 0$, and so the map ϕ, and hence also v, would drop rank along a surface, whereas in reality v drops rank exactly on the curve Y. Hence $s_j \ne 0$ for some $j \le \ell$, as claimed.

If Q is a general form of degree $b_\ell - b_j$, then the section

$$F = s_\ell + Qs_j \in H^0(\mathbb{P}^3, I_X(b_\ell))$$

is non-zero. Let X_2 be the curve obtained from $X = X_1$ by a basic double linkage using F and a general surface H of degree δ_ℓ. Then by (0.4), X_2 and Y (trivially) are realized via

$$0 \longrightarrow \bigoplus_{i=1}^{r} \mathcal{O}(-a_i) \oplus \mathcal{O}(-a_\ell - \delta_\ell) \longrightarrow E \oplus \mathcal{O}(-a_\ell) \longrightarrow I_{X_2}(\delta_\ell) \longrightarrow 0$$

$$0 \longrightarrow \bigoplus_{i=1}^{r} \mathcal{O}(-a_i - \delta_i) \oplus \mathcal{O}(-a_\ell) \longrightarrow E \oplus \mathcal{O}(-a_\ell) \longrightarrow I_Y(\delta) \longrightarrow 0.$$

These sequences satisfy the conditions stated at the beginning of the proof, hence the existence of the desired sequence of curves follows by induction. Finally, the last assertion of the proposition follows from the fact that if X lies on no surfaces of degree $e(X) + 4$, and if $X \neq Y$, then at least one δ_i is non-zero.

Corollary 1.5. If X lies on no surface of degree $e(X) + 4$ or less, and Y is evenly linked to X, then either $X = Y$ or $\deg(Y) > \deg(X)$ and $p_a(Y) > p_a(X)$.

We now show that under a small additional hypothesis, a similar picture applies to odd linkage:

Proposition 1.6. Let $X \subseteq \mathbb{P}^3$ be a curve not lying on any surface of degree $e(X) + 3$ or less. Choose a system of minimal generators

$$F_i \in I(X) \quad (1 \leq i \leq \ell)$$

of the homogeneous ideal of X, with $c_i = \deg(F_i)$ non-decreasing in i. Assume that F_1 and F_2 meet properly, so that they link X to a curve Z. Then Z does not lie on any surface of degree $e(Z) + 3$ or less.

Remarks.
(1) If X is reduced and irreducible, it is automatic that F_1 and F_2 meet properly.
(2) The curve Z may depend on the choice of the surfaces F_1 and F_2. However if Z' is obtained via F_1' and F_2', then Z' is a deformation of Z through curves with fixed deficiency module (thanks to (1.4)).

Proof. Consider a minimal free resolution of $H^0_*(\mathbb{P}^3, N)$, where N is the vector bundle constructed from X in Lemma 1.1.

Such a resolution has length 2, and sheafifying yields an exact sequence which we use to define a bundle E as shown:

(1.7)
$$0 \longrightarrow L_2 \longrightarrow L_1 \longrightarrow L_0 \longrightarrow N \longrightarrow 0$$
with E fitting in via $0 \to E \to L_1$ and $E \to L_0 \to 0$.

Thus each L_i is a sum of line bundles, and in particular $H^1_*(\mathbb{P}^3, E) = 0$. On the other hand, we have from (1.1) an exact sequence

$$0 \longrightarrow P \longrightarrow N \longrightarrow I_X \longrightarrow 0,$$

where

$$P = \bigoplus_{i=1}^{s} \mathcal{O}(-d_j), \text{ with all } d_j \leq e(X) + 4.$$

Observe that our hypothesis on X implies that

(*) $d_j \leq c_i$ $(1 \leq j \leq s, \; 1 \leq i \leq \ell)$.

It follows in particular from (*) that any minimal generator of $H^0_*(\mathbb{P}^3, P)$ must be a minimal generator of $H^0_*(\mathbb{P}^3, N)$. Therefore

$$L_0 = \bigoplus_{i=1}^{\ell} \mathcal{O}(-c_i) \oplus P,$$

and then one sees that the bundle E defined above is isomorphic to the kernel of the natural map $\mathcal{O}(-c_i) \longrightarrow I_X$:

(**) $$0 \longrightarrow E \longrightarrow \bigoplus_{i=1}^{\ell} \mathcal{O}(-c_i) \xrightarrow{(F_1, \ldots, F_\ell)} I_X \longrightarrow 0$$

We next use (0.3) to read off from (**) a locally free resolution of I_Z. Cancelling redundant terms, one finds:

$$(***) \quad 0 \longrightarrow \bigoplus_{i=3}^{\ell} O(c_i - c_1 - c_2) \longrightarrow E^v(-c_1 - c_2) \longrightarrow I_Z \longrightarrow 0$$

Since $H_*^2(\mathbb{P}^3, E^v) = 0$, it follows that $H_*^1(\mathbb{P}^3, O_Z)(=H_*^2(\mathbb{P}^3, I_Z))$ injects into

$$\bigoplus_{i=3}^{\ell} H_*^3(\mathbb{P}^3, O(c_i - c_1 - c_2)).$$

Recalling that the c_i are non-decreasing, we deduce that

$$e(Z) \leq c_1 + c_2 - (c_3 + 4) \leq c_1 - 4.$$

On the other hand $H_*^1(\mathbb{P}^3, N^v) = 0$, so it follows from (1.7) that $H_*^0(\mathbb{P}^3, L_0^v(-c_1-c_2))$ surjects onto $H_*^0(\mathbb{P}^3, E^v(-c_1-c_2))$, and hence, thanks to (***), onto $H_*^0(\mathbb{P}^3, I_Z)$. Therefore, generators of the homogeneous ideal $I(Z) = H_*^0(\mathbb{P}^3, I_Z)$ can occur only in degrees

$$c_1, \ c_2, \ \text{and } c_1 + c_2 - d_j \quad (1 \leq j \leq s).$$

But $c_1 + c_2 - d_j \geq c_2 \geq c_1$ by (*), so $H_*^0(\mathbb{P}^3, I_Z)$ vanishes in degrees $c_1 - 1$ and less. Since $e(Z) + 3 \leq c_1 - 1$, this proves the Proposition.

<u>Corollary 1.8.</u> <u>Let X and Z be as in the statement of Proposition 1.6, and let Y be any curve oddly linked to X. Then Y is obtained from Z by a succession of basic double linkages, and then a deformation, as described in (1.4). In particular, $\deg(Y) \geq \deg(Z)$, and $p_a(Y) \geq p_a(Z)$.</u>

§2. Proof of Lemma 1.2.

We keep the notation introduced in §1, so that we have exact sequences

$$0 \longrightarrow P \xrightarrow{u} N \longrightarrow \mathcal{I}_X \longrightarrow 0$$

and

$$0 \longrightarrow B \xrightarrow{v} N \oplus F \longrightarrow \mathcal{I}_Y(\delta) \longrightarrow 0.$$

Here B, F and P are direct sums of line bundles, every summand of P having degree $\geq -f$, where $f = e(X) + 4$. We assume that X lies on no surface of degree $e(X) + 3$ or less, i.e., that

(2.1) $\qquad H^0(\mathbb{P}^3, P(t)) \xrightarrow{\sim} H^0(\mathbb{P}^3, N(t)) \quad \text{for } t \leq f - 1.$

Note that this implies that N cannot itself split as a sum of line bundles (at least if $X \neq \emptyset$). We wish to compare the degrees of the summands of $A = P \oplus F$ to those of B.

To begin with, we rephrase the desired statement (1.2). If H is any bundle on $\mathbb{P}^3$ splitting as a direct sum of line bundles, and ℓ is any integer, set

$$H_{\geq \ell} = \bigoplus (\text{summands of H of degree} \geq \ell),$$

and define $H_{< \ell}$ similarly,[*] so that $H = H_{< \ell} \oplus H_{\geq \ell}$. A moment's thought shows that the first statement of Lemma 1.2 is equivalent to

---

[*] More formally, for any $\ell \in \mathbb{Z}$ there is a natural map

$$H^0(\mathbb{P}^3, H(-\ell)) \otimes_k \mathcal{O}(\ell) \longrightarrow H.$$

$H_{\geq \ell}$ is its image, and $H_{< \ell} = H/H_{\geq \ell}$.

(2.2) For every $\ell \in \mathbb{Z}$,

$$rk(P_{\geq \ell} \oplus F_{\geq \ell}) \geq rk(B_{\geq \ell}).$$

Proof of (2.2). We proceed in two cases: (i) $\ell \leq -f$; and (ii) $\ell > -f$.

Case (i): $\ell \leq -f$.

We have $B = B_{\geq \ell} \oplus B_{< \ell}$ and $F = F_{\geq \ell} \oplus F_{< \ell}$, and the given map $v \in \text{Hom}(B, N \oplus F)$ is an injection of sheaves. Since $\text{Hom}(B_{\geq \ell}, F_{< \ell}) = 0$, we get an injection $B_{\geq \ell} \hookrightarrow N \oplus F_{\geq \ell}$. Hence

(*) $\qquad rk(B_{\geq \ell}) \leq rk(N) + rk(F_{\geq \ell})$

Now suppose that equality holds in (*). Then $B_{\geq \ell}$ and $N \oplus F_{\geq \ell}$ are vector bundles of the same rank, and since N is not a sum of line bundles, the map $B_{\geq \ell} \to N \oplus F_{\geq \ell}$ (thought of as a homomorphism of vector bundles) must drop rank along a surface. Hence so too must the given homomorphism v. But v in fact drops rank only along the curve Y, and therefore strict inequality must hold in (*), i.e.,

(**) $\qquad rk(B_{\geq \ell}) \leq rk(N) - 1 + rk(F_{\geq \ell}).$

Finally, since $\ell \leq -f$, one has $P_{\geq \ell} = P$. But $rk(P) = rk(N) - 1$, so (2.2) follows from (**) in the case at hand.

Case (ii): $\ell > -f$.

Let $N_{< \ell}$ denote the cokernel of the composition of the natural inclusion $P_{\geq \ell} \hookrightarrow P$ with the given map $u: P \to N$:

$$0 \longrightarrow P_{\geq \ell} \longrightarrow N \longrightarrow N_{< \ell} \longrightarrow 0$$

In view of (2.1) and the hypothesis $\ell > -f$, one has

$$H^0(\mathbb{P}^3, P_{\geq \ell}(t)) \xrightarrow{\simeq} H^0(\mathbb{P}^3, P(t)) \xrightarrow{\simeq} H^0(\mathbb{P}^3, N(t))$$

when $t \leq -\ell$ ($\leq f-1$). Hence

(*) $\qquad H^0(\mathbb{P}^3, N_{<\ell}(t)) = 0 \quad$ for any $t \leq -\ell$.

Consider now the diagram

$$\begin{array}{c} 0 \\ \downarrow \\ B_{\geq \ell} \oplus B_{<\ell} \\ \downarrow v \\ 0 \longrightarrow P_{\geq \ell} \oplus F_{\geq \ell} \longrightarrow N \oplus F \longrightarrow N_{<\ell} \oplus F_{<\ell} \longrightarrow 0, \end{array}$$

where horizontally we've just formed the sum of two exact sequences. Evidently $\text{Hom}(B_{\geq \ell}, F_{<\ell}) = 0$, and it follows from (*) that similarly

$$\text{Hom}(B_{\geq \ell}, N_{<\ell}) = 0.$$

Hence v induces an injection of sheaves $B_{\geq \ell} \hookrightarrow P_{\geq \ell} \oplus F_{\geq \ell}$, and taking ranks gives (2.2) when $\ell > -f$. ∎

To complete the proof of Lemma 1.2, it remains to show that if X lies on no surfaces of degrees $f = e(X) + 4$ or less, and if $B = P \oplus F$, so that one has an exact sequence

$$0 \longrightarrow P \oplus F \xrightarrow{v} N \oplus F \longrightarrow I_Y \longrightarrow 0,$$

then $Y = X$. To this end, observe that v gives rise via the decomposition $F = F_{\geq -f} \oplus F_{<-f}$ to a homomorphism

$$\alpha: F_{<-f} \longrightarrow F_{<-f},$$

and since $\text{Hom}(P \oplus F_{\geq -f}, F_{<-f}) = 0$ for reasons of degree, $\text{coker } \alpha$ is

a quotient of coker $v = I_Y$. Since in any event coker α is
locally free, this implies that α is an isomorphism. Hence α
splits off from v, i.e., I_Y arises as the cokernel of an injective
sheaf homomorphism $w: P \oplus F_{\geq -f} \longrightarrow N \oplus F_{\geq -f}$. But then consider the
diagram

(*)
$$\begin{array}{ccccccccc} & & & & P \oplus F_{\geq -f} & & & & \\ & & & \nearrow & \downarrow w & & & & \\ 0 & \longrightarrow & P \oplus F_{\geq -f} & \xrightarrow{u \oplus 1} & N \oplus F_{\geq -f} & \longrightarrow & I_X & \longrightarrow & 0. \end{array}$$

Since $H^0(\mathbb{P}^3, I_X(t)) = 0$ for $t \leq f$, (*) shows that w factors through
an isomorphism as indicated, whence $X = Y$.

The proof of Lemma 1.2 is now complete.

§3. <u>The liaison class of a general curve of large degree.</u>

We now apply the results of §1 to a smooth irreducible curve
$X \subseteq \mathbb{P}^3$ of genus g and degree d>>g. To begin with, we assume that
$d \geq 2g - 1$, so that $e(X) \leq 0$. Moreover, when $d \geq 2g - 1$ the
family of all such curves is irreducible, so it makes sense to
speak of the properties enjoyed by a general smooth curve of genus
g and degree d.

All that is needed at this point is to bound from below the
degrees of the surfaces on which X lies. For our purposes, the
following elementary estimate is sufficient:

Lemma 3.1. <u>Fix an integer</u> $g \geq 0$. <u>Then there exists a constant</u>
$C(g) \geq 2g-1$ <u>such that a sufficiently general curve of genus</u> g <u>and</u>

degree $d \geq C(g)$ lies on no surfaces of degree $\sqrt{5d}$ or less.

Proof. Recall first of all that Hirschowitz [Hi] has shown that a general rational curve $D \subseteq \mathbb{P}^3$ of degree f lies on a surface of degree n if and only if

$$\binom{n+3}{3} > nf + 1,$$

i.e.,

$$n > -3 + \sqrt{6f - 2}.$$

Now let $C \subseteq \mathbb{P}^3$ be any smooth curve of genus g and degree $d_0 \geq 2g-1$. Choose a smooth rational curve $D \subseteq \mathbb{P}^3$ of degree f for which Hirschowitz's theorem holds. Translate D by an automorphism of $\mathbb{P}^3$ so that it meets C at a single point with distinct tangents, and let

$$X_0 = C \cup D.$$

Thus X_0 has degree $d = d_0 + f$ and arithmetic genus g, and by construction X_0 lies on no surfaces of degree $\leq \sqrt{6(d-d_0) - 2} - 3$. But X_0 moves in an irreducible flat family of curves in $\mathbb{P}^3$ whose general member is smooth (cf [T]). Therefore a generic smooth curve of degree d and genus g lies on no surfaces of degree $\leq \sqrt{6(d-d_0) - 2} - 3$, and letting $f \to \infty$ the lemma follows.

Thus all the results of §1 apply in the case at hand, and in summary we have established

Theorem 3.2. Let $X \subseteq \mathbb{P}^3$ be a general smooth irreducible curve of genus g and degree d>>g, and let Z be the curve directly linked to X by two irreducible surfaces of lowest degree through X. (Thus

for $d > C(g)$, Z has degree $> 4d$ and arithmetic genus $\geq g + 3d(\sqrt{5d}-2)$.)
Then:

(a) X is the only curve with deficiency module $M(X)$ (with the given grading), and for $n > 0$ there is no curve with module $M(X)(n)$.

(b) If $Y \subseteq \mathbb{P}^3$ is evenly linked to X, then Y is a deformation, through curves with fixed deficiency module, of a curve obtained from X by a sequence of basic double linkages. In particular, if $Y \neq X$, then

$$\deg(Y) > \deg(X) \quad \text{and} \quad p_a(Y) > p_a(X).$$

(c) If $Y \subseteq \mathbb{P}^3$ is oddly linked to X, then Y is a deformation of a curve obtained from Z by a sequence of basic double linkages, and in particular

$$\deg(Y) \geq \deg(Z) > \deg(X)$$

and

$$p_a(Y) \geq p_a(Z) > p_a(X).$$

Remark. In case (a) one has the estimates

$$\deg(Y) \geq d + \sqrt{5d}$$

$$p_a(Y) \geq g + (7d - 3\sqrt{5d})/2.$$

One can replace $\sqrt{5d}$ in the lemma by $\sqrt{(6-\varepsilon)d}$ ($\varepsilon > 0$), and this leads to somewhat sharper bounds on the degree and genus of Y.

References

[A]   - Apéry, R.: Sur les courbes de première espèce de l'espace à trois dimensions, C.R.A.S. Vol 220, 271-272 (1945,1)

[G]   - Gaeta, F.: Quelques progrès récents dans la classification des variétés algèbriques d'un espace prójectif, Deuxièume Collogue de Géométrie Algèbrique Liège, C.B.R.M., 1952

[GLP]   - Gruson, L., Lazarsfeld, R., Peskine, C.: On a theorem of Castelnuovo, and the equations defining space curves (to appear)

[H]   - Harris, J.: Curves in projective space, Les Presses de L'Université de Montréal, 1982

[Hi]   - Hirschowitz, A.: Sur la postulation générique des courbes rationelles, Acta Math. 146 No. 3-4, 209-230 (1981)

[M]   - Migliore, J.: Ph.D. thesis, Brown University (1983)

[P-S]   - Peskine, C., Szpiro, L.: Liaison des variétés algèbriques I. Inventiones Math. 26, 271-302 (1974)

[R]   - Rao, P.: Liaison among curves in $\mathbb{P}^3$, Inventiones Math. 50, 205-217 (1979)

[S]   - Schwartau, P.: Ph.D. thesis, Brandeis University (1982)

[Se]   - Sernese, E.: L'unirazionalità della varietà dei moduli delle curve di genere dodici, Annali della Scuola Normale Superiore di Pisa, Vol 8, 3, 405-440, (1981)

[T]   - Tannenbaum, A.: Deformations of space curves, Arch. Math. Vol. 34, 37-42 (1980)

R. Lazarsfeld
Harvard University
Cambridge, Massachusetts

Prabhakar Rao
Northeastern University
Boston, Massachusetts

SOME PROBLEMS AND RESULTS ON FINITE SETS OF POINTS IN $\mathbb{P}^n$

Paolo Maroscia
Istituto Matematico "G. Castelnuovo"
Università di Roma
00100 Roma, Italy

0. INTRODUCTION

Let $P_1,\ldots, P_s$ be distinct points in $\mathbb{P}^n(k)$, where k is an algebraically closed field and $n \geq 2$, and let $A = \oplus A_i$ ($i \geq 0$) denote their homogeneous coordinate ring. Also, let $S = \{b_i\}_{i \geq 0}$, with $b_i = \dim_k A_i$, denote the Hilbert function of A or, equivalently, the Hilbert function or the "postulation" of the s points.

In this paper (which is an expanded version of a talk given at the Conference), we investigate some relevant properties of finite sets of points in $\mathbb{P}^n$, by making use of a careful study of their Hilbert functions, which turns out to be very helpful in various cases.

More precisely, in Section 1 we characterize (Theorem 1.8) those sequences of integers that are the Hilbert function of a finite set of points in $\mathbb{P}^n$; moreover, if S is a sequence as above, the proof we give provides an explicit set of points in $\mathbb{P}^n$ having Hilbert function S. This problem has been studied recently by L. Roberts in [R], where a solution has been given in the case $n = 2$. We present here the complete solution, for any $n \geq 2$, in a somewhat simpler form than the original one given in [Mar]. Also, a generalization of this result is proved in [G-M-R].

Section 2 contains one basic result (Theorem 2.3) that gives some sufficient conditions for a sequence S as above which force any set of s points in $\mathbb{P}^n$ with Hilbert function S to contain at least a fixed number of points lying in a

---

This work has been partially supported by Consiglio Nazionale delle Ricerche.

subspace of $\mathbb{P}^n$ or even on a rational normal curve embedded in a subspace of $\mathbb{P}^n$. Also, we give a few applications of this result to the study of a problem of unicity for certain complete linear series on subcanonical curves in $\mathbb{P}^n$. In particular, we show (Theorem 2.11) that "a non-singular irreducible complete intersection of two surfaces in $\mathbb{P}^3$ of respective degrees m,n with m = 3,4 and n ⩾ m , admits a unique simple g_{mn}^3 without fixed points".

The last section is devoted to the discussion of two open problems, closely related to our study. The first one regards a characterization of a complete intersection zero-cycle in $\mathbb{P}^n$, in terms of the Hilbert function and the so-called Cayley-Bacharach property (cf. [G-H]$_2$). The second one relates the "generic" Hilbert function of a finite set of points in $\mathbb{P}^n$ with the minimal number of generators of the homogeneous ideal of a set of points in $\mathbb{P}^n$ having that Hilbert function. Finally, in connection with a remark made in [St], we show the existence of 1-dimensional reduced (Cohen-Macaulay) standard G-algebras (Def. 1.1) that have the Hilbert function of a complete intersection and nevertheless have any prescribed Cohen-Macaulay type.

Throughout the paper, unless otherwise specified, k denotes an algebraically closed field. Also, if $\{g_i\}_{i \geqslant 0}$ is a sequence of integers such that for some $t \geqslant 0$, we have $g_t = g_{t+1} = \ldots = g_{t+j}$ for all $j \geqslant 1$, then we shall use the following notation

$$g_0 \quad g_1 \quad \ldots \quad g_t \rightarrow .$$

1. THE HILBERT FUNCTION OF A FINITE SET OF POINTS IN $\mathbb{P}^n$

Our main objective in this section is to prove a characterization of the Hilbert function of a finite set of points in $\mathbb{P}^n$ ($n \geq 2$). First we recall some general notions and results.

DEFINITION 1.1 ([St]): Let $R = \bigoplus R_i$ ($i \geq 0$) be a Noetherian commutative ring with identity graded by the non-negative integers and let R_0 be a field k, so that R is a k-algebra. Then we say that R is a G-algebra. If moreover, R is generated as a k-algebra by R_1, we say that R is a standard G-algebra.

DEFINITION 1.2 ([G-K]): Let h, i be positive integers. Then h can be written uniquely in the form

$$h = \binom{m_i}{i} + \binom{m_{i-1}}{i-1} + \ldots + \binom{m_j}{j} \quad \text{with} \quad m_i > m_{i-1} > \ldots > m_j \geq j \geq 1 \quad,$$

and this expression for h is called the i-binomial expansion of h. Also, we define

$$h^{<i>} = \binom{m_i + 1}{i+1} + \binom{m_{i-1}+1}{i} + \ldots + \binom{m_j+1}{j+1} \quad \text{and} \quad 0^{<i>} = 0 \quad .$$

DEFINITION 1.3 ([St],[R]): A sequence of non-negative integers $\{c_i\}_{i \geq 0}$ is called an O-sequence, if

$$c_0 = 1 \quad \text{and} \quad c_{i+1} \leq c_i^{<i>} \quad \text{for all} \quad i \geq 1 \quad .$$

Now we state a simple property of the function $h \to h^{<i>}$ (for a proof, see [Mar] or [G-M-R]).

PROPOSITION 1.4 : Let a, b be positive integers, with $a < b$. Then

$$a^{<i>} < b^{<i>} \quad \text{for any} \quad i > 0 \quad .$$

COROLLARY 1.5 : Let $\{b_i\}_{i \geq 0}$ be an O-sequence, with $b_1 = m$. Then

$$b_{i+1} \leq \binom{m+i}{i+1} \quad \text{for all } i \geq 0 .$$

The first basic property of an O-sequence was discovered by Macaulay [Mac]; for a modern proof, see [St] or [T].

THEOREM 1.6 : Let $\{c_i\}_{i \geq 0}$ be a sequence of non-negative integers and let k be any field. Then the following are equivalent:

(1) $\{c_i\}_{i \geq 0}$ is an O-sequence ;

(2) $\{c_i\}_{i \geq 0}$ is the Hilbert function of a standard G-algebra, say $R = \bigoplus R_i$ $(i \geq 0)$, with $R_0 = k$.

Before proving our main result, we state a simple but useful lemma.

LEMMA 1.7 : Let B be a finite set of points in $\mathbb{P}^n$ $(n \geq 2)$ with Hilbert function $\{c_i\}_{i \geq 0}$ and let B' be a finite set of points lying on a hyperplane L in $\mathbb{P}^n$ such that $L \cap B = \emptyset$, with Hilbert function $\{e_i\}_{i \geq 0}$. Also, let $\{b_i\}_{i \geq 0}$ denote the Hilbert function of $B \cup B'$. Then

(1) $b_i \geq e_i + c_{i-1}$ for any $i \geq 1$;

(2) if $c_j = c_{j-1}$ for some $j \geq 1$, then $b_i = e_i + c_{i-1}$ for all $i \geq j$;

(3) if $e_j = \binom{j+n-1}{j}$ for some $j \geq 1$, then $b_i = e_i + c_{i-1}$ for $1 \leq i \leq j$

Proof : (1) Let W_i, W_i' and W_i'' denote, respectively, the linear system of all hypersurfaces of degree i of $\mathbb{P}^n$ through $B \cup B'$, the linear system cut out on L by W_i and the linear system consisting of all hypersurfaces in W_i that contain L as a component. Looking at them as k-vector spaces, we get:

$\dim W_i = \binom{i+n}{n} - b_i$, $\dim W_i' = \binom{i+n-1}{n-1} - e_i'$ $(e_i' \geq e_i)$, $\dim W_i'' = \binom{i-1+n}{n} - c_{i-1}$.

Also, we have $\dim W_i = \dim W_i' + \dim W_i''$, whence the conclusion.

(2) Let $W'_{B,i}$ denote the linear system cut out on L by the linear system of all hypersurfaces of degree i of $\mathbb{P}^n$ through B. Then, for any $i \geq j$,

we get $\dim W'_{B,i} = \binom{i+n-1}{n-1}$; hence, with the notation introduced in (1) above, $\dim W'_i = \binom{i+n-1}{n-1} - e_i$, and so we are done.

(3) This statement follows directly from the proof of (1) above, since by Proposition 1.4, we have: $e_t = \binom{t+n-1}{t}$ for $1 \leq t \leq j$. This concludes the proof of Lemma 1.7.

THEOREM 1.8 : Let $S = \{b_i\}_{i \geq 0}$ be a sequence of integers of the form

$$b_o(=1) \quad b_1(\geq 2) \ldots b_d = s \to \quad \text{with } b_d \neq b_{d-1} \quad .$$

Then the following are equivalent:

(1) The difference sequence $\{b'_i\}_{i \geq 0}$ of S :

$$b'_o = 1 , b'_1 = b_1 - b_o , \ldots , b'_i = b_i - b_{i-1} , \ldots$$

is an O-sequence ;

(2) There exist s points in $\mathbb{P}^{b_1 - 1}(k)$ such that their homogeneous co-ordinate ring has Hilbert function S .

REMARK 1.9 : It is worth observing that statement 1.8(1) above automatically implies that S is an O-sequence. In fact, following [St] , if R is a standard G-algebra with Hilbert function $\{b'_i\}_{i \geq 0}$, it is easy to check that S is the Hilbert function of $R[X]$, where X is an indeterminate; hence the conclusion, in view of Theorem 1.6.

<u>Proof</u> : (2) $\Rightarrow$ (1) follows from Theorem 1.6, since the homogeneous co-ordinate ring of the s points contains a non-zero divisor of degree 1 .

(1) $\Rightarrow$ (2) : The proof goes by induction on $b_1 \geq 2$, the case $b_1 = 2$ being trivial, since then we have $b_i = i+1$ for $0 \leq i \leq d$, i.e., S is the Hilbert function of $s = d+1$ collinear points (cf. also [G-M]$_2$).

So, write $b_1 = n+1$ (> 2) and assume the result true for $b_1 \leq n$. We will divide the proof into three steps.

Step 1 : Let $\{c_i\}_{i \geq 0}$ be the sequence obtained from S by subtracting the Hilbert function of a hyperplane in $\mathbb{P}^n$:

1	b_1	...	b_h	b_{h+1}	b_{h+2}	...	$b_d = s$	$\to$
1	$\binom{n}{1}$	...	$\binom{n+h-1}{h}$	$\binom{n+h}{h+1}$	$\binom{n+h+1}{h+2}$	...	$\binom{n+d-1}{d}$	...

$$c_0 \quad ... \quad c_{h-1} \quad c_h \quad c_{h+1} \quad ... \quad c_{d-1} \quad ...$$

Then there exists an integer h, with $0 \leq h \leq d-1$, such that

$$1 = c_0 \leq c_1 \leq ... \leq c_h \quad \text{and} \quad c_h > c_{h+1} \ .$$

Let S_1 denote the sequence $c_0 \ c_1 \ ... \ c_h \to$.

CLAIM : The difference sequence $\{c_i'\}_{i \geq 0}$ of S_1, where $c_0' = 1$, $c_1' = c_1 - c_0$, ..., $c_i' = c_i - c_{i-1}$, ... is an O-sequence.

Proof (of the Claim) : First we observe that, if $1 \leq r \leq h$, then :

(α) $\qquad \binom{r+n-1}{r+1} \leq b_{r+1}' \leq \binom{r+n}{r+1}$,

(β) $\qquad b_{r+1}'^{<r+1>} = [b_{r+1}' - \binom{r+n-1}{r+1}]^{<r>} + \binom{r+n}{r+2}$.

In fact, since $c_0 \leq c_1 \leq ... \leq c_h$, we get: $b_{r+1}' \geq \binom{r+n-1}{r+1}$ for $1 \leq r \leq h$; the other inequality in (α) follows from Corollary 1.5. Then, from (α) we get $b_{r+1}' \leq \binom{r+n}{r+1}$, $(1 \leq r \leq h)$. If we have equality, then (β) is clear. If not, the $(r+1)$-binomial expansion of b_{r+1}' is of the form: $b_{r+1}' = \binom{r+n-1}{r+1} + [\ ...\ ]$, hence the parenthesis $[\ ...\ ]$ gives precisely the r-binomial expansion of $b_{r+1}' - \binom{r+n-1}{r+1}$, whence (β).

Now, the Claim follows immediately. In fact, by our hypotheses, we have $b_{r+2}' \leq b_{r+1}'^{<r+1>}$ for any $r \geq 0$; hence, from (β) we get

$$b'_{r+2} \leq [b'_{r+1} - \binom{r+n-1}{r+1}]^{<r>} + \binom{r+n}{r+2} , \quad \text{for } 1 \leq r \leq h-1 ,$$

and so we are done.

Step 2 : Define a new sequence $T_1 = \{e_i\}_{i \geq 0}$ by putting :

$$e_0 = 1$$

$$e_1 = n$$

$$\ldots\ldots\ldots\ldots$$

$$e_h = \binom{n+h-1}{h}$$

$$e_{h+1} = b_{h+1} - c_h$$

$$e_{h+2} = b_{h+2} - c_h$$

$$\ldots\ldots\ldots\ldots$$

$$e_d = b_d - c_h = s - c_h$$

$$e_{d+j} = e_d \quad \text{for any } j \geq 1 .$$

It is easy to check that the difference sequence of T_1 is an 0-sequence. Hence, by induction, there exists a set of $s - c_h$ points in $\mathbb{P}^n$ lying on a hyperplane, with Hilbert function T_1 .

Step 3 : Now we distinguish two cases :

(I) $c_1 \leq n$. In this case, if we denote by B_1 a set of c_h points in $\mathbb{P}^n$ with Hilbert function S_1 (such a set certainly exists, by induction) and by L a hyperplane of $\mathbb{P}^n$ such that $L \cap B_1 = \emptyset$ and containing also a set of $s - c_h$ points defined as in Step 2 , say B'_1 , then, by Lemma 1.7 , the Hilbert function of $B_1 \cup B'_1$ is precisely $S = \{b_i\}_{i \geq 0}$, which proves the theorem.

(II) $c_1 = n+1$. In this case, we start from S_1 and repeat the construction that gave us S_1 starting from S (see Step 1) . So we get a new sequence, say S_2 :

$$c_0^{(2)} \quad c_1^{(2)} \quad \ldots \quad c_j^{(2)} \quad \to \quad (j \leq h , c_j^{(2)} < c_h) .$$

Then, if $c_1^{(2)} \leq n$, we are done (see the argument given in case (I) above). If not, we repeat the construction starting from S_2. Clearly, after a finite number of steps, we get a sequence, say S_t:

$$c_0^{(t)} \quad c_1^{(t)} \quad \ldots \quad c_m^{(t)} \to \qquad \text{with } c_1^{(t)} \leq n.$$

Now, by using the induction hypothesis and Lemma 1.7 again, we are done. This completes the proof of the theorem.

Next we state a simple consequence of Theorem 1.8.

If $S = \{b_i\}_{i \geq 0}$ is a sequence of integers, with $b_0 = 1$, and d is any positive integer, we shall denote by $S^{(d)}$ the dth-difference sequence of S, defined in the obvious way : $S^{(1)} = \{1, b_1' = b_1 - b_0, \ldots, b_i' = b_i - b_{i-1}, \ldots\}$, $S^{(2)} = \{1, b_1'' - b_0'', \ldots, b_i'' - b_{i-1}'', \ldots\}$ and so on.

<u>COROLLARY 1.10</u> : Let $S = \{b_i\}_{i \geq 0}$ be a sequence of integers with $b_0 = 1$ such that for a given integer $d \geq 1$, $S^{(d)}$ is an O-sequence of the form

$$1 \quad g_1 \quad g_2 \quad \ldots \quad g_m = 0 \to \qquad .$$

Then S is the Hilbert function of a reduced Cohen-Macaulay standard G-algebra of dimension d.

Proof : We proceed by induction on $d \geq 1$, the case $d = 1$ being clear, in view of Theorem 1.8.

So, let $d > 1$ and assume the result true for $d' \leq d-1$. Then, by induction, $S^{(1)}$ is the Hilbert function of a reduced Cohen-Macaulay standard G-algebra of dimension $d-1$, say R. Hence S is precisely the Hilbert function of $R[X]$, where X is an indeterminate, and so we are done.

We conclude this section with a few general remarks. In particular, Remark 1.11(c) provides some useful geometric information that can be extracted directly from the Hilbert function of a finite set of points in $\mathbb{P}^n$.

REMARK 1.11 : (a) It is easily seen that Theorem 1.8 holds more generally for any infinite field k . Also, Corollary 1.10 slightly improves an analogous result by Macaulay (cf. [St, Corollary 3.11]).

(b) We note that the construction given in the proof of Theorem 1.8 in order to obtain a set of s points with the required properties stops exactly after $m-1$ steps, where m is the least integer such that $b_m < \binom{m+b_1-1}{m}$. Hence the set of s points we get at the end lies on a reducible hypersurface of degree m consisting of m distinct hyperplanes of $\mathbb{P}^{b_1-1}(k)$.

(c) Finally, with the notation introduced in the proof of Theorem 1.8, given any set of s points in $\mathbb{P}^{b_1-1}(k)$ with Hilbert function $S = \{b_i\}_{i \geq 0}$, it follows from Lemma 1.7 that at most $s - c_h$ of the s points can lie on a hyperplane. Moreover, this maximum is achieved in our construction of a set of s points with Hilbert function S . Similarly, one can determine explicitly the maximum number of points that can lie on a subspace of any given dimension of $\mathbb{P}^{b_1-1}(k)$. In particular, the maximum number of collinear points is $d + 1$, which can also be proved by using a straightforward argument (cf. [G-M-R]) .

2. HILBERT FUNCTIONS AND GEOMETRIC PROPERTIES OF SETS OF POINTS IN $\mathbb{P}^n$

In this section we prove some results concerning geometric properties of finite sets of points in $\mathbb{P}^n$ ($n \geqslant 2$) having certain given Hilbert functions. Also, we give a few applications of these results to a problem of unicity for certain complete linear series on subcanonical curves in $\mathbb{P}^n$.

We start with two lemmas before stating our main result.

LEMMA 2.1 : Let $P_1, \ldots, P_s$ be distinct points in $\mathbb{P}^n$ ($s > n \geqslant 2$) with Hilbert function $\{b_i\}_{i \geqslant 0}$, and let d be the least integer such that $b_d = s$. Suppose there exists some index j ($1 \leqslant j \leqslant d-1$) such that $b_j - b_{j-1} = n-h$ ($h \geqslant 1$).

Then at least $n-h+2$ of the s points lie in a subspace $\mathbb{P}^{n-h}$ of $\mathbb{P}^n$. In particular, if the s points are in general position (i.e., no $n+1$ of them are linearly dependent), then $b_t \geqslant tn+1$, for $t = 1, \ldots, d-1$.

Proof : Let $P_{r+1}, P_{r+2}, \ldots, P_s$ with $r = s-b_j$, impose independent conditions to the linear system $|0_{\mathbb{P}^n}(j)|$. Then, we may assume (without loss of generality) that $P_r, P_{r+1}, \ldots, P_{s-(n-h+1)}$ impose b_{j-1} independent conditions to $|0_{\mathbb{P}^n}(j-1)|$. Now take a form F of degree $j-1$ vanishing at $P_{r+1}, \ldots, P_{s-(n-h+1)}$ but not at P_r and let H denote any hyperplane through $P_{s-(n-h)}, \ldots, P_s$. Then, by our assumptions, $P_r \in FH$. Hence $P_r \in H$, which proves Lemma 2.1.

The next lemma is due to Castelnuovo (see $[H_2]$).

LEMMA 2.2 : Let $P_1, \ldots, P_s$ be distinct points in $\mathbb{P}^n$ ($s > n \geqslant 2$) in general position, with Hilbert function $\{b_i\}_{i \geqslant 0}$ and satisfying one of the following conditions :

(1) $b_t = tn+1$ for some $t \geqslant 3$ and $s \geqslant tn+2$,
(2) $b_2 = 2n+1$ and $s \geqslant 2n+3$.

Then $P_1, \ldots, P_s$ lie on a rational normal curve.

THEOREM 2.3 : Let $P_1, \ldots, P_s$ be distinct points in $\mathbb{P}^n$ ($s > n \geq 2$) with Hilbert function $\{b_i\}_{i \geq 0}$ and let d be the least integer such that $b_d = s$. Suppose that

(a) $b_{d-1} - b_{d-2} = n-h$, with $h \geq 1$, and

(b) no $n-h+1$ of the s points lie in a $\mathbb{P}^{n-h-1}$ of $\mathbb{P}^n$.

Then we have :

(1) at least $(n-h)(d-1)+2$ of the s points lie in a $\mathbb{P}^{n-h}$ of $\mathbb{P}^n$;

(2) if moreover, $b_{d-n+1} - b_{d-n} = \ldots = b_{d-1} - b_{d-2} = n-h$, then at least $(n-h)(d-1)+2$ of the s points lie on a rational normal curve in a $\mathbb{P}^{n-h}$ of $\mathbb{P}^n$.

Proof : (1) Let $P_{r+1}, \ldots, P_s$ with $r = s - b_{d-1}$, impose independent conditions to $|\mathcal{O}_{\mathbb{P}^n}(d-1)|$. Then, we may assume (without loss of generality) that $P_r, P_{r+1}, \ldots, P_{s-(n-h+1)}$ impose independent conditions to $|\mathcal{O}_{\mathbb{P}^n}(d-2)|$. Hence, if L denotes the $(n-h)$-dimensional subspace of $\mathbb{P}^n$ spanned by $P_{s-(n-h)}, \ldots, P_s$ (cf. (b)), we get $P_r \in L$ (see the proof of Lemma 2.1).

Now, by our assumptions, the points $P_r, P_{r+1}, \ldots, P_{s-(n-h+1)}, P_{s-(n-h)}$ impose b_{d-2} independent conditions to $|\mathcal{O}_{\mathbb{P}^n}(d-2)|$. Hence there exists some index i, with $r+1 \leq i \leq s-(n-h+1)$, such that the b_{d-2} points[1] $P_r, P_{s-(n-h)}, P_{r+1}, \ldots, \hat{P}_i, \ldots, P_{s-(n-h+1)}$ impose independent conditions to $|\mathcal{O}_{\mathbb{P}^n}(d-2)|$. Hence, arguing as above, we get $P_i \in L$. Also, we can repeat the argument. Thus, there exists some index j ($j \neq i$; $r+1 \leq j \leq s-(n-h+1)$) such that the b_{d-2} points $P_r, P_{s-(n-h)}, P_{r+1}, \ldots, \hat{P}_j, \ldots, P_{s-(n-h+1)}$ impose independent conditions to $|\mathcal{O}_{\mathbb{P}^n}(d-2)|$, and so on. Then, if we iterate the argument above, it follows from Lemma 2.1 that the set $\{P_r, P_{r+1}, \ldots, P_{s-(n-h+1)}, P_{s-(n-h)}\}$ contains at least $(n-h)(d-2)+1$ points which lie in L and moreover impose independent conditions to $|\mathcal{O}_{\mathbb{P}^n}(d-2)|$. Hence L contains at least $[(n-h)(d-2)+1] + (n-h+1)$ points (among the $P_i's$), which proves (1).

(2) This statement follows from Lemma 2.2, by using an argument quite similar to the one developed in the proof of (1) above.

---

[1] Here the notation $\hat{P}_i$ means omit P_i.

The following result is an immediate consequence of Theorem 2.3.

COROLLARY 2.4 : Let $P_1, \ldots, P_s$ be distinct points in $\mathbb{P}^n$, with $s > 2n$ and $n \geq 2$, and let $\{b_i\}_{i \geq 0}$ denote their Hilbert function. Suppose that

(1) $b_2 \leq 2n$, and

(2) no n of the s points lie in a $\mathbb{P}^{n-2}$ of $\mathbb{P}^n$.

Then, either the s points lie in a hyperplane (of $\mathbb{P}^n$) or all but one lie in a hyperplane, in which case these $s-1$ points actually lie on a rational normal curve.

REMARK 2.5 : (a) It is easily seen that, in view of Lemma 2.2, statement 2.3(1) can be completed as follows : "If the maximum number of points lying in a $\mathbb{P}^{n-h}$ is exactly $(n-h)(d-1)+2$ and also $d \geq 4$, then these points actually lie on a rational normal curve".

(b) We note that conditions (a),(b) in Theorem 2.3 are by no means necessary conditions too. Take for instance three lines in $\mathbb{P}^2$ all passing through a point O, say L_1, L_2, L_3, and choose one point on L_1, three points on L_2 and five points on L_3, all distinct from O. Then, these nine points have Hilbert function 1 3 6 8 9 → ; also, they contain the maximum number of collinear points (cf. Remark 1.11(c)). Yet, the Hilbert function above does not satisfy 2.3(a).

(c) Theorem 2.3 gives an answer to a problem raised in [G-M-R], where the first case : $b_{d-2} = s-2 = b_{d-1} - 1$ is also settled. In this case, with the notation of Theorem 2.3, we have $h = n-1$, and our "minimum" number of collinear points is $d+1$, which is also equal to the maximum number of collinear points for a set of s points in $\mathbb{P}^n$ with the given Hilbert function (cf. Remark 1.11(c)). It is worth observing that such an equality is no longer true, in general, for $h \leq n-2$. For example, choose 12 points in $\mathbb{P}^3$ in the following manner: take 10 points on a non-singular conic lying in a plane L and any 2 points outside L in such a way that no 3 of the 12 points are collinear. Then, the Hilbert function of these points is 1 4 7 9 11 12 → ; hence (cf. Remark 1.11(c)), in this case, the maximum number of coplanar points is 11.

We now turn to the promised applications to subcanonical curves.

First recall that a non-singular irreducible curve $C \hookrightarrow \mathbb{P}^n$ $(n \geq 2)$ is called a <u>subcanonical curve of type</u> t if there exists an integer $t \geq 1$ such that $|tH| = |K|$, where H, K denote, respectively, a hyperplane section of C and a canonical divisor on C. For example, any non-singular plane curve of degree $d \geq 4$ is subcanonical of type $d-3$. More generally, a non-singular irreducible curve in $\mathbb{P}^n$ $(n \geq 3)$ which is the complete intersection of $n-1$ hypersurfaces of respective degrees $r_1, \ldots, r_{n-1}$, is subcanonical of type $t = r_1 + \ldots + r_{n-1} - n - 1$.

We shall now prove a few results related to the following problem (cf. also [Se]).

PROBLEM 2.6 : Let $C \hookrightarrow \mathbb{P}^n$ $(n \geq 2)$ be a subcanonical curve of degree d, which is linearly normal, i.e., such that the linear system of hyperplane sections is complete. Does then C admit a <u>unique</u> simple complete linear series g_d^n without fixed points ?

Clearly the above problem has an affirmative answer when $t = 1$. Also, a classical result, due to H. Weber, provides a positive answer for non-singular plane curves. We next give a slight generalization of Weber's result, after recalling a well-known basic fact. (For the rest of this section, we shall assume char $k = 0$.)

LEMMA 2.7 : Let g_d^n be a simple linear series without fixed points on an irreducible projective curve. Then the "general" divisor D of g_d^n satisfies the following

(1) D consists of d distinct points, and

(2) any n points of D impose independent conditions on g_d^n.

THEOREM 2.8 : Let C be a non-singular plane curve of degree $d > 5$ and let G denote a simple complete linear series g_m^h without fixed points on C, with $h \geq 1$ and $m \leq d+1$. Then only the following possibilities for G can occur :

(a) G is a g_{d-1}^1, each of whose divisors consists of $d-1$ collinear points,

(b) G is the complete g_d^2 cut out on C by all lines of $\mathbb{P}^2$.

Proof: Let D be a general divisor of our g_m^h. Then, in view of Lemma 2.7, we can write $D = P_1 + \ldots + P_m$ where the P_i's are all distinct, and let $\{b_i\}_{i \geq 0}$ denote their Hilbert function. By Riemann-Roch, our series is special; hence we get $b_{d-3} = m-h$. It follows that $d-2 \leq m-h \leq m-1$, hence $m \geq d-1$. We now discuss all possible cases:

(1) $m = d-1$. This implies $b_{d-3} = d-2$, hence (cf. Theorem 2.3(1)) we are in case (a).

(2) $m = d$. In this case, we have necessarily $b_{d-3} = d-2$; otherwise (in view of (1) above) our series would be a g_d^1 with one fixed point: contradiction. Hence D consists of d collinear points, i.e., we are in case (b).

(3) $m = d+1$, which clearly implies $b_{d-3} > d-2$, hence $d-1 \leq b_{d-3} \leq d$. Now, in the light of Theorem 2.3(1), if $b_{d-3} = d-1$ (resp. $b_{d-3} = d$) our series would be a g_{d+1}^2 (resp. a g_{d+1}^1) with one fixed point (resp. with two fixed points): contradiction. This completes the proof of the theorem.

Our next result gives an affirmative answer to Problem 2.6 in the case of a subcanonical curve of "low" degree. Indeed, if C is a subcanonical curve in $\mathbb{P}^n$ of degree d and type t which is linearly normal, then (cf. [A-S]) we have:
$$d \geq (t+1)(n-1) + 2.$$

THEOREM 2.9: Let C be a subcanonical curve in $\mathbb{P}^n$ ($n \geq 3$) of type t (≥ 2) and degree $d \leq (t+1)(n-1)+3$ (which is linearly normal).

Then C admits a unique simple g_d^n without fixed points.

Proof: Let $D = P_1 + \ldots + P_d$ be a general divisor of a simple linear series g_d^n without fixed points on C, and let $\{b_i\}_{i \geq 0}$ denote the Hilbert function of these points. By Riemann-Roch, our g_d^n is special, hence $b_t \leq d-n \leq tn$. Hence, by Lemma 2.7 and Lemma 2.1, we get $b_2 \leq 2n$. Now we wish to show that the P_i's are coplanar. Suppose not. Then, by Corollary 2.4, $d-1$ of the d points lie on a hyperplane L and the remaining point is outside L. Hence, if we compare the t-multiple of the divisor cut out on C by L with tD, we get a g_t^1 on C, which is a contradiction. So we are done.

REMARK 2.10 : We recall from [A-S] that if C is a linearly normal subcanonical curve in $\mathbb{P}^n$ ($n \geq 3$) of type t and degree $d = (t+1)(n-1)+2$, then C is a Castelnuovo curve, i.e., C is a curve of maximal genus with respect to its degree. Hence, in this case, Theorem 2.9 also follows from a more general result proved by Accola [A]. Furthermore, it is worth observing that Theorem 2.9 covers, in particular, the case of a complete intersection lying on a quadric surface in $\mathbb{P}^3$ and also the case of the complete intersection of two cubic surfaces in $\mathbb{P}^3$.

We can say a bit more in the case of a complete intersection in $\mathbb{P}^3$.

THEOREM 2.11 : Let C be a non-singular irreducible curve in $\mathbb{P}^3$ which is the complete intersection of two surfaces S,T of respective degrees m,n with $m = 3,4$ and $n \geq m$.

Then C admits a unique simple complete g^3_{mn} without fixed points.

Proof : Let $D = P_1 + \ldots + P_{mn}$ be a general divisor of a simple complete g^3_{mn} without fixed points on C, and let $\{b_i\}_{i \geq 0}$ denote the Hilbert function of these mn points. Then we discuss separately the two cases :

(I) $m = 3$. In this case, since $|D|$ is special and C is subcanonical of type $n-1$, we get $b_{n-1} = 3n-3$. Now, suppose $b_1 = 4$. Then we have necessarily : $b_{n-1} - b_{n-2} = 2$ (since the difference sequence of $\{b_i\}_{i \geq 0}$ is still an O-sequence) and also $b_n = 3n-1$ (in view of Theorem 2.3(1)). Hence, by Theorem 2.3(2) at least $2n+2$ of the $3n$ points of D lie on a non-singular plane conic, say F. Hence F lies on both S and T : contradiction. So $b_1 = 3$, and we are done.

(II) $m = 4$. In this case, since $|D|$ is special and C is subcanonical of type n, we get $b_n = 4n-3$. Hence $b_{n-1} = 4n-6$. In fact, first observe that $b_{n-1} \geq 4n-6$ (since the difference sequence of $\{b_i\}_{i \geq 0}$ is still an O-sequence). Now, if $b_{n-1} > 4n-6$, we have necessarily $b_{n-1} = 4n-5$, hence $b_{n+1} = 4n-1$. Then, applying Theorem 2.3(2) and arguing as in (I) above, we get a contradiction.

We wish to show that $b_1 = 3$, i.e., $P_1, \ldots, P_{4n}$ lie on a plane.

Suppose not. Then $b_1 = 4$, which implies $b_{n-2} = 4n-9$. Let $P_4, \ldots, P_{4n}$ impose independent conditions to $|\mathcal{O}_{\mathbb{P}^3}(n)|$ and let $P_3, P_4, \ldots, P_{4n-7}$ impose independent conditions to $|\mathcal{O}_{\mathbb{P}^3}(n-2)|$. Then every quadric surface through the 7 points $P_{4n-6}, \ldots, P_{4n}$ also contains P_3.

CLAIM : At least $2n+3$ of the $4n$ points of D lie on a plane.

Proof (of the Claim) : We first observe that $P_{4n-6}, \ldots, P_{4n}$ are not in general position in $\mathbb{P}^3$. Otherwise, by our hypotheses, the points $P_3, P_{4n-6}, \ldots, P_{4n}$ are in general position. Then, if we replace P_{4n-6} by a suitable P_i ($4 \leq i \leq 4n-7$) (see the proof of Theorem 2.3(1)), we get that the 9 points $P_3, P_i, P_{4n-6}, \ldots, P_{4n}$ are also in general position, and their Hilbert function is 1 4 7 9 →; hence, by Lemma 2.2, they lie on a rational normal curve, say E. Now, by repeating an argument similar to the one developed in the proof of Theorem 2.3(1), it follows that at least $3n+2$ of the $4n$ points of D lie on E; hence E lies on both S and T : contradiction.

On the other hand, in view of Lemma 2.7 and Theorem 2.3, the Hilbert function of the set $U = \{P_{4n-6}, \ldots, P_{4n}\}$ must be one of the following

(1)   1 4 7 →
(2)   1 4 6 7 →
(3)   1 3 5 7 →
(4)   1 3 6 7 →

We now discuss all possibilities for the set U.

(1a) U contains exactly 4 coplanar points, say $P_{4n-6}, \ldots, P_{4n-3}$, lying on a plane L with the remaining points outside L and spanning a plane L_1. Then, it follows from our hypotheses that $P_3 \in L_1$. Now, if we replace (in the usual way) P_{4n-2} by P_i ($4 \leq i \leq 4n-7$), we have $P_i \in L_1$; hence, iterating the argument, we get that at least $2n$ of the $4n$ points of D lie on L_1. Indeed, L_1 contains at least $2n+5$ points of D; otherwise, by Lemma 2.7(2), Theorem 2.3(1) and Lemma 1.7, we get $b_{n-1} \geq (2n-1)+(2n-4) = 4n-5$: contradiction, whence the Claim.

(1b) U contains 5 points on a non-singular conic F lying in a plane L, with the remaining 2 points outside L. Then, it is easily seen that this possibility can be ruled out.

(1c) U consists of 6 points on a plane L and not lying on a conic, say $P_{4n-6},\ldots, P_{4n-1}$, with $P_{4n} \notin L$. Hence $P_3 \in L$. Now, we may assume (without loss of generality) that P_3 does not lie on the conic through $P_{4n-5},\ldots, P_{4n-1}$. Then, replacing P_{4n-6} by a suitable P_i (as usual) and iterating the argument, we get that at least $2n+3$ of the $4n$ points of D lie on L, which concludes the proof of the Claim in case (1).

Now, a similar discussion can be made in each of the other cases (2), (3), (4), and so we are done.

Going back to the proof of the theorem, let M denote the maximum number of coplanar points in the set $\{P_1,\ldots, P_{4n}\}$; then, by the Claim above, we have: $M \geq 2n+3$. Also, let $\{d_i\}_{i \geq 0}$ denote the Hilbert function of these M points lying, say, on a plane L. Since $4n-M \leq 2n-3$, we get (cf. Lemma 2.7(2), Theorem 2.3(1), Lemma 1.7): $b_{n-1} = d_{n-1}+(4n-M) = 4n-6$, $b_n = d_n+(4n-M) = 4n-3$. Hence $d_n - d_{n-1} = 3$, which implies $d_n \geq 3n$ (since the difference sequence of $\{d_i\}_{i \geq 0}$ is still an O-sequence). Hence $M = d_n+3 \geq 3n+3$.

Now, we may assume that $S \cap L$ is an irreducible plane quartic. Then we easily get: $d_2 = 6$, $d_3 = 10$ and $d_i = 4i-2$ for $i = 4,\ldots, n-1$. Hence $M = 4n$, which contradicts the assumption $b_1 = 4$, made at the beginning.

This completes the proof of the theorem.

It is worth observing that the argument given in the proof of Theorem 2.11 could be shortened, by using the Uniform Position Lemma (cf.[H_1]). Also, we hope that the argument above can be rearranged in such a way to provide an affirmative answer to Problem 2.6 for any complete intersection curve in $\mathbb{P}^3$.

Finally, C. Ciliberto informed me that he proved both Theorem 2.9 and Theorem 2.11, in a different way (and in a more general context): he is writing a paper which will contain, in particular, these results.

3. OPEN PROBLEMS AND FINAL REMARKS

Let S be the Hilbert function of a finite set of points in $\mathbb{P}^n$ ($n \geq 2$); then it is clear that there exist, in general, quite different sets of points with Hilbert function S. So, a first naive problem is the following: "Are there any geometric properties shared by all sets of points in $\mathbb{P}^n$ having a given Hilbert function?". Such a question is doubtless too vague, yet we already saw in the previous sections some expressive results in this direction: in particular, Theorem 2.3 and Remark 1.11(c). In this connection, a very natural question about a finite set of points in $\mathbb{P}^n$ is to characterize complete intersection zero-cycles in $\mathbb{P}^n$ (i.e., to give a converse of the Bézout theorem), assuming that the cycles in question have the Hilbert function of a complete intersection. Clearly this assumption is not enough to get the desired result (cf., e.g., Remark 2.5(b)).

Now, let us start with the case $n = 2$, for simplicity. Then one way we can approach the problem is to appeal to a classical result due to Cayley and Bacharach (cf. [S-R], [G-H]$_1$). First recall, following [G-H]$_1$, that a zero-cycle Z in $\mathbb{P}^2$ consisting of mn distinct points, with $3 \leq m \leq n$, is said to have the Cayley-Bacharach property, if any curve of degree $m+n-3$ passing through all but one point of Z necessarily contains Z.

PROBLEM 3.1 : Let Z be a zero-cycle in $\mathbb{P}^2$ consisting of mn distinct points, with $3 \leq m \leq n$, such that

(a) Z has the Hilbert function of the complete intersection of two curves of respective degrees m, n (with no common components), and

(b) Z has the Cayley-Bacharach property.

Is then Z a complete intersection ?

It is fairly clear that condition (b) alone does not imply necessarily that Z is a complete intersection. However, it is shown in [G-H]$_2$ (and in a more general context) that, when $m = n \geq 3$, the Cayley-Bacharach property is "in general" sufficient that Z be a complete intersection. On the other hand, if we assume moreover that Z lies on an irreducible curve of degree m , then the Cayley-Bacharach property certainly implies that Z is a complete intersection (cf. also [S]).

We feel that Problem 3.1 admits an affirmative answer. Yet, we are able to prove that, only in the case m = 3 .

PROPOSITION 3.2 : With the notation introduced in 3.1 above, let Z be a zero-cycle in $\mathbb{P}^2$ consisting of 3n distinct points (n ⩾ 3) and satisfying conditions (a), (b) . Then Z is the complete intersection of a plane cubic and a curve of degree n , with no common components .

Proof : In view of (a) , our cycle Z lies on two curves, say F, G, of respective degrees 3, n, with the property that G does not contain F as a component. Now, suppose Z is not a complete intersection. Then we consider the two possible cases :

(1) F and G have exactly one line L in common ; so we can write F = LF' and G = LG' . Now consider the set , Z' , of all points of Z lying on the conic F' and let $\{c_i\}_{i \geq 0}$ denote their Hilbert function. It follows from (b) that $c_{n-1} \leq c_n - 1$. Also, the line L contains at least n+2 points of Z (otherwise F' and G' would have a common component); hence $c_n \leq 2n-2$. This implies that $c_{n-1} = c_n - 1$; hence, by Theorem 2.3(1) , Z' contains at least n+1 collinear points: contradiction .

(2) F and G have one conic C in common . Hence C contains at least 2n+2 points of Z (otherwise F would be a component of G). Then, using (b) , we get a contradiction also in this case, which concludes the proof .

Finally, we point out that Problem 3.1 can be stated also in $\mathbb{P}^n$, with n ⩾ 3, using the generalized Cayley-Bacharach property defined in $[G-H]_2$.

Another problem that we would like to discuss here regards the relationship between a given Hilbert function S of a finite set of points in $\mathbb{P}^n$ (n ⩾ 2) and the minimal number of generators for the homogeneous ideal of a set of points with

Hilbert function S. Instead of trying to state the problem in its full generality, we will stick to the "generic" case, corresponding to the following definition.

DEFINITION 3.3 ([G-O]$_1$, [G-O]$_2$, [G-M]$_2$) : Let $P_1, \ldots, P_s$ be distinct points in $\mathbb{P}^n(k)$ ($s > n \geq 2$) with Hilbert function $\{b_i\}_{i \geq 0}$. Then we say that $P_1, \ldots, P_s$ are in <u>generic position</u> (or also in generic s-position) if

$$b_i = \min\left(\binom{i+n}{n}, s\right) \quad \text{for any } i \geq 0 \; .$$

It is shown in [G-O]$_1$ that the sets of s points in $\mathbb{P}^n$ (considered as points in $(\mathbb{P}^n)^s$) which are in generic position form a non-empty open set of $(\mathbb{P}^n)^s$. Next we recall the first basic properties of a set of points in generic position.

PROPOSITION 3.4 : Let $P_1, \ldots, P_s$ be points in $\mathbb{P}^n(k)$ in generic position and let $I = I_d \oplus I_{d+1} \oplus \ldots$, with $I_d \neq 0$, be the homogeneous ideal of these points in $k[X_o, \ldots, X_n]$. Also, let $\nu(I)$ denote the minimal number of generators of I. Then

(a) the ideal I is generated by forms of degree $\leq d+1$;

(b) $\nu(I) = \dim_k I_d + \dim_k I_{d+1} - \dim_k W$, where W denotes the k-subspace of I_{d+1} generated by I_d ;

(c) if we write $s = \binom{(d-1)+n}{n} + h$, with $d \geq 2$ and $0 \leq h \leq \binom{(d-1)+n}{n-1} - 1$, we have :

(1) $\quad \left[\binom{d+n-1}{n-1} - h\right] - \min\left\{n\left[\binom{d+n-1}{n-1} - h\right] - \binom{d+n}{n-1}, 0\right\} \leq \nu(I) \leq \binom{d+n}{n-1} - (n-1)$.

Proof : (a) follows directly from some general properties of a 1-dimensional Cohen-Macaulay standard G-algebra (cf., e.g., [G-M]$_2$) and (b) is an immediate consequence of (a). Also, in view of (b), (1) is equivalent to

(1)' $\quad 2 \dim_k I_d + (n-1) \leq \dim_k W \leq \min\{(n+1)\dim_k I_d, \dim_k I_{d+1}\}$.

We shall now show the first inequality in (1)' in the case $n = 2$, for simplicity (the general argument being clear after that), while the other inequality is immediate. So, let $F_1, \ldots, F_m$ be a basis of I_d (over k). Then we may assume (without loss of generality) that the co-ordinate axes of $\mathbb{P}^2$ contain none of the s points in question and furthermore $[0:0:1] \notin F_1$. Hence, $X_0F_1,\ldots,X_0F_m, X_1F_1,\ldots,X_1F_m, X_2F_1$ are linearly independent forms (over k) in W, and so we are done.

Now we are able to state our problem.

<u>PROBLEM 3.5</u> : Let s be an integer of the form $s = \binom{(d-1)+n}{n} + h$, with $d \geq 2$, $n \geq 2$ and $0 \leq h \leq \binom{(d-1)+n}{n-1} - 1$. Does there exist, for any integer r satisfying the inequalities (for $\nu(I)$) given in (1) above, a set of s points in generic position in $\mathbb{P}^n$, say $P_1,\ldots, P_s$ such that $\nu(I(P_1,\ldots, P_s)) = r$?

<u>REMARK 3.6</u> : It was conjectured in $[G-O]_2$ that there exists a non-empty Zariski open set in $(\mathbb{P}^n)^s$ consisting of s-tuples of points in generic position in $\mathbb{P}^n$ for which the minimal number of generators (of the corresponding homogeneous ideal) is equal to the lower bound given for $\nu(I)$ in Proposition 3.4(c). This conjecture has been proved in $[G-M]_2$, while only a few (positive) results are known when $n \geq 3$ (see $[G-M]_1$). Also, we point out that in order to prove this conjecture (which expresses an open condition), it would be enough to produce, for any $s > n$, a set of s points in generic position in $\mathbb{P}^n$ whose homogeneous ideal has the required minimal number of generators. Further, the upper bound for $\nu(I)$ given in Proposition 3.4(c) improves similar bounds shown in $[G-O]_2$ and $[G]$; moreover, when $n = 2$, it coincides with the classical upper bound given by Dubreil in $[D]$ (see also $[G]$, $[D-G-M]$) for the minimal number of generators of the ideal of a finite set of points in $\mathbb{P}^2$ such that d is the least degree of a curve containing them.

Next we show that Problem 3.5 admits an affirmative answer when $n = 2$, i.e., the minimal number of generators of the ideal of a set of points in generic position in $\mathbb{P}^2$ can take all possible values (allowed by Proposition 3.4(c)). However, if moreover such points are in uniform position in $\mathbb{P}^2$ ($[G-O]_1$, $[G-O]_2$, $[H_1]$), this is no longer true (see $[G-M]_2$).

We first state a simple lemma.

LEMMA 3.7 : Let t be an integer of the form $t = \binom{m+2}{2} + q$, with $m \geq 2$, $1 \leq q \leq m+1$, and let $t' = t-(m+2)$. Also, let $\nu_{o,t}$ (resp. $\nu_{o,t'}$) denote the lower bound for the minimal number of generators of the ideal of t (resp. t') points in generic position in $\mathbb{P}^2$. Then

(1) given any set of t' points in generic position in $\mathbb{P}^2$, say $P_1,\ldots,P_{t'}$, there exist t points in generic position in $\mathbb{P}^2$, say $Q_1,\ldots,Q_t$ such that $\nu(I(Q_1,\ldots,Q_t)) = \nu(I(P_1,\ldots,P_{t'}))+1$.

(2) if we write $u = [(m+2) - \nu_{o,t}]+1$, $u' = [(m+1) - \nu_{o,t'}]+1$, then we have $u' \leq u \leq u'+1$.

Proof : (1) Let L be a line in $\mathbb{P}^2$ missing $P_1,\ldots,P_{t'}$ and let $R_1,\ldots,R_{m+2}$ denote any $m+2$ distinct points on L. Then the t points $P_1,\ldots,P_{t'}$, $R_1,\ldots,R_{m+2}$ are in generic position in $\mathbb{P}^2$. Now, applying Proposition 3.4(b), we are done.

(2) This follows essentially from statement (1), which concludes the proof.

THEOREM 3.8 : Let s be an integer of the form $s = \binom{(d-1)+2}{2} + h$, with $d \geq 2$ and $0 \leq h \leq d$. Then, given any integer r with

$$(d+1-h) - \min\{d-2h, 0\} \leq r \leq d+1,$$

there exist s points in generic position in $\mathbb{P}^2$, say $P_1,\ldots,P_s$ such that

$$\nu(I(P_1,\ldots,P_s)) = r.$$

Proof : We proceed by induction on the integer $d \geq 2$, the case $d = 2$ being immediate. So, we assume the result true for any integer $\leq d-1$ and we wish to prove it for $s = \binom{(d-1)+2}{2} + h$, with $d \geq 3$ and $1 \leq h \leq d$ (clearly, if $h = 0$, there is nothing to be shown). Now, write $s' = s-(d+1) = \binom{d}{2} + (h-1)$. Then, in view of Lemma 3.7(2) (and with the notation introduced there), we have only two possible cases:

(a) $\nu_{o,s} = \nu_{o,s'} + 1$, in which case the theorem follows by induction, in the light of Lemma 3.7(1);

(b) $\nu_{o,s} = \nu_{o,s'}$. Now, the first case $r = \nu_{o,s}$ is settled in $[G-M]_2$. Then, by using induction and Lemma 3.7(1) again, we are done.

Finally, we would like to make a brief comment about the Hilbert function of a complete intersection.

In [St], Stanley raised the following question : "If a G-algebra R has the Hilbert function of a complete intersection, under what circumstances can we conclude that R actually is a complete intersection ?" . Furthermore, he gave an example of a 2-dimensional reduced Cohen-Macaulay standard G-algebra that has the Hilbert function of a Gorenstein ring and yet is not a Gorenstein ring (cf. also [G-M-R]) . We shall now show that there exist 1-dimensional reduced (Cohen-Macaulay) standard G-algebras that have the Hilbert function of a complete intersection and nevertheless have any prescribed Cohen-Macaulay type.

EXAMPLE 3.9 : Let $d \geq 1$ be any positive integer and let $L_1, \ldots, L_d$ be distinct lines in $\mathbb{P}^2$, all passing through a point O . Then choose d^2 points in $\mathbb{P}^2$ (all distinct from O) in the following way: take 1 point on L_1 , 3 points on $L_2, \ldots,$ $2i-1$ points on $L_i, \ldots,$ $2d-1$ points on L_d . It is easy to check that these points have the Hilbert function of the complete intersection of two plane curves of degree d , with no common components . Also, the minimal number of generators of the homogeneous ideal of these points is d+1 , hence (cf., e.g., [G-O]$_2$), the Cohen-Macaulay type of the homogeneous co-ordinate ring of these points is precisely d .

REFERENCES

[A] R.D.M. Accola, On Castelnuovo's inequality for algebraic curves, I, Trans. Amer. Math. Soc. 251 (1979), 357-373.

[A-S] E. Arbarello and E. Sernesi, Petri's approach to the study of the ideal associated to a special divisor, Inventiones math. 49 (1978), 99-119.

[D] P. Dubreil, Sur quelques propriétés des systèmes de points dans le plan et des courbes gauches algébriques, Bull. Soc. Math. France 61 (1933), 258-283.

[D-G-M] E.D. Davis, A.V. Geramita and P. Maroscia, Perfect homogeneous ideals of height 2 in polynomial rings: Dubreil's theorems revisited, I, (to appear).

[G] A.V. Geramita, Remarks on the number of generators of some homogeneous ideals, Bull. Soc. Math. France (to appear).

[G-H]$_1$ P. Griffiths and J. Harris, Principles of algebraic geometry, Wiley, New York, 1978.

[G-H]$_2$ _____, Residues and zero-cycles on algebraic varieties, Annals of Math. 108 (1978), 461-505.

[G-K] C. Greene and D.J. Kleitman, Proof techniques in the theory of finite sets, Studies in Mathematics, Vol. 17, Mathematical Association of America, 1978, 22-79.

[G-M]$_1$ A.V. Geramita and P. Maroscia, The ideal of forms vanishing at a finite set of points in $\mathbb{P}^n$, C.R. Math. Rep. Acad. Sci. Canada 4 (1982), 179-184.

[G-M]$_2$ _____, The ideal of forms vanishing at a finite set of points in $\mathbb{P}^n$, Queen's University Mathematical Preprint No. 1981-5.

[G-M-R] A.V. Geramita, P. Maroscia and L.G. Roberts, The Hilbert function of a reduced K-algebra, Queen's Papers in Pure and Applied Mathematics, No. 61, Kingston, Ontario, 1982, pp. C1-C63.

[G-O]$_1$ A.V. Geramita and F. Orecchia, On the Cohen-Macaulay type of s-lines in A^{n+1}, J. Algebra 70 (1981), 116-140.

[G-O]$_2$ _____ , Minimally generating ideals defining certain tangent cones, J. Algebra 78 (1982), 36-57.

[H$_1$]   J. Harris, The genus of space curves, Math. Ann. 249 (1980), 191-204.

[H$_2$]   _____ , A bound on the geometric genus of projective varieties, Ann Scuola Norm. Sup. Pisa 8 (1981), 35-68.

[Mac]   F.S. Macaulay, Some properties of enumeration in the theory of modular systems, Proc. London Math. Soc. 26 (1927), 531-555.

[Mar]   P. Maroscia, The Hilbert function of a finite set of points in $\mathbb{P}^n$, Queen's University Mathematical Preprint No. 1982-14.

[R]   L.G. Roberts, The Hilbert function of some reduced graded k-algebras, Queen's University Mathematical Preprint No. 1982-15.

[S]   B. Segre, Sui teoremi di Bézout, Jacobi e Reiss, Ann. Mat. Pura Appl. (4) 26 (1947), 1-26.

[S-R]   J.G. Semple and L. Roth, Introduction to algebraic geometry, Clarendon Press, Oxford, 1949.

[Se]   E. Sernesi, On the problem of uniqueness for certain linear series (talk given at the Conference).

[St]   R.P. Stanley, Hilbert functions of graded algebras, Advances in Math. 28 (1978), 57-82.

[T]   B. Teissier, Variétés toriques et polytopes, Sém. Bourbaki 1980/81, Lecture Notes in Mathematics, Vol. 901, Springer, Berlin 1981, 71-84.

P.S. After this paper was typed, an affirmative answer has been given to Problem 3.1 ; the proof will appear in [D-G-M] .

"Homogeneous Bundles in characteristic p"

V.B. Mehta and A. Ramanathan

Introduction

Let G be a reductive group, P a parabolic subgroup and $X = G/P$. In [4] Ramanan proved, when P is maximal, that for any representation $\sigma : P \to GL(n)$, which is trivial on the radical of P, the associated bundle V_σ on X is semistable, in fact a direct sum of stable bundles. This result was extended by Umemura [7] for arbitrary parabolic subgroups. Their methods use the vanishing theorems of Borel-Weil-Bott and hence are restricted to characteristiz zero. It is natural to ask whether the results hold in characteristic p. In this note we prove the result in characteristic p.

We make essential use of the notion of a Harder-Narasimhan filtration for a nonsemistable principal G-bundle on X [See Thm.3.1]. We also need the fact that on a homogeneous space the Frobenius morphism preserves semistability and consequently, associated bundles of semistable bundles are again semistable.

We would like to thank S. Ramanan and M. Nori for discussions and suggestions. Ramanan also has a different proof of our result.

During the preparation of this paper, the first-named author was a Visiting Professor at the University of Naples, supported by the Consiglio Nazionale delle Ricerche of Italy. He is grateful to the University of Naples and the C.N.R. for their hospitality.

§ 2.

In this section we do the preliminaries needed for the main theorem. We deal with the following situation : X is a nonsingular projective variety over an algebraically closed field k of char p > 0, and H is an ample line bundle on X. For the definitions of H stability and semi-stability and for the notion of the Harder-Narasimhan filtration of an unsemistable vector bundle on X, we refer to [2]. Now let $\pi : X \to X$ be the Frobenius morphism. Then we have

Theorem 2.1. 1) Let $T = T_X$ be the tangent bundle of X. Assume that if $0 = T_0 \subset T_1 \subset T_r = T$ is the Harder-Narasimhan filtration of T, then $\mu_r(T) \geq 0$. Then for any semistable vector bundle V on X, $\pi^*(V)$ is also semistable.

2) If $\mu_r(T) > 0$, then, for stable V on X, $\pi^*(V)$ is also stable.

Proof. We induct on the rank of V for 1). Suppose for all semistable V of rank less than n, $\pi^*(V)$ is semistable. Then it follows from [5, Thm.3.23] that whenever V_1 and V_2 are semistable, each of rank less than n, that $V_1 \otimes V_2$ is also semistable. Now let V be of rank n and semistable. If $\pi^*(V)$ is not semistable, let $B \subset \pi^*(V)$ be the β - subbundle of $\pi^*(V)$. We get a canonical map f : $T_X \to \underline{\text{Hom}}(B, \pi^*(V)/B)$. By induction we get that $\underline{\text{Hom}}(B, \pi^*(V)/B)$ is filtered by semistable bundles of negative degree. By our assumption on T_X it follows that f must be zero.

By purely inseparable descent theory it follows that B descends to a subsheaf of V, contradicting the semi-stability of V. Hence $\pi^*(V)$ is semistable. Now assume that $\mu_r(T) > 0$ and let V be a stable sheaf on X. Suppose $W \subset \pi^*(V)$ with $\mu(W) = \mu\pi^*(V)$. Now both W and $\pi^*(V)/W$ are semistable and again we get a map $f : T_X \to \underline{\text{Hom}}(W, \pi^*(V)/W)$. By 1) $W \otimes \pi^*(V)/W$ is semistable, and of degree zero. Hence f=0, and W descends to a subsheaf of V, contradicting the stability of V.

Q.E.D.

Remark 2.2. If T_X is generated by $H^0(X, T_X)$ (e.g. homogeneous spaces) then we always have $\mu_r(T) \geq 0$. Hence the above Theorem is a slight improvement of Prop.1.1 in [3].

For example, let X be a complete intersection in $\mathbb{P}^n$, dim $X \geq 3$, and if X is defined by $(f_1 \ldots f_t)$, we assume that $\sum_{i=1}^{t} \deg f_i \leq n+1$. Then T_X is semistable of nonnegative degree, although T_X may not have any sections.

Remark 2.3. Applications of the above to uniformization of semistable bundles on simply-connected varieties and other related questions will appear separately. There is also a partial converse, which enables one to construct bundles whose Frobenius pull-back is not semistable.

§3.

Let G be a reductive group over k. We fix a maximal torus T and a Borel subgroup B ⊃ T. Let E → X be a principal G-bundle over X. We call E a <u>stable</u> (resp. semistable) G-bundle if for any reduction of structure group of E, restricted to any open subset whose complement has codimension ≥ 2, the line bundle associated to any dominant character has degree < 0 (resp. ≤ 0). See [6].

We then have the following result.

<u>Theorem 3.1.</u> Let E → X be a nonsemistable G-bundle where X is a projective nonsingular variety. Then there is a unique reduction of structure group of E to a proper parabolic subgroup such that

1) the P/U - bundle (U being the unipotent radical of P) obtained by the extension of structure group P → P/U from the reduced P-bundle is a semistable P/U - bundle.

2) for any nontrivial character of P which can be expressed as a positive combination of simple roots the associated line bundle has degree strictly greater than zero (simple roots taken with respect to any Borel subgroup contained in P). See [6].

We make use of this result to extend a theorem of Ramanan and Umemura [4,7] on homogeneous bundles on G/P to the case when the base field has characteristic p.

<u>Theorem 3.2.</u> Let P be a parabolic subgroup of G. Let U be its unipotent radical and M a Levicomponent so that P = M.U. Then the M-bundle obtained from the P-bundle G → G/P by the extension of structure group P → M is semistable. For any irreducible representation of M

the associated vector bundle on G/P is semistable.

Proof. Let $E \to G/P$ be the associated P/U - bundle obtained from $G \to G/P$ by extension of structure group. Then it is a homogeneous bundle i.e. G acts on E as a bundle automorphism group compatible with its action on G/P. If $E \to G/P$ is not semistable then by Theorem 3.1 there will be a unique reduction of structure group to a parabolic subgroup P' of M over a suitable open subset of G/P. Because of the uniqueness of this reduction it is invariant under the action of G on E. As is easily seen this implies that the parabolic subgroup P' is invariant under the adjoint action of M which is impossible since P' is a proper non-normal subgroup of M.

The last statement of the theorem now follows from Theorem 2.1, Remark 2.2 and the result proved in [5, Theorem 3.22] that if a bundle as well as all its Frobenius twists are semistable then any associated bundle is semistable.

Remark 3.3. More generally if H is a reductive group and $M \to H$ is a homomorphism such that the connected component of the center of M goes into the center of H then the extended H-bundle obtained from E is semistable.

Remark 3.4. Note that our proof goes through in characteristic zero also, making use of [5, Theorem 3.18]. See also [1].

REFERENCES

1) Kobayashi, S. : Curvature and Stability of vector bundles. Proc. Japan Acad. Ser. A, 58 (1982).

2) Maruyama, M. : Boundedness of semistable sheaves of small ranks. Nagoya Math. J. 78 (1980) 65-94.

3) Mehta, V.B., Nori, M.V. : Semistable sheaves on homogeneous spaces and abelian varieties. (preprint)

4) Ramanan, S. : Holomorphic vector bundles on homogeneous spaces. Topology, 5 (1966).

5) Ramanan, S., Ramanathan, A. : Some remarks on the instability flag. (preprint)

6) Ramanathan, A. : Moduli for principal bundles. In : Algebraic Geometry Proceedings, Copenhagen 1978. Lecture Notes 732. Springer 1979.

7) Umemura, H. : On a theorem of Ramanan. Nagoya Math. J. 69 (1978), 131-138.

A. Ramanathan
School of Mathematics
Tata Institute of Fundamental Research
Homi Bhabha Road
Bombay 400 005
I N D I A

V.B. Mehta
Dept. of Mathematics
University of Bombay
Bombay 400 098
I N D I A

The Group of Sections on a Rational Elliptic Surface

by

Ian Morrison* (Department of Mathematics,
University of Toronto,
Toronto, Canada)

and

Ulf Persson** (Institut Mittag-Leffler,
Djursholm, Sweden)

The diophantine problem

(1)
$$n^2 = \sum_{i=1}^{k} n_i^2 - 1$$

$$3n = \sum_{i=1}^{k} n_i + 1$$

has for $k \leq 8$ only a finite number of solutions. Indeed

$$n^2 = \sum_{i=1}^{k} (n_i)^2 - 1$$

$$\geq \frac{1}{k} (\sum_{i=1}^{k} n_i)^2 - 1 = \frac{9}{k} n^2 + 0(1).$$

*Research partially supported by the National Science and Engineering Research Council of Canada

**Research partially supported by the Swedish Research Council of Natural Sciences

As is well-known, however, if $f:X \longrightarrow \mathbb{P}^1$ is a relatively minimal rational elliptic surface then every section S of X gives rise to a solution of these equations with $k = 9$. The choice of an origin section S_0, makes the set of sections of X into an abelian group under the operation of fibre-by-fibre addition. If X is generic in a suitable sense any $S \neq S_0$ will have infinite order; hence for $k = 9$, the equations have infinitely many solutions. This construction was first used by Nagata [6] to give examples of relatively minimal surfaces with infinitely many exceptional curves of the first kind : the exceptional curves of the first kind on X are exactly the sections.

This paper grew out of the authors' attempts to answer two natural problems. Let Φ denote the set of solutions to (1) with $k = 9$. First, enumerate Φ. Second, describe numerically the group law on Φ induced by transport-of-structure. While we were writing the paper, Igor Dolgachev kindly pointed out to us that the first of these problems and part of the second had been solved by Manin [3]. Our method enables us to simplify Manin's solutions to the first problem and to complete his results on the second.

We first recall some facts about linear systems of curves in the plane defined by base point conditions, in particular, the system of curves of degree n passing n_i-fold through each of nine points P_i. We illustrate the special properties such systems have when the nine points are themselves the base points of a pencil of cubics. Next, we study the class of rational elliptic surfaces X whose sections, $\Phi(X)$, correspond bijectively to elements of Φ. These are the X all of whose fibres are irreducible; we then call X irreducible. (Most of this material is well-known). We then determine the principal homogeneous space structure of $\Phi(X)$ in terms of the intersection form on X (16), and obtain numerical formulae (17) for the corresponding structure induced on Φ by identifying Φ with $\Phi(X)$ for (any) irreducible X. These formulae, which are the crucial novelty in our treatment, are then used to recover Manin's enumeration of Φ; in particular, they greatly simplify the key step of identifying the subgroup generated by the standard exceptional divisors of a blowdown $\pi: X \longrightarrow \mathbb{P}^2$. We give two additional applications. The first is the determination of the group $Aut_0(X)$ of fibre preserving automorphisims of X and the computation of its representation on NS(X). The second is the identification in NS(X) of the divisors of n-torsion points on X.

We plan to take up in a subsequent paper [5] a number of generalizations and extensions of these results.

We take great pleasure in acknowledging a number of debts. Manin's enumeration of Φ is based on a geometric computation of the action on NS(X) of translation by the standard exceptional divisors. Although our method recovers this as a corollary of the numerical formulae for the group law, our insight was helped at an early stage by seeing this result from Joe Harris who had obtained it independently with Rick Miranda some years ago. We wish to thank Igor Dolgachev for helpful discussions and for bringing Manin's paper to our attention and Henry Pinkham for suggesting some useful references. This work was begun while the authors were C.N.R. visitors at the University of Rome. We are grateful to both institutions for their hospitality. Most of all, we would like to thank Enrico Arbarello for making Rome such a delightful place to work and live.

LINEAR SYSTEMS IN $\mathbb{P}^2$

A classical problem in the theory of plane curves is to study linear systems defined by base point conditions. Given a finite set $\{P_1,\ldots,P_k\}$ of points in the plane (some possibly infinitely near) and a $(k+1)$-tuple $(n;n_1,\ldots,n_k)$ with $n > 0$ and $n_i \geq 0$, one seeks the linear system L of curves of degree n in the plane passing through each P_i with multiplicity at least n_i. In this paragraph, which is primarily motivational, we adopt the simplifying assumption that the P_i are distinct: i.e. none is infinitely near.

Nowadays we let $\pi: X \to \mathbb{P}^2$ be the blow-up of the plane at the points P_i and carry on the analysis on the rational surface X. If H is the pullback of the class of a line in $\mathbb{P}^2$ and $E_i = \pi^{-1}(P_i)$, then pullback via π gives a bijection of L with the linear system $|nH - \sum_{i=1}^{k} n_i E_i|$ on X. The classes $H, E_1,\ldots,E_k$ are a basis for $NS(X)$, the Neron-Severi group of X. If $C \in |nH - \sum_{i=1}^{k} E_i|$, then following Nagata [6], we refer to the $(k+1)$-tuple $(n; n_1,\ldots,n_k)$ as the numerical data or numerical character of C. Since

$$(2) \quad K_X \cong -3H + \sum_{i=1}^{k} E_i$$

we obtain from Riemann-Roch that

$$(3) \quad h^0(nH - \sum_{i=1}^{k} n_i E_i) = \tfrac{1}{2}((n^2 - \sum_{i=1}^{k} n_i^2) + (3n - \sum_{i=1}^{k} n_i) + 2)$$
$$+ h^1(nH - \sum_{i=1}^{k} n_i E_i).$$

The h^1 term, classically called the superabundance to reflect the postulation that for generic choices of the P_i it is zero, is a rather subtle invariant measuring the failure of the sets of linear conditions imposed at each of the P_i to be independent.

The first example every student sees is the linear system of cubics passing simply through nine base points. By (3), this linear system is never empty. We say $P_1, \ldots, P_9$ are associated if any of the equivalent conditions below holds.

(4)
i) $h^0(3H - \sum_{i=1}^{9} E_i) = 2$

ii) $h^1(3H - \sum_{i=1}^{9} E_i) = 1$

iii) The $\{P_i\}$ are the base points of a pencil of plane cubics.

iv) If $E \in |3H - \sum_{i=1}^{9} E_i|$, then $3H|_E \equiv \sum_{i=1}^{9} P_i$

v) There is a map $f: X \longrightarrow \mathbb{P}^1$ making X a relatively minimal elliptic surface.

(Recall that X relatively minimal means there are no exceptional curves of the first kind lying in any fibre of f.) In this case, the fibres F_λ of f are the curves in $|3H - \sum_{i=1}^{9} E_i|$ so by (2) are anti-canonical: $F = -K_X$.

Most of this paper is concerned with the geometry on X when $k = 9$ and $P_1,\ldots P_9$ are associated. In the rest of this section we wish to motivate this choice by comparing the associated and generic cases when $k = 9$. To avoid a number of special arguments, we assume that there is an ireducible curve E in $|3H - \sum_{i=1}^{9} E_i|$.

We denote by Φ the set of 10-tuples $N=(n;n_1,\ldots,n_9)$ that satisfy

(5)
i) $n^2 - \sum_{i=1}^{9} n_i^2 = -1$

ii) $3n - \sum_{i=1}^{9} n_i = 1$

iii) $(n^2 - \sum_{i=1}^{9} n_i^2) - (3n - \sum_{i=1}^{9} n_i) = -2$

We denote by L_N the linear system $|nH - \sum_{i=1}^{9} n_i E_i|$ on X, and if $C \in L_N$ we call $N = N(C)$ the numerical character of C. In terms of a curve $C \in L_N$, the conditions (5) may be reexpressed as

(6)
i) $C^2 = -1$

ii) $(-K_X) C = 1$

iii) $p_a(C) = \frac{C.C + K.C}{2} + 1 = 0$

Condition (6.ii) is equivalent to $F.C = 1$ for associated P_i. Clearly, any two of these conditions imply the third.

PROPOSITION 7 If $N \in \Phi$, then the linear system L_N is not empty. If L_N contains an irreducible curve C, then

i) C is the unique curve in L_N.

ii) C is smooth and rational

iii) $\pi(C)$ has multiple points of order n_i at P_i and no other singularities.

REMARK. The condition in i) means, in this case, that the superabundance is zero. That in iii) was classically expressed by saying that "the virtual data are effective."

PROOF: The non-emptyness is immediate from (3). If C is irreducible and $C' \sim C$, then since $C\, C' = -1$, we must have $C' = C + D$ with D effective and $D \sim 0$. But then $D = 0$ so $C = C'$. This gives i). A singular irreducible curve has arithmetic genus at least one which yields ii) and iii) is immediate from ii).

Now if $N \in \Phi$ and $C \in L_N$, write $C = \sum_{j=0}^{\ell} m_j C_j$ where the C_j are distinct irreducible curves with numerical characters $R_j = (r_j; r_{j1}, \ldots, r_{j9})$. Then (5.ii) implies $\sum_{j=0}^{\ell} m_j (r_j - \sum_{i=1}^{9} r_{ji}) = 1$. But $E \cdot C_j = 3r_j - \sum_{i=1}^{9} r_{ji} \geq 0$ for each j. Hence, after possibly reordering the C_j, we have $m_0 = 3r_0 - \sum_{i=1}^{9} r_{0i} = 1$ and $3r_j - \sum_{i=1}^{9} r_{ji} = 0$ for $j \geq 1$.

Now suppose in addition that $\{P_1, \ldots, P_9\}$ are associated. If S is any irreducible numerical section of X, then the map $\sigma: \mathbb{P}^1 \to S$ by $\lambda \longrightarrow S \cdot F_\lambda$ is an isomorphism so S is a section. In particular S is rational so its numerical character is in Φ. Hence,

LEMMA 8   i) If $N \in \Phi$ and $C \in L_N$ then $C = C_0 + D$ where C_0 is irreducible and satisfies (5, ii) and $E \cdot C_j = 0$ for every component C_j of D.

ii) If in addition $\{P_1, \ldots, P_q\}$ are associated, then $N(C_0) \in \Phi$ and C_0 is a section of X.

In fact, one can show that if the P_i are associated the curve C_0 above depends only on N and not on the choice of a curve $C \in L_N$. (cf. [5]).

We conclude this section with an example which illustrates the pathologies which arise. Let C be an irreducible quintic with five nodes $P_1,\ldots,P_5$ – for example, a generic projection of an elliptic normal curve in $\mathbb{P}^4$. Let D be an irreducible quartic with nodes at P_1, P_2 and P_3 and passing through P_4 and P_5. (An easy count of constants shows that not all quartics satisfying these conditions are reducible.) Let L be the line through P_4 and P_5 and let P_6 be its residual intersection with C. Let P_7, P_8, P_9 and P_{10} be the residual intersections of C and D. (Again it is easily checked that C and D can be chosen so P_1 to P_{10} are distinct). Consider the data (6; 2,2,2,3,3,2,1,1,1) posed at $P_1,\ldots,P_9$. This data is in Φ but there are (at least) two solutions: C+L and D+2L. Moreover the data of C, (5;2,2,2,2,2,1,1,1,1) satisfies (1.4.ii) but not (1.4.i) so the proper transform of C is an irreducible formal section whose data is <u>not</u> in Φ. Of course, the cubic E through $P_1,\ldots,P_q$ is unique. (It is also irreducible: use Bezout).

In the next section, we will define a class of elliptic surfaces X for which the superabundance of linear systems with data in Φ is always zero. We do not know of any set of non-associated points with this property though of course one suspects most sets will have it. The difficulty is that for non-associated sets, one must apparently rule out $E \cdot I = 0$ for an infinite set of types I. For another examination of this problem see Nagata [6].

IRREDUCIBLE RATIONAL ELLIPTIC SURFACES.

Let $f: X \longrightarrow C$ be a relatively minimal elliptic surface with a section. We wish to describe a class of group structures on the set of sections of X. Roughly speaking once one fixes a section and uses it as the origin on each fibre then one can add any two sections "fibre by fibre". Formally let $(F_\lambda)_{n.s.}$ be the set of non-singular points on the fibre F_λ of X over $\lambda \in C$. This is naturally a principal homogeneous space for a commutative algebraic group. (cf. [1]). If S is a section of X let $S_\lambda = S \cdot F_\lambda$. Since S is a section $S_\lambda \in (F_\lambda)_{n.s.}$ so taking S_λ as the origin fixes a group structure on $(F_\lambda)_{n.s.}$ whose operation we denote $+_{S_\lambda}$. Then define a group operation on the sections of S, denoted $\oplus_S$ by requiring that $S_1 \oplus_S S_2$ solve

$$(9) \qquad (S_1 \oplus_S S_2)_\lambda = (S_1)_\lambda +_{S_\lambda} (S_2)_\lambda .$$

This uniquely determines a section $S_1 \oplus_S S_2$. Thus the set of sections of X becomes a principal homogeneous space for a commutative group which we denote by $\Phi(X)$.

There is a second well-known interpretation of this group in case the j-invariant of the fibres of X is non-constant. Let $f_\eta : X_\eta \longrightarrow \eta = \text{Spec}(k(C))$ be the generic fibre of f. Specialization then gives a bijection between $k(C)$-valued points S of the genus 1 curve X_η over η and sections S of X. If we choose an origin S for $X_\eta(k(C))$ and let the specialization S of S be the origin section, as above, then this bijection becomes a group isomorphism. For more details, see Manin [3].

For the remainder of this paper, let $f:X \to \mathbb{P}^1$ be a relatively minimal rational elliptic surface with a section. Then X can be realized by resolving the base locus of a pencil G_λ of cubic curves in $\mathbb{P}^2$ whose generic member is smooth. (Some of the base points will, in general, be infinitely near. For facts about X which we quote without proof a good reference is Miranda [4]). We will say that X is <u>irreducible</u> if either of the following equivalent conditions hold.

(10')
    i) All fibres F_λ of X are irreducible.

    ii) The base points of the pencil G_λ of cubic curves giving rise to X are distinct and no three lie on a line.

To see that these are equivalent, first observe that there is an infinitely near base point if and only if some fibre contains the proper transform of an exceptional divisor. Hence we may as well assume the base points $P_1, \ldots, P_9$ of the pencil are distinct. Let G_λ be a smooth element of the pencil and suppose $G_\mu = Q.L$ is a reducible member with L linear. Then Bezout's theorem implies $\#(Q \cap G_\lambda) \le 6$ and $\#(L \cap G_\lambda) \le 3$. But $\#(G_\mu \cap G_\lambda) = 9$ so in fact $\#L \cap G_\lambda = 3$ and three base points are collinear. Conversely, if three P_i lie on a line L and Q is the quadric through some five of the remaining G, then Q.L contains 8 of the P_i and again Bezout's theorem implies Q.L is a member of the pencil $\{G_\lambda\}$.

We remark that irreducible X exist. In fact since the reducible cubics have codimension 2 in the space of all cubics such X are generic. We shall give some other characterizations of the irreducible rational elliptic surfaces in (14) and (26). The interest of the hypothesis of irreducibility from the point of view of the linear systems considered in the preceding section is expressed by,

LEMMA 11. <u>If X is irreducible and C is an effective curve on X then the following are equivalent</u>:

    i) C is a section of X

    ii) C is an exceptional curve of the first kind on X.

    iii) The numerical character $N(C)$ is in Φ.

PROOF: A section is an irreducible curve satisfying (6.ii) and (6.iii) and an exceptional curve of the first kind is an irreducible curve satisfying (6.i) and (6.iii). Since two of the conditions in (6) imply the third it is immediate that i) ii) and that both imply iii). The critical implication iii) i) follows from

LEMMA 12. <u>If X is irreducible and C is an effective numerical section of X (i.e. $C \cdot F = 1$), then C is irreducible if and only if $C^2 = -1$. Moreover in this case C is a section.</u>

PROOF: Write $C = \sum_{i=0}^{k} C_i + \sum_{j=1}^{\ell} F_j$ where each F_j is a fibre and each C_i is an irreducible curve not equal to a fibre. Since X is irreducible, $C_i \cdot F > 0$ for each i. Hence $C \cdot F = 1$ implies $i = 1$ and $C_0 \cdot F = 1$. Since C_0 is irreducible, it is a section hence $c_0^2 = -1$. But then $c^2 = 2\ell - 1$ so $c^2 = -1$ if and only if $\ell = 0$; this in turn is equivalent to the irreducibility of C.

COROLLARY 13. <u>If X is irreducible the map $C \longrightarrow N(C)$ gives a bijection between $\Phi(X)$ and Φ.</u>

In view of this we shall often speak of the curve $N = N(C)$ when X is irreducible and $C \in \Phi(X)$. We conclude this section by giving another characterization of irreducible X.

LEMMA 14 <u>X is irreducible if and only if rank $\Phi(X) = 8$.</u>

PROOF: This is a consequence of a general formula of Tate [9] and Shioda [8] for elliptic surfaces $f: X \longrightarrow C$.

(15) $\quad \text{rank}_Z(NS(X)) = \text{rank}_Z \Phi(X) + 2 + \sum_{\lambda \in C} (m_\lambda - 1)$.

where m_λ is the number of components in the fibre F_λ. When X is rational, of course, $\text{rank}(NS(X)) = 10$.

THE GROUP LAW ON Φ.

Our first goal is to write down the group structures $\oplus_S$ on $\Phi(X)$ in terms of the addition $+$ in $NS(X)$.

LEMMA 16 $\underline{S_1 \oplus_S S_2 \simeq (S_1+S_2-S+\alpha F)}$ where $\alpha = (S_1+S_2)S - S_1 \cdot S_2 + 1$.

PROOF: Denote by $+$ the sum of divisors on F_λ and by $\sim_\lambda$ linear equivalence of divisors on F_λ. Then the condition of (9)

$$(S_1 \oplus_S S_2)_\lambda = (S_1)_\lambda +_{S_\lambda} (S_2)_\lambda \quad \text{means that}$$

$(S_1 \oplus_S S_2)_\lambda - S_\lambda \sim_\lambda ((S_1)_\lambda - S_\lambda) + ((S_2)_\lambda - S_\lambda)$ or equivalently
$(S_1 \oplus_S S_2)_\lambda \sim_\lambda (S_1)_\lambda + (S_2)_\lambda - S_\lambda$. But this means that on X, $S_1 \oplus_S S_2$ is linearly equivalent to $S_1 + S_2 - S + \alpha F$ for some α. Since $S_1 \oplus_S S_2$ is a section, $(S_1 \oplus_S S_2)^2 = -1$ by Lemma 11, and a short calculation yields the value of α.

An immediate corollary of this lemma is that the group structure induced on Φ by the choice of an irreducible surface X and a section S depends only on the numerical character of S, not on X. On the other hand for fixed X, the group structures corresponding to different S are isomorphic. Hence we may fix once and for all as the origin section the exceptional divisor E_1 with character $(0; -1,0,0,0,0,0,0,0)$ and then speak of the group structure on Φ or on $\Phi(X)$. We denote the corresponding operation by $\oplus$ in both cases. More precisely, we can now write down the "universal" group law on Φ.

THEOREM 17 i) $(n;n_1,\ldots,n_9) \oplus (m;m_1,\ldots,m_9)$
$= (n+m+3\alpha, n_1+m_1+1-\alpha, n_2+m_2-\alpha, n_3+m_3-\alpha,\ldots,n_9+m_9-\alpha)$.
where $\alpha = n_1 + m_1 - nm + \sum_{i=1}^{9} n_i m_i + 1$.

ii) $\ominus (n;n_1,\ldots,n_9) = (6+6n_1-n ; n_1, 2+2n_1-n_2,\ldots,2+2n_1-n_9)$.

PROOF: i) Simply set $S_1 = (n;n_1,\ldots,n_9)$, $S_2 = (m;m_1,\ldots,m_9)$ and $S = (0;1,0,\ldots,0)$ in Lemma 16, compute α and evaluate the formula of the lemma for $S_1 \oplus_S S_2$.

ii) Let $S_2 = \ominus S_1$ with the notation above. Then $S_1 \oplus S_2 = E_1$ Hence i) implies that $(m;m_1,\ldots,m_9)$ is the unique solution in Φ of

$$n+m+3\alpha = 0$$
$$n_1+m_1+1+\alpha = -1$$
$$n_i+m_i+\alpha = 0, \text{ for } i = 2,\ldots,9$$

But the right hand side of the formula is easily checked to be such a solution using (5) for $(n;n_1,\ldots,n_9)$.

COROLLARY 18 i) The map $\Phi \longrightarrow Z/3Z$ by $(n;n_1,\ldots,n_9) \longrightarrow n \pmod{3}$ is a surjective group homomorphism.

ii) The action of S_8 on Φ by permuations of $(n_1,\ldots,n_9)$ commutes with $\oplus$.

EXAMPLE 19

$\ominus (0;0,-1,0,0,0,0,0,0,0) = (6;0,3,2,2,2,2,2,2,2)$
$\ominus (1;0,1,1,0,0,0,0,0,0) = (5;0,1,1,2,2,2,2,2,2)$
$\ominus (2;0,1,1,1,1,1,0,0,0) = (4;0,1,1,1,1,1,2,2,2)$
$\ominus (3;0,2,1,1,1,1,1,1,0) = (3;0,0,1,1,1,1,1,2)$

The orbits of these solutions under the action of S_8 yield all 240 elements of Φ with $n_1 = 0$. (cf. Manin [2], Proposition 26.1])

With a little more effort we can list the elements of Φ.

PROPOSITION 20   (Manin [3], Theorem 5])
Fix an 8-tuple of integers $A = (a_2, \ldots, a_9)$ and set

$$N_A = \sum_{j=2}^{9} a_j$$

$$M_A = \sum_{j=2}^{9} (a_j)^2 + \sum_{2 \le j < k \le 9} a_j a_k - N_A.$$

$$E_A = (3 M_A); M_A + N_A - 1, M_A - a_2, \ldots, M_A - a_9)$$

Then if we let $C^k = \underbrace{C \oplus C \oplus \ldots \oplus C}_{k \text{ times}}$, we have

$$E^A = E_2^{a_2} \oplus \ldots \oplus E_9^{a_9}$$

PROOF: The formula is easily checked when E^A is one of the exceptional divisors E_2 to E_9. Let $A' = (a_2 + 1, a_3, a_4, \ldots, a_9)$ By induction and symmetry it suffices to check that the assertion of the Proposition for E^A is equivalent to the assertion for $E^{A'}$ or equivalently that $E^A \oplus E_2 = E^{A'}$. If we add E^A and E_2, then in the notation of Theorem 17 $\alpha = m_A + N_A - 1 + 0 - 0 - (M_A - n_2) + 1 = N_A + n_2$, so that $E^A \oplus E_2$ is given by

$(3(M_A + (N_A + n_2)); (M_A + N_A + n_2) + (N_A + 1 - 1), (M_A + N_A + n_2) - (n_2 + 1),$

$(M_A + N_A + n_2) - n_3, \ldots, (M_A + N_A + n_2) - n_9)$.

But $N_{A'} = N_A + 1$ and $M_{A'} = M_A + N_A + n_2$, hence $E^A \oplus E_2 = E^{A'}$.

This Proposition gives us a numerical description of the subgroup $\xi \cong \mathbb{Z}^8$ of Φ generated by the exceptional divisors $E_2, \ldots, E_9$.

COROLLARY 21   i) If $N(C) = (m; m_1, \ldots, m_9) \in \Phi$, then $C \in \xi$ if and only if $m \equiv 0 \pmod{3}$.

ii) $\Phi/\xi \cong \mathbb{Z}/3\mathbb{Z}$.

PROOF: Part ii) is immediate from i) and (18.i)) and the necessity in i) is immediate from the addition formula (17.i). On the other hand, suppose $m = 3m'$. If we are to have $C = E^A$, then we must have

$$m' = M_A$$
$$m_1 = M_A + N_A - 1$$
$$m_i = M_A - a_i, \quad i \geq 2$$

Hence we must have,

(22) $a_i = m' - m_i, \quad i \geq 2$

Let us fix these a_i and prove that $m' = M_A$ and $m_1 = M + N - 1$. Since $N(C) \in \Phi$, we have by (5)

$$9(m')^2 = \sum_{i=1}^{9} m_i^2 - 1$$
$$9m' = \sum_{i=1}^{9} m_i + 1$$

From the second we find

$$m_1 = 9m' - \sum_{j=2}^{9} m_j - 1$$

$$= m' - \sum_{j=2}^{9} (m' - m_j) - 1$$

$$= m' + \sum_{j=2}^{9} a_j$$

$$= m' + N - 1$$

Hence it will suffice to show that $m' = M$. Now rewrite the first equality as

$$9(m')^2 = m_1^2 + \sum_{j=2}^{9} (m_j)^2 - 1$$

$$= (m' + N - 1)^2 + \sum_{j=2}^{9} (m' - a_j)^2 - 1$$

After expanding the right hand side and simplifying the resulting expression, we obtain

$$2m' = N^2 - 2N + \sum_{j=2}^{9} a_j^2$$

$$= (\sum_{j=2}^{9} a_j)^2 + \sum_{j=2}^{9} a_j^2 - 2N$$

$$= 2M$$

as required

The simplest section which has degree n not congruent to zero modulo three and which is symmetric in $(n_2, \ldots, n_9)$ is $G = (4; 3, 1, 1, 1, 1, 1, 1, 1, 1)$. Using (20.i)

we find $G^2 = (23; 15, 10,10,10,10,10,10,10,10)$ and
$G^3 = (84; 35, 27, 27, 27, 27, 27, 27, 27, 27) = E_2 \oplus E_3 \oplus \ldots \oplus E_9$.
(The last equality can be read off from (22))

COROLLARY 23 **Let** $N(C) = (n; n_1, \ldots, n_9) \in \Phi$. Then there is a unique 8-tuple of integers $A = (a_2, \ldots, a_9)$ such that

i) if $n \equiv 0 \pmod{3}$, then $C = E^A$ and
$(n; n_1, \ldots, n_9) = (3M_A; M_A + N_A - 1, M_A - a_2, \ldots, M_A - a_9)$.

ii) if $n \equiv 1 \pmod{3}$, then $C = G \oplus E^A$ and
$(n; n_1 \ldots, n_9) = (3M_A + 4 + 9N_A; M_A + N_A + 3 - 3N_A, M_A - a_2 + 1 - 3N_A, \ldots, M_A - a_9 + 1 - 3N_A)$.

iii) if $n \equiv 2 \pmod{3}$, then $C = G^2 \oplus E^A$ and
$(n; n_1, \ldots, n_9) = (3M_A + 32 + 18N_A; M_A + N_A + 15 - 6N_A, M_A - a_2 + 10 - 6N_A, \ldots, M_A - a_9 + 10 - 6N_A)$.

PROOF: The assertions about the form of C follow from (20.i); those about the form of $(n; n_1, \ldots, n_9)$ from (20) and the addition formula (17.i)

A description of Φ differing only on the choice of coset representatives of Φ/ξ is given by Manin [3, Theorem 6]. The preceding enumeration has the effect of linearizing Φ. In [5], we obtain a linear parametrization of Φ which is independent of its diophantine description and of which the results above are formal consequences.

RELATIONS WITH $\mathrm{Aut}_0(X)$

Fix a rational elleptic surface $f: X \to \mathbb{P}^1$ (not necessarily irreducible) with non-constant j-invariant and a section S of X and denote by $\oplus$ and $\ominus$ the group operations this induces on $\Phi(X)$ and by $+_\lambda$ the group operation this induces on the set of non-singular points of the fibre F_λ.

Let $\mathrm{Aut}_0(X)$ be the set of fibre-preserving automorphisms α of X i.e. $f \circ \alpha = f$. If C is a section of X, let C_λ be the point $C \cdot F_\lambda$. If Q is a non-singular point of F_λ, we set $i(C)(Q) = Q +_\lambda C_\lambda$. Then $i(C)$ extends by continuity to an automorphism of X. Informally, $i(C)$ is just translation by C fibre-by-fibre. Since $i(C)$ is clearly fibre preserving we obtain an inclusion of groups $i: \Phi(X) \to \mathrm{Aut}_0(X)$. Let σ denote the fibre-preserving involution on X induced by taking fibrewise inverses: if Q is non-singular on F_λ, $\sigma(Q) = -_\lambda Q$. Then,

PROPOSITION 24 $\mathrm{Aut}_0(X) \cong i(\Phi(X)) \, \alpha \, \mathbb{Z}/2$ where the $\mathbb{Z}/2$ factor is generated by σ and the action of σ on $i(\Phi(x))$ is by $\sigma i(C) \sigma^{-1} = i(\ominus C)$.

PROOF: Given any $\alpha \in \mathrm{Aut}_0(X)$, let $C = \alpha(S)$ which is again a section of X and let $\beta = i(\ominus C) \alpha$. Then β is an element of $\mathrm{Aut}_0(X)$ fixing S, hence induces automorphisms of the fibres of X fixing the origin. Since we assume $j_\lambda = j(F_\lambda)$ is not constant, this means that for generic λ, $\beta|_{F_\lambda}$ is given by the identity or by $-_\lambda$. Hence β itself is either the identity or σ. So $\alpha = i(C)$ or $\alpha = i(C) \circ \sigma$. Moreover if $Q \in F_\lambda$, $\sigma \circ i(C) \sigma (Q) = -_\lambda (C_\lambda -_\lambda Q) = (\Theta C)_\lambda +_\lambda Q = i(\Theta C)(Q)$.

This result implies that the natural map $M: \text{Aut}(X) \longrightarrow \text{Aut}(NS(X))$ is injective when restricted to $\text{Aut}_0(X)$. (The map M will <u>not</u> be faithful on all of $\text{Aut}(X)$). Observe that if $C' \in \Phi(X) \subseteq NS(X)$, then $M(i(C))(C') = C \oplus C'$. Now let us assume that X is an irreducible surface and that our origin section $S = E_1$, as in §3. The next proposition determines the representation of $\text{Aut}_0(X)$ on $\text{Aut}(NS(X))$ given by M.

PROPOSITION 25 Fix the ordered basis $(H; -E_1, \ldots, -E_9)$ of $NS(X)$.

1) If $C = (n; n_1, \ldots, n_9) \in \Phi$ and $\mathcal{C} = M(i(C))$ then the matrix coefficients of $\mathcal{C}$ are

$\mathcal{C}_{HH} = 10 + 9n_1 \qquad \mathcal{C}_{HE_1} = 6n_1 + 6 - n \qquad \mathcal{C}_{HE_i} = 3(n_1 + n_i + 1) - n$
$ i \geq 2$

$\mathcal{C}_{E_1 H} = n \qquad \mathcal{C}_{E_1 E_1} = n_1 \qquad \mathcal{C}_{E_1 E_1} = n_i, \quad i \geq 2$

$\mathcal{C}_{E_j H} = 3(n_1 - n_j - 1) + n \qquad \mathcal{C}_{E_j E_1} = n_1 + 1 + (n_1 - n_j + 1)$

$ j \geq 2 j \geq 2$

and for $i \geq 2$ and $j \geq 2$

$$c_{E_j E_i} = \begin{cases} n_i + (n_1 - n_j + 1), & i \neq j \\ n_i, & i = j \end{cases}$$

ii) <u>The matrix coefficients of $\sum = M(\sigma)$ are</u>

$\sum_{HH} = 17$ $\sum_{H \cdot E_1} = 0$ $\sum_{H \cdot E_i} = 6$, $i \geq 2$

$\sum_{E_1 H} = 0$ $\sum_{E_1 E_1} = -1$ $\sum_{E_1 E_i} = 0$, $i \geq 2$

$\sum_{E_j H} = 6$ $\sum_{E_j E_1} = 0$ $\sum_{E_j E_i} = \begin{cases} 2, i \neq j & i \geq 2, j \geq 2 \\ 3, i = j \end{cases}$

The addition formula (17.i) gives the action of $M(i(\Phi))$ on $E_1, \ldots, E_9$ and the inversion formula (17.ii) does the same for $M(\sigma)$. Since elements of $\text{Aut}_0(X)$ fix the class of a fibre, this reduces the proposition to a straightforward changed-of-basis computation.

REMARK: 1) The matrices associated to the exceptional divisions E_i were obtained by Manin [3] who based his proof of Proposition 20 on them. Rick Miranda and Joe Harris (unpublished) later rediscovered these matrices and the matrix $\sum$. By diagonalizing the matrices of the E_i's they were able to compute the matrices of all sections $C \in \xi$. In both cases, the matrices are obtained by direct geometric arguments.

2) If we do not suppose X irreducible, but continue to assume that E_1 is a section, then the matrix of $C = M(i(C))$ is still given by the proposition for any $C \in \Phi(X)$. For more details see [5].

§7 CURVES OF TORSION POINTS

We continue to fix a rational elliptic surface $f: X \to \mathbb{P}^1$ with section, and an origin section S on X and to denote by σ the correspoinding minus-one involution on X. Let

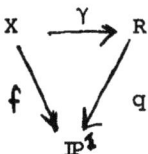

be the quotient of X by this involution. Then R is a ruled surface (not necessarily minimal) and the branch locus on R has the form $\bar{B} + \bar{S}$ where $\bar{S} = \gamma(S)$ is a section of R and where $\bar{B}$ is a smooth trisection of R disjoint from $\bar{S}$. (Observe that two generically distinct torsion points Q_λ and Q'_λ can coalesce at λ_0 only if Q_{λ_0} is singular on F_{λ_0}.) The curve $B = \gamma^{-1}(\bar{B})$ is just the locus of primitive 2-torsion elements on X. There is a bijection between pairs of sections $(C, \ominus C)$ on X and sections of R everywhere tangent to B given by associating to each pair their common image $\bar{C} = \gamma(C)$ on R. (For more details the geometry of double covers see [7]).

Define a rational morphism Γ which fits in a diagram

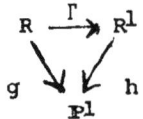

by blowing down all exceptional divisors on R which are disjoint from $\bar{S}$. Then r is an isomorphism in a neighbourhood of $\bar{S}$ and R' is the relatively minimal model of R. Since $\bar{S}^2 = -2 = 2(S^2)$ and $r(\bar{S})$ is again a section of R' of self-intersection -2, $R' = \mathbb{F}_2$.

LEMMA 26⁻ <u>The map r is an isomorphism if and only if X is irreducible.</u>

PROOF: Clear from the construction.

As generators of $NS(\mathbb{F}_2)$, we take as usual the class of a generic section T and of a fibre G (so $T^2 = 2$, $T \cdot G = 1$ and $G^2 = 0$). Since $r(\bar{S})^2 = -2$, $r(\bar{S})$ is the unique section of $\mathbb{F}_2$ of negative self intersection and $r(\bar{S}) \in |T - 2G|$. Since $r(\bar{B})$ is still a trisection and is disjoint from $r(\bar{S})$, $r(\bar{B}) \in |3T|$. While $r(\bar{B})$ is no longer in general a smooth trisection, there is a dictionary between the singularities of $r(\bar{B})$ and the Kodaira types of the degenerate fibres of X. We will content ourselves here with the example that an I_2 fibre produces an ordinary double point on $r(\bar{B})$.

Now let X be irreducible and let $S = E_1$ as usual Define T_n to be the locus of n-torsion points of X. Then $T_n = T_n' + E_1$ where T_n' is the curve of non-trivial n-torsion points and $T_n' \cdot E_1 = 0$. (A point of $T_n \cap E_1$ would be singular in its fibre, hence could not lie on the section E_1, as was remarked above.)

LEMMA 27 i) $T'_n(X) \in |(n^2-1)(3H - \sum_{j=2}^{9} E_i)|$

ii) If C is any section disjoint from E_1, $T'_n(X) \cdot C = (n^2-1)$

PROOF: 1) Clearly $T'_n(X) \cdot F = n^2 - 1$, the number of non-trivial n-torsion points on an elliptic curve. As above, $T'_n(X) \cdot E_1 = 0$ and if $i > 1$, $T'_n(X) \cdot E_i = E_1 E_i^n = n^2 - 1$ by (20).

ii) Immediate from 1) and $C \cdot F = 1$.

We remark that part i) of the Lemma remains valid if we suppose that X is any rational elliptic surface, $\pi: X \longrightarrow \mathbb{P}^2$ is a blow-down, $P_1, \ldots, P_9$ are the base points (some possibly infinitely near) of the associated pencil of cubics and $E_i = \pi^{-1}(P_i)$.

COROLLARY 28 Let $T^*_n(X)$ be the locus of primitive n-torsion points of X. Then $T^*_n(X) \in |\gamma(n)(3H - \sum_{j=2}^{9} E_i)|$ where $\gamma(n) = 2n^2 - \sum_{d|n} d^2$

REFERENCES

1. Kodaira, K.: On Compact Analytic Surfaces, II,III, Ann. of Math. 77 (1963), 563-626; ibid 78(1963), 1-40.

2. Manin, Yu. I. : Cubic Forms., Amsterdam, North-Holland(1974).

3. Manin, Yu. I. : The Tate Height of Points on an Abelian Variety. Its Variants and Applications. A.M.S. Translations, Ser. 2,59 59 (1966), 82-110.

4. Miranda, R. : On the Stability of Rational Elleptic Surfaces with Section , Thesis (1979) M.I.T.

5. Morrison, I. and Persson, U. : Numerical Sections on Elliptic Surfaces (to appear).

6. Nagata, M. : On Rational Surfaces, I, II. Mem. Coll. Sci. Univ. Kyoto, Ser. A. Math. 32 (1960), 351-370; ibid. 33(1960-1961), 271-293.

7. Persson, U. : Double Coverings and Surfaces of General Type, 168-195 in : Algebraic Geometry, Springer L.N.M. 687 (1977).

8. Shioda, T. : On Elliptic Modular Surfaces, J. Math. Soc. Japan, 24 (1972), 20-59.

9. Tate, J. : On the Conjecture of Birch and Swinnerton-Dyer and a Geometric Analogue, Sem. Bourbaki, Exp. 306, (Feb. 1966), 1-26.

On the Kodaira Dimension of the Siegel Modular Variety

by David Mumford

Let A_g represent the quotient of Siegel's upper half-space $\mathfrak{h}_g$ of rank g by the full integral symplectic group $Sp(2g, \mathbb{Z})$: this is known as Siegel's modular variety, or as the moduli space of g-dimensional principally polarized abelian varieties (called p.p.a.v. below). A_g has been shown to be a variety of general type (i.e., Kodaira dimension = dimension) for various g's: Freitag [F1] proved this first if $24|g$; Tai [T] proved this recently for all $g \geq 9$. On the other hand, A_g is known to be unirational for $g \leq 5$: Donagi [D] for $g = 5$, Clemens [C] for $g = 4$, classical for $g \leq 3$. The purpose of this paper is to refine Tai's result, showing:

Theorem: A_g is of general type if $g \geq 7$.

Note that this leaves only the Kodaira dimension of A_6 still to be determined. We shall use results of Freitag and Tai in a crucial way, but the idea of the proof is a direct adaption of the proof [H-M] by Harris and the author that $\mathfrak{M}_g$ is of general type if $g \geq 25$, g odd. In that proof the divisor D_k of curves which are k-fold covers of $\mathbb{P}^1$, $k = \frac{g+1}{2}$, is shown to be linearly equivalent to

$$nK-(\text{ample divisor})-(\text{effective divisor}).$$

Here we prove the same thing except that the role of D_k is taken by the components of N_o, where

$$N_k = \left[\text{locus of p.p.a.v. where dim(sing. locus of } \Theta) \geq k.\right]$$

These sets N_k were introduced by Andreotti and Mayer [A-M], and studied recently by Beauville [B]. I want to thank Beauville very much for stimulating discussions which led me to this result. At the same time, I would like to raise the question which seems very interesting to me: is there an explicit polynomial in theta constants, or other modular forms constructed from theta series (with quadratic forms and pluri-harmonic coefficients) whose zeroes give N_0 with suitable multiplicities? Although important steps are taken in this direction in Andreotti-Mayer [A-M] and Beauville [B], this is not answered because the "theta nulls" $C(r,\mu,z)$ are not in general modular forms — they are theta series whose coefficients are not pluri-harmonic; esp. you cannot form a modular form out of the $\partial^2 \vartheta / \partial u_k^2$'s alone without using mixed derivatives $\partial^2 \vartheta / \partial u_k \partial u_\ell$ too. Finally, I want to mention the related results of Stillman [S] (based on earlier ideas of Freitag [F2]) which prove A_g carries holomorphic $(4g-6)$-forms for $g \geq 7_{,k}^{ij \neq 8}$. These results are directly based on the use of theta series.

§1. A partial compactification of the Siegel modular variety.

Satake's compactification A_g^* of A_g consists, set-theoretically, in the union of $(g+1)$-strata:

$$A_g^* = A_g \amalg A_{g-1} \amalg \cdots \amalg A_0.$$

The Kodaira dimension of A_g is based on pluri-canonical differentials on a desingularization $\tilde{A}_g$ of A_g^*. However, Tai has shown that a pluri-canonical differential form with "no poles above $A_g \amalg A_{g-1}$", is everywhere regular, so we do not have to study the full $\tilde{A}_g$. We will make this precise in a minute. The space we want to work with is a blow-up of $A_g \amalg A_{g-1}$ first introduced by Igusa [I] and studied by the author [M] and by Namikawa [N]. To describe this space geometrically, let us define a <u>rank 1 degeneration of a p.p.a.v.</u> as follows: it is a pair $(\overline{G}, D)$ where $\overline{G}$ is a complete g-dimensional variety and D is an ample divisor (i.e., $\overline{G}$ is to be the limit of a g-dimensional abelian variety and D the limit of its theta divisor). $\overline{G}$ is constructed as follows:

1) let B^{g-1} be a (g-1)-dimensional p.p.a.v., $\Xi \subset B$ its theta divisor

2) let G be an algebraic group which is an extension of B by $\mathbb{C}_m$:
$$0 \longrightarrow \mathbb{C}_m \longrightarrow G \longrightarrow B \longrightarrow 0.$$

3) Considering G as a $\mathbb{C}_m$-bundle over B, let $\tilde{G}$ be the associated $\mathbb{P}^1$-bundle:

$$\begin{array}{c} G \subset \tilde{G} \\ \mathbb{C}_m \searrow \pi \swarrow \mathbb{P}^1 \\ B \end{array}$$

Then $\tilde{G}-G$ equals $\tilde{G}_0 \amalg \tilde{G}_\infty$, the union of 2 sections of $\tilde{G}$ over B.

4) Then $\bar{G}$ is to be the non-normal variety obtained by glueing $\tilde{G}_0, \tilde{G}_\infty$ with a translation by a point $b \in B$.

5) Note that on $\tilde{G}$

$$\tilde{G}_0 - \tilde{G}_\infty \equiv \pi^{-1}(E), \quad E \text{ algebraically equivalent to 0 on B}$$
$$\equiv \pi^{-1}(\Xi - \Xi_{b_1}), \text{ for a unique } b_1 \in B.$$

Thus

$$\tilde{G}_0 + \pi^{-1}(\Xi_{b_1}) \equiv \tilde{G}_\infty + \pi^{-1}(\Xi).$$

Let $\tilde{L} = \mathcal{O}_{\tilde{G}}(\tilde{G}_\infty + \pi^{-1}(\Xi))$. Via the Leray spectral sequence for π, we see that $h^0(\tilde{L}) = 2$ and that $\tilde{G}_0 + \pi^{-1}(\Xi_{b_1}), \tilde{G}_\infty + \pi^{-1}(\Xi)$ span the linear system $|\tilde{L}|$. Then $|\tilde{L}|_{\tilde{G}_0} \cong \mathcal{O}_B(\Xi)$ and $\tilde{L}|_{\tilde{G}_\infty} \cong \mathcal{O}_B(\Xi_{b_1})$, so if b is chosen to be b_1 (and only then) the line bundle $\tilde{L}$ can be descended to a line bundle L on $\bar{G}$. Choose such an L and let

$$D = \text{the unique divisor in } |L|.$$

We now define

(1.1) $\bar{A}_g^{(1)} = \begin{cases} \text{coarse moduli space of p.p.a.v.}(A,\Theta) \text{ of} \\ \text{dimension g and their rank 1 degenerations} \end{cases}.$

As first shown by Igusa, this space exists, is a quasi-projective variety, and is essentially the blow-up of the open set $A_g \amalg A_{g-1}$ in A_g^* along its boundary A_{g-1}. $\bar{A}_g^{(1)}$ is the union of A_g and a divisor Δ parametrizing the rank 1 degenerations. Via the map

$$(\bar{G}, D) \longmapsto (B, \Xi)$$

the divisor Δ is seen to be fibred:

(1.2)
$$\begin{matrix} \Delta \\ \delta \downarrow & \text{fibres } B/\text{Aut}(B,\Xi) \\ A_{g-1} \end{matrix}$$

Analytically, we may consider $\bar{A}_g^{(1)}$ to represent precisely the degenerations of the abelian variety $A_{\Omega(t)}$ with period matrix $\Omega(t)$ when:

$$\left. \begin{matrix} \text{Im } \Omega_{11} \longrightarrow \infty \\ \text{and } \Omega_{ij}, \; i > 1 \; \text{ or } \; j > 1, \text{ have finite limits} \end{matrix} \right\} \text{ as } t \longrightarrow 0 \; .$$

Then $B = B_{\Omega^{(1)}}$, where $\Omega^{(1)}$ is the lower right block of the limit

$$\Omega(0) = \left(\begin{array}{c|c} i\infty & \omega \\ \hline {}^t\omega & \Omega^{(1)} \end{array} \right)$$

and b is the image of the vector $\vec{\omega} = (\Omega_{12}(0), \Omega_{13}(0), \cdots, \Omega_{1g}(0))$ in $B_{\Omega^{(1)}}$. To find D, we must translate $\Theta_{\Omega(t)} \subset A_{\Omega(t)}$ as $t \longrightarrow 0$.

Thus

$$\Theta_{\Omega(t)} = \left\{ \text{zeroes of } \vartheta(z,\Omega) = \sum_{n \in \mathbb{Z}^g} e^{\pi i {}^t n \Omega(t) n + 2\pi i {}^t n \cdot z} \right\}.$$

Translate $\Theta_{\Omega(t)}$ by $b(t)$, the image of $(\frac{\Omega_{11}(t)}{2}, 0, \cdots, 0)$:

$$T_{b(t)}(\Theta_{\Omega(t)}) = \left\{ \text{zeroes of } \sum e^{\pi i (n_1^2 - n_1)\Omega_{11}(t)} \cdot e^{\left\{ \pi i \sum_{i,j \neq 1,1} n_i n_j \Omega_{ij}(t) + 2\pi i {}^t nz \right\}} \right\}.$$

Then $e^{\pi i (n_1^2 - n_1)\Omega_n(t)} \longrightarrow 0$ unless $n_1 = 0$ or 1, hence the limit is

(1.13)
$$\left\{ \text{zeroes of } \sum_{n_2, \cdots, n_g \in \mathbb{Z}} e^{\left\{ \pi i \sum_{i,j \geq 2} n_i n_j \Omega_{ij}(0) + 2\pi i \sum_{j \leq 2} n_j z_j \right\}} \cdot \left(1 + e^{2\pi i z_1} \cdot e^{2\pi i \sum_{j \leq 2} n_j \Omega_{1j}} \right) \right\}$$

$$= \left\{ \text{zeroes of } \vartheta(z^{(1)}, \Omega^{(1)}) + e^{2\pi i z_1} \cdot \vartheta(z^{(1)} + \omega, \Omega^{(1)}) \right\}$$

where $z^{(1)} = (z_2, \cdots, z_g)$ is the analytic coordinate on $B_{\Omega^{(1)}}$. Interpreting $e^{2\pi i z_1}$ as the algebraic coordinate in the fibre $\mathbb{G}_m$ of G, and Ξ as the zeroes of $\vartheta(z^{(1)}, \Omega^{(1)})$, this is immediately seen to be D if L is suitably defined.

Next, let $\overline{A}_g^{(1),0}$ be the open set in $\overline{A}_g^{(1)}$ parametrizing those pairs (A,Θ) or $(\overline{G},D)$ whose automorphism group is the minimal one, $\{\pm 1\}$. More precisely, the only non-trivial automorphism of A (or $\overline{G}$) mapping Θ (resp. D) to itself is of the form $x \longmapsto -x+a$, some a^*. Then $\overline{A}_g^{(1),0}$ is locally isomorphic

---
*We have not normalized Θ and D to be symmetric. On the other hand, we have not fixed an origin either, so the pairs (A,Θ) and (A,Θ_c) are isomorphic by translation by c, and define the same point of $\overline{A}_g^{(1)}$.

to the universal deformation space of (A,Θ) (or $(\bar{G},D)$), hence is a smooth of dimension $g(g+1)/2$. Analytically, A_g^0 is the open subset of A_g of points which are images of $\Omega \in \mathcal{H}_g$ whose stabilizer in $Sp(2g,\mathbb{Z})$ are just $(\pm I)$. Likewise, using the analytic description of $\bar{A}_g^{(1)}$ in Ash et al [A-M-R-S], $\bar{A}_g^{(1),0}$ is the open subset of $\bar{A}_g^{(1)}$ of points which are images of points in $(\mathcal{H}_g/U_{\mathbb{Z}})_{\{\sigma_\alpha\}}$ whose stabilizer in the normalizer of the first boundary component is just $(\pm I)$. (Compare Tai [T], §). This set includes, in particular, those $\bar{G}$ constructed from a $(B,\Xi) \in A_{g-1}^0$ and a point $b \in B$ <u>not</u> of order 2. We are now in a position to state one of the main results of Tai's paper [T], in the form in which we need it:

<u>Theorem 1.4</u> (Tai). <u>If</u> $g \geq 5$, <u>then</u>

a) $\operatorname{codim} (\bar{A}_g^{(1)} - \bar{A}_g^{(1),0}) \geq 2$

<u>and</u>

b) $\Gamma(\tilde{A}_g, \mathcal{O}(nK)) = \Gamma(\bar{A}_g^{(1),0}, \mathcal{O}(nK))$, <u>if</u> $n \geq 1$.

This means that a pluri-canonical differential with no poles on $\bar{A}_g^{(1),0}$ is everywhere regular on a full desingularization $\tilde{A}_g$ of A_g^*.

The second result we need is the calculation of $\operatorname{Pic}(A_g^0)$. This follows from the theory of Matsushima, Borel, Wallach and others on the low cohomology groups of discrete subgroups of Lie groups. In particular, the results of Borel [Bo] imply that for any subgroup $\Gamma \subset Sp(2g,\mathbb{Z})$ of finite index:

$$H^*(\Gamma,\mathbb{Q}) \equiv \mathbb{Q}[C_2, C_6, C_{10}, \ldots] \text{ , in degrees } \leq g-2 \text{ .}$$

In particular:

$$H^2(A_g,\mathbb{Q}) \cong H^2(Sp(2g,\mathbb{Z}),\mathbb{Q}) \cong \mathbb{Q} \quad \text{if } g \geq 4 \text{ .}$$

An immediate corollary* is:

Theorem 1.5 (Borel et al): $\text{Pic}(A_g^0) \otimes \mathbb{Q} \cong \mathbb{Q} \cdot \lambda$, <u>if $g \geq 4$</u>,
where λ <u>is the line bundle on</u> A_g^0 <u>defined by the co-cycle</u> $\det(C\Omega+D)$.

Corollary 1.6: $\text{Pic}(\bar{A}_g^{(1),0}) \otimes \mathbb{Q} \cong \mathbb{Q}\lambda + \mathbb{Q} \cdot \delta$
where δ <u>is the divisor class of the boundary</u> Δ.

In terms of these generators, a standard result is:

Proposition 1.7. $K_{\bar{A}_g^{(1),0}} \equiv (g+1)\lambda - \delta$.

For a proof, see for instance Tai [T], §1. Another fairly standard result that we need is:

Proposition 1.8. <u>Let</u> (B,Ξ) <u>be a</u> $(g-1)$<u>-dimension p.p.a.v. whose automorphism group is</u> (± 1). <u>Consider the</u> 2-1 <u>map</u>

$$\phi: (B - B_2) \longrightarrow \bar{A}_g^{(1),0}$$

<u>defined by</u> $\phi(b) =$ <u>the pair</u> $(\bar{G}, D)$ <u>constructed from</u> (B, Ξ) <u>with</u>

---

* If $\tilde{A}_g$ is a smooth compactification of A_g^0, then use:

$$\begin{array}{ccccc}
\oplus \mathbb{Q}\delta_i & \longrightarrow & \text{Pic}\tilde{A}_g \otimes \mathbb{Q} & \xrightarrow{\text{res}} & \text{Pic}A_g^0 \otimes \mathbb{Q} \\
\| & & \updownarrow & & \downarrow \\
\oplus \mathbb{Q}\delta_i & \longrightarrow & H^2(\tilde{A}_g, \mathbb{Q}) & \twoheadrightarrow & H^2(A_g^0, \mathbb{Q})
\end{array}$$

plus $H^2(A_g^0, \mathbb{Q}) \cong H^2(\tilde{A}_g, \mathbb{Q}) \cong \mathbb{Q}$.

glueing via b.   Then

$$\phi^*(\mathcal{O}_{\overline{A}_g}(1),_0(\Delta)) \cong \mathcal{O}_B(-2\Xi).$$

Proof: Let's construct over B the family of $(\overline{G},D)$'s made up with all possible b's. To do this, let P be the Poincaré bundle over B×B, trivial on e×B, B×e. Then P* = P-(0-section) serves as the universal family of G's. Let $\overline{P} \supset P$ be the associated $\mathbb{P}^1$-fibre bundle, and

$$\mathfrak{P} = \overline{P}/(b_1,b_2,0) \sim (b_1,b_1+b_2,\infty).$$

Then the projection on the first factor:

$$p_1: \mathfrak{P} \longrightarrow B$$

is the universal family of $\overline{G}$'s. The deformation theory of such a $\overline{G}$ gives an exact sequence:

$$0 \longrightarrow H^1(\overline{G},T^0(\mathcal{O}_{\overline{G}})) \longrightarrow T^1(\overline{G}) \longrightarrow H^0(\text{Sing } \overline{G}, T^1(\mathcal{O}_{\overline{G}}))$$
$$\parallel$$
$$H^0(B,N_0 \otimes N_\infty)$$

where N_0, N_∞ are the normal bundles to the locus of double points of $\overline{G}$. For one $\overline{G}$, made up starting from a line bundle L over B, completed at ∞ and glued by translation by $b \in B$,

$$N_0 \otimes N_\infty \cong L \otimes T_b^*(L^{-1}).$$

Note that L must be algebraically equivalent to 0, hence $T_b^* L^{-1} \cong L^{-1}$, hence $N_0 \otimes N_\infty \cong \mathcal{O}_B$. Thus $H^0(B, N_0 \otimes N_\infty) \cong k$. This one-dimensional vector space represents the normal bundle to Δ in $\bar{A}_g$ at the point $(\bar{G}, D)$. Doing this now for the whole family $\mathcal{Y} \longrightarrow B$, $N_0 \otimes N_\infty$ is the line bundle on $B \times B$ given by

$$P \otimes T^*(P^{-1})$$

where $T(x,y) = (x, x+y)$.

Then the normal bundle to Δ, pulled back to this family, is

$$p_{1,*}(P \otimes T^*(P^{-1}))$$

which is the same as the restriction of $P \otimes T^* P^{-1}$ to $B \times e$, i.e., $\delta^*(P^{-1})$, where $\delta(x) = (x,x)$. Since P, along the diagonal of $B \times B$ is $\mathcal{O}(2\Xi)$, this proves the Proposition.   QED

§2. The divisor N_0 and its class in $\mathrm{Pic}(\bar{A}_g^{(1)})$.

Andreotti-Mayer [A-M] defined the important subsets N_k in A_g:

(2.1) $N_k = \{(A, \Theta) \mid \mathrm{Sing}\ \Theta \neq \emptyset$ and $\dim(\mathrm{Sing}\ \Theta) \geq k\}$.

Andreotti and Mayer prove by using the Heat equation for that $N_0 \subsetneq A_g$, but it is not easy to estimate the dimension of N_k in general. However, we are interested only in codimension 1 and we must at least check that none of the N_k, $k \geq 1$, have codimension 1 components. This follows by an elaboration of

Andreotti-Mayer's arguments using the heat equation:

Lemma 2.2. The codimension of N_1 (hence of $N_2, N_3, \cdots$) in A_g is greater than 1.

Proof: We use the heat equation

$$(2\pi i)(1+\delta_{\alpha\beta})\frac{\partial \vartheta}{\partial \Omega_{\alpha\beta}} = \frac{\partial^2 \vartheta}{\partial z_\alpha \partial z_\beta}.$$

If the lemma were false, we could find a matrix $\bar{\Omega}$, a smooth analytic hypersurface $g(\Omega) = 0$ defined in a neighborhood of $\bar{\Omega}$ and containing $\bar{\Omega}$, and a vector-valued function

$$\vec{f}(\Omega, t) \in \mathbb{C}^g$$

defined in a neighborhood of $\bar{\Omega}$ and for $|t|$ small, such that

$$\vartheta(\vec{f}(\Omega, t), \Omega) \equiv 0$$
$$\frac{\partial \vartheta}{\partial z_k}(\vec{f}(\Omega, t), \Omega) \equiv 0, \quad 1 \leq k \leq g$$

whenever $g(\Omega) = 0$.

We may assume that for each Ω, $t \longmapsto \vec{f}(\Omega, t)$ is part of an algebraic curve $C_\Omega \subset A_\Omega$. Note that the lemma is obvious if $g = 2$ and if $g \geq 3$, then the codimension of the locus of non-simple abelian varieties is greater than 1. Therefore we can also assume that the abelian variety $A_{\bar{\Omega}}$ is simple. It follows that the set of differences $x-y, x,y \in C_{\bar{\Omega}}$ generates $A_{\bar{\Omega}}$, hence the set of differences $x-y, x,y \in C_\Omega$, generates A_Ω for Ω near $\bar{\Omega}$. Therefore, for no Ω near $\bar{\Omega}$ is there a vector $\vec{a}$ such that

$$\frac{\partial}{\partial t}(\vec{a}\cdot\vec{f}) = (\vec{a}\cdot\frac{\partial \vec{f}}{\partial t}) = 0, \quad \text{all } t.$$

We prove **by induction on d** that:

(*)$_d$ If $|\alpha| = d$, then $\left(\dfrac{\partial^\alpha \vartheta}{\partial z_1^{\alpha_1}\cdots\partial z_g^{\alpha_g}}\right)(\vec{f}(\Omega,t),\Omega) \equiv 0$ whenever $g(\Omega) = 0$.

Since $(z,\overline{\Omega})$ does not vanish identically as a function of z, this is a contradiction. In fact, to prove this it will suffice to apply:

If $\eta(\Omega,z)$ satisfies the heat equation and

(**) $\left.\begin{array}{l}\eta(\vec{f}(\Omega,t),\Omega) \equiv 0 \\[6pt] \dfrac{\partial \eta}{\partial z_k}(\vec{f}(\Omega,t),\Omega) \equiv 0\end{array}\right\}$ whenever $g(\Omega) = 0$

then

$$\frac{\partial^2 \eta}{\partial x_k \partial z_\ell}(\vec{f}(\Omega,t),\Omega) \equiv 0 \qquad \text{whenever } g(\Omega) = 0$$

to all the partial derivatives of ϑ in turn. To prove (**), differentiate the first relation with respect to Ω. We find that if $\omega_{k\ell}$ satisfies $\sum \omega_{k\ell}\partial g/\partial \Omega_{k\ell}(\Omega) = 0$, then $\Omega+\varepsilon\omega$ is tangent to the hypersurface $g(\Omega) = 0$, hence

$$\begin{aligned}0 &= \eta(\vec{f}(\Omega+\varepsilon\omega,t),\Omega+\varepsilon\omega) \\ &= \varepsilon\Bigl\{\sum_{k,a,b}\frac{\partial \eta}{\partial z_k}(\vec{f}(\Omega,t),\Omega)\cdot\frac{\partial f_k}{\partial \Omega_{ab}}\cdot\omega_{ab} + \sum_{a\le b}\frac{\partial \eta}{\partial \Omega_{ab}}(\vec{f}(\Omega,t),\Omega)\cdot\omega_{ab}\Bigr\} \\ &= \frac{\varepsilon}{4\pi i}\sum_{a,b}\frac{\partial^2 \eta}{\partial z_a \partial z_b}(\vec{f}(\Omega,t),\Omega)\cdot\omega_{ab}\ .\end{aligned}$$

Therefore

$$\frac{\partial^2 \eta}{\partial z_a \partial z_b}(\vec{f}(\Omega,t),\Omega) = \emptyset(\Omega,t) \cdot (1+\delta_{ab}) \cdot \frac{\partial g}{\partial \Omega_{ab}}(\Omega)$$

with some factor $\emptyset$, for all Ω near $\bar{\Omega}$, all small t. Now differentiate the second relation in (**) with respect to t. We find:

for all a, $\quad \sum_b \frac{\partial^2 \eta}{\partial z_a \partial z_b}(\vec{f}(\Omega,t),\Omega) \cdot \frac{\partial f_b}{\partial t}(\Omega,t) \equiv 0 \quad$ whenever $g(\Omega) = 0$.

If $\emptyset(\Omega,t) \equiv 0$ when $g(\Omega) = 0$, we are done. If not, we find by substitution that

for all a, $\quad \sum_b (1+\delta_{ab})\frac{\partial g}{\partial \Omega_{ab}}(\Omega) \cdot \frac{\partial f_b}{\partial t}(\Omega,t) \equiv 0 \quad$ whenever $g(\Omega) = 0$,

i.e.,

(***) $\qquad\qquad (\vec{c}(a) \cdot \frac{\partial \vec{f}}{\partial t}) = 0$

where

$$c(a)_b = (1+\delta_{ab})\frac{\partial g}{\partial \Omega_{ab}}(\Omega).$$

For some a, $\vec{c}(a) \neq 0$ since $g(\Omega) = 0$ is a smooth hypersurface. But we saw that (***) did not occur, so thus completes the proof.

In the other direction, Beauville [B], Remark 7.7 proved*:

<u>Proposition 2.3</u> (Beauville): N_0 <u>has codimension 1 in</u> A_g.

---

*The result is stated only for g = 4; however the argument works without any modification for all g.

His proof also uses an elaboration of the techniques of Andreotti-Mayer — in this case their technique for deriving "explicit" equations for the N_k. (It might be thought that this Proposition could be proven from general principles, but I don't see how, without specific information, one could have excluded the possibilities that some component of some N_k, $k > 1$, was not in the closure of $N_0 - N_1$.)

We want now to consider the closure $\overline{N}_0$ of N_0 in $\overline{A}_g^{(1)}$, and to give multiplicities to its components. To do this, we would like to use the "universal family" of pairs $(A, \Theta), (\overline{G}, D)$ over $\overline{A}_g^{(1)}$. However, even generically these pairs still have an automorphism group of order 2, so a universal family need not exist. However, $\overline{A}_g^{(1)}$ admits a "covering" $U_\alpha \longrightarrow \overline{A}_g^{(1)}$ such that over U_α there are flat, proper families

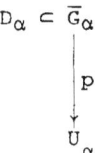

consisting of abelian varieties and rank 1 degenerations thereof, and such that p is locally the universal deformation space of its fibre $(\overline{G}_s, D_s)$. Outside $\Delta \cap U_\alpha$, $\overline{G}_\alpha$ will be smooth over U_α; over points of $\Delta \cap U_\alpha$, $\overline{G}_\alpha$ itself will still be smooth, but at the double points of the fibres, p will look like the universal local deformation space:

$$\hat{\mathcal{O}}_{\overline{G}_\alpha} \cong \mathbb{C}[[z_1, z_1', z_2, \ldots, z_{g-1}, t_2, \ldots, t_{g(g+1)/2}]]$$

$$\hat{\mathcal{O}}_{U_\alpha} \cong \mathbb{C}[[t_1, t_2, \ldots, t_{g(g+1)/2}]]$$

$$t_1 = z_1 \cdot z_1'.$$

On $\overline{G}_\alpha$, define the subsheaf of the tangent sheaf T_{vert} to be the kernel:

$$0 \longrightarrow T_{vert} \longrightarrow T_{\overline{G}_\alpha} \longrightarrow p^* T_{U_\alpha}.$$

Note that T_{vert} is locally free of rank g (at double points of the fibres, T_{vert} is spanned by $z_1 \partial/\partial z_1 - z_1' \partial/\partial z_1'$, $\partial/\partial z_2, \ldots, \partial/\partial z_g$). Using a local equation $\delta = 0$ of D_α, and interpreting sections of T as derivations, define:

$$T_{vert} \xrightarrow{\alpha} \mathcal{O}(D_\alpha)/\mathcal{O}$$

$$D \longmapsto D\delta/\delta \qquad \text{(independent of } \delta\text{)}.$$

Let

$$\text{Sing}_{vert} D_\alpha = \text{subscheme of } D_\alpha \text{ where } \alpha \text{ is zero}.$$

Thus $\text{Sing}_{vert} D_\alpha$ is defined locally by g equations and has codimension at most g. Set-theoretically:

(2.4) $p(\text{Sing}_{\text{vert}} D_\alpha)$ = set of points whose fibres are of 3 types

1) fibre is (A,Θ), A abelian variety, and Θ singular

2) fibre is $(\bar{G}, D)$ and D has a singularity in G

3) fibre is $(\bar{G}, D)$ and the divisor
$\bar{D} = D.(\bar{G}-G)$ on $\bar{G}-G$ is singular.

To see this at fibres of type $(\bar{G},D)$, at points of $\bar{G}-G$, expand δ in a power series in $z_1, z_1', z_2, \ldots, z_g$, t's: then the origin lies in $\text{Sing}_{\text{vert}} D_\alpha$ if and only if

$$\delta \in (z_1, z_1', z_i z_j \ (2 \leq i,j \leq g), t_i) ,$$

i.e., if and only if $\delta = 0$ is singular in $\mathbb{C}[[z_2, \ldots, z_g]]$. The sets $p(\text{Sing}_{\text{vert}} D_\alpha)$ patch together into a subset $\tilde{N}_0$ of $\bar{A}_g^{(1)}$. (We shall see shortly that $\tilde{N}_0 = \bar{N}_0$.)

Let us work out which $(\bar{G},D)$ arise in cases (2) and (3). Let G be the extension:

$$0 \longrightarrow \mathbb{C}_m \longrightarrow G \longrightarrow B \longrightarrow 0.$$

Then $\bar{G}-G \cong B$ and $D.(\bar{G}-G)$ is the theta divisor of B, called Ξ at the beginning of this section. Thus if $\pi: \Delta \longrightarrow A_{g-1}$ is the natural projection, case (3) contributes $\pi^{-1}(N_0(A_{g-1}))$ to $\tilde{N}_0$. As for case (2), if translation by $b \in B$ is used in glueing together $\bar{G}$, then a local equation of D at any point of G is of the form

$$f(x,z) = \delta_p(x) + z \cdot \delta_{p+b}(x+b)$$

Here δ_p (resp. δ_{P+b}) are local functions on B near p (resp. P+b) which define the non-zero section of $\mathcal{O}_B(\Xi)$ near P (resp. P+b), and z is a vertical coordinate on G in a local splitting $G \cong \mathbb{C}_m \times B$. (We may use the analytic equation (1.13) if we want.) Taking derivatives of f, we see that:

$f(x,z) = 0$ is singular at $x = P$, some $z \in \mathbb{C}^*$ $\iff$ P, P+b $\in \Xi$ and either Ξ has the same tangent plane at P,P+b, or is singular at both pts.

Looking at points $(\bar{G},D)$ not already covered in case (3), this shows that $\tilde{N}_0$ contains the set of pairs $(\bar{G},D)$ such that $\Xi \subset B$ is smooth and Ξ, Ξ_b are tangent somewhere. If Ξ is smooth, let

$$\gamma_B : \Xi \longrightarrow \mathbb{P}^{g-2}$$

be the "Gauss map" associating to each $P \in \Xi$, the tangent plane $T_{P,\Xi}$, as a point of $\mathbb{P}(T^*_{0,B})$. Then Ξ and Ξ_b are tangent at P if and only if $\gamma_B(P) = \gamma_B(P+b)$. Thus for any principally polarized abelian variety (B,Ξ) with smooth Ξ we may define

$$c(B,\Xi) = \text{locus of points x-y, where } \gamma_B(x) = \gamma_B(y)$$
$$= \text{locus of points x such that } \Xi, \Xi_x \text{ are tangent somewhere.}$$

Then in the description (2.4):

$$\tilde{N}_0 \cap \Delta \cong \left[\bigcup_{(\bar{G},D)} c(B,\Xi) \right] \cup \left[\delta^{-1}(N_0 \text{ for } \mathcal{A}_{g-1}) \right].$$

Next, the method of Andreotti-Mayer-Beauville extends to rank 1 degenerations, to prove that $\tilde{N}_0$ is a divisor. For abelian varieties A, their technique is to map A to $\mathbb{P}^{2^g-1}$ by $|2\Theta|$, i.e., explicitly by the theta functions

$$\vartheta_\mu(z,\Omega) = \sum_{n\in\mathbb{Z}^g} e^{2\pi i\,^t(n+\mu)\Omega(n+\mu)+4\pi i\,^t(n+\mu)\cdot z}.$$

Call this $\phi: A \longrightarrow \mathbb{P}^{2^g-1}$.

They define a linear subspace $L_\Omega \subset \mathbb{P}^{2^g-1}$ of codimension g+1 by

(2.5)
$$\sum \vartheta_\mu(0,\Omega)\cdot X_\mu = 0$$
$$\frac{\partial^2 \vartheta_\mu}{\partial z_i^2}(0,\Omega)\cdot X_\mu = 0, \quad 1 \le i \le g$$

and prove

(2.6) $$\phi^{-1}(L_\Omega) = \text{Sing } \Theta,$$

hence

$$(A,\Theta) \in N_0 \iff L_\Omega \cap \phi(A) \neq \emptyset$$
$$\iff \text{Chow form of } \phi(A) \text{ varieties at Plücker Coord of } L_\Omega$$

Now if Im $\Omega_{11} \longrightarrow \infty$, the limit of $\phi(A)$ is $\phi(\bar{G})$, where ϕ is defined by the 2^g "theta functions"

$$\left.\begin{array}{c}\vartheta_\mu(z^{(1)},\Omega^{(1)}) + u^2\vartheta_\mu(z^{(1)}+\omega,\Omega^{(1)}) \\ u\vartheta_\mu(z^{(1)}+\frac{1}{2}\omega,\Omega^{(1)})\end{array}\right\} \quad \mu \in \tfrac{1}{2}\mathbb{Z}^{g-1}/\mathbb{Z}^{g-1}$$

(where, as above, G is a $\mathbb{C}_m$-bundle over B, $\Omega^{(1)}$ = period matrix of B, $\bar{G}$ is glued via ω, $z^{(1)}$ is the coordinate on B, u the coordinate on $\mathbb{C}_m$). The basic theta identity on which the proof of (2.6) is based becomes

(2.7)
$$[\vartheta(x+y)+uw\,\vartheta(x+y+\omega)] \cdot [\vartheta(x-y)+\frac{u}{w}\vartheta(x-y+\omega)] =$$
$$\sum_{\mu \in \frac{1}{2}\mathbb{Z}^{g-1}/\mathbb{Z}^{g-1}} [\vartheta_\mu(x)+u^2\vartheta_\mu(x+b)] \cdot \vartheta_\mu(y)+uw\vartheta_\mu(x+\frac{\omega}{2}) \cdot [\vartheta_\mu(y+\frac{\omega}{2})+\frac{1}{w^2}\vartheta_\mu(y-\frac{\omega}{2})]$$

The limit of L_Ω is the linear space

(2.8)
$$\sum \vartheta_\mu(0,\Omega^{(1)}) \cdot X_\mu + 2\sum \vartheta_\mu(\frac{\omega}{2},\Omega^{(1)}) \cdot Y_\mu = 0$$
$$\sum \frac{\partial^2 \vartheta_\mu}{\partial z_i^2}(0,\Omega^{(1)}) \cdot X_\mu + 2\sum \frac{\partial^2 \vartheta_\mu}{\partial z_i^2}(\frac{\omega}{2},\Omega^{(1)}) \cdot Y_\mu = 0$$
$$\sum \vartheta_\mu(\frac{\omega}{2},\Omega^{(1)}) \cdot Y_\mu = 0 \ .$$

(The last equation comes from the $2^{\underline{nd}}$ derivative of (2.7) with respect to $w\,\partial/\partial w$; these equations are not the exact analogs of the (2.5) because, in passing to the limit, we have renormalized the origin.) Then it follows from (2.7) exactly as in Andreotti-Mayer-Beauville that

$$\emptyset^{-1}(L_\Omega) = \begin{pmatrix} \text{singularities of D in G plus singularities} \\ \text{of } \bar{D} \cdot (\bar{G}-G) \text{ in } \bar{G}-G \end{pmatrix}$$

hence

$$\begin{pmatrix} \text{Chow form of } \emptyset(\bar{G}) \text{ is} \\ \text{zero at } L_\Omega \end{pmatrix} \iff (\bar{G},D) \in \tilde{N}_0 \ .$$

This proves that $\tilde{N}_0$ is a divisor.

On the other hand, it is clear that for all B, $c(B,\Xi) \subsetneq B$ and for generic B, Ξ is smooth: hence $\tilde{N}_0 \cap \Delta \subsetneq \Delta$. Thus $\tilde{N}_0$ must be the closure $\overline{N}_0$ of N_0. Incidentally, this proves that $c(B,\Xi)$ is always a divisor in B. At the same time, we can now give multiplicities to the components of $\overline{N}_0$. I think the Andreotti-Mayer-Beauville equation gives artificially large multiplicities, and want, instead, to assign multiplicities via the local description of $\overline{N}_0$ in U_α as $p(\text{Sing}_{\text{vert}} D_\alpha)$. Let $\overline{N}_0'$ be the maximal open set of points of $\overline{N}_0$ such that for all α

$$p: \text{Sing}_{\text{vert}} D_\alpha \longrightarrow (\overline{N}_0 \cap U_\alpha)$$

is <u>finite</u> over $\overline{N}_0'$. Because N_1 has codimension at least 2, $\overline{N}_0'$ is dense in $\overline{N}_0$. Then over $\overline{N}_0'$

$$\dim(\text{Sing}_{\text{vert}} D_\alpha) = \dim N_0$$

hence

$$\text{codim}(\text{Sing}_{\text{vert}} D_\alpha) = g+1 = \text{\# of equations defining Sing}_{\text{vert}} D_\alpha$$

hence $\text{Sing}_{\text{vert}} D_\alpha$ is Cohen-Macauley. Therefore, over $\overline{N}_0'$, $p_*(\mathcal{O}_{\text{Sing}_{\text{vert}} D_\alpha})$ has a locally free resolution:

$$0 \longrightarrow \mathcal{E}_1 \xrightarrow{f} \mathcal{E}_0 \longrightarrow p_*(\mathcal{O}_{\text{Sing}_{\text{vert}} D_\alpha}) \longrightarrow 0$$

and det f gives a local equation for $\overline{N}_0' \cap D_\alpha$, and this assigns multiplicities to $\overline{N}_0$. Next, we want to break $\overline{N}_0$ up into 2 pieces: the first piece is

(2.9) $\quad \vartheta_{null} = \left\{ (A,\Theta) \;\middle|\; \begin{array}{l} \text{if } \Theta \text{ is normalized to be symmetric about } e, \\ \text{then } \Theta \text{ has a singularity at a point of order } 2 \end{array} \right\}$.

It is easy to see that:

$$\vartheta_{null} \cap \Delta = \left[\bigcup_{\text{all } \overline{G},D} 2_B(\Xi) \right] \cup \left[\delta^{-1}(\text{ null for } A_{g-1}) \right]$$

where we note that (assuming Ξ is symmetric too) $c(B,\Xi)$ contains the "obvious" component:

$$2_B(\Xi) = \{2x \mid x \in \Xi\}$$

because $\gamma(-x) = \gamma(x)$, all $x \in \Xi$.

If a symmetric Θ has a singularities at a point x not of order 2, it is also singular at $-x$. Thus $\overline{N}_0$ breaks up:

$$\overline{N}_0 = \vartheta_{null} + 2 \cdot \overline{N}_0^*$$

where all multiplicities in the $2^{\underline{nd}}$ piece are divisible by 2. We can now state the main result of this paper:

<u>Theorem (2.10)</u>: <u>The divisor classes of</u> $\overline{N}_0, \vartheta_{null}, \overline{N}_0^*$ <u>are given by</u>:

$$[\overline{N}_0] = (\tfrac{(g+1)!}{2} + g!)\lambda - \tfrac{(g+1)!}{12}\delta$$
$$[\vartheta_{null}] = 2^{g-2}(2^g+1)\lambda - 2^{2g-5}\cdot\delta$$
$$[\overline{N}_0^*] = \left[\tfrac{(g+1)!}{4} + \tfrac{g!}{2} - 2^{g-3}(2^g+1)\right]\lambda - \left[\tfrac{(g+1)!}{24} - 2^{2g-6}\right]\delta \quad .$$

Here is a table for low degrees:

g	$[\overline{N}_0]$	$[\vartheta_{null}]$	$[N_0^*]$	slope
2	$5\lambda - \frac{1}{2}\delta$	$5\lambda - \frac{1}{2}\delta$	0	—
3	$18\lambda - 2\delta$	$18\lambda - 2\delta$	0	—
4	$84\lambda - 10\delta$	$68\lambda - 8\delta$	$8\lambda - \delta$	8
5	$480\lambda - 60\delta$	$264\lambda - 32\delta$	$108\lambda - 14\delta$	7.71
6	$3,240\lambda - 420\delta$	$1,040\lambda - 128\delta$	$1,100\lambda - 146\delta$	7.53
7	$25,200\lambda - 3,360\delta$	$4,128\lambda - 512\delta$	$10,536\lambda - 1,424\delta$	7.40

Note that the figures imply $\overline{N}_0^* = \emptyset$ for $g = 2,3$ as is well known. We also see that the divisor class of $\overline{N}_0^*$ is the same as that of the Jacobian locus for $g = 4$, confirming Beauville's results. The last column, "slope", refers to the ratio of the coefficient of λ to the coefficient of δ. As soon as this drops below the same ratio for K, A_g is of general type:

<u>Corollary (2.11)</u>. $\frac{(g+1)!}{12} K_{\overline{A}_g}(1) = [\overline{N}_0] + g!(g^2 - 4g - 17)\lambda$.

Proof: Combine 1.7 and 2.10.

<u>Corollary (2.12)</u>. <u>If</u> $g \geq 7$, A_g <u>is of general type</u>.

Proof: Combine 1.4 and 2.11.

§3. **Proof of the Theorem.**

Now how are we going to prove the Theorem? The formula for $[\vartheta_{null}]$ is immediate, because we know the modular form that cuts out this divisor, viz.:

$$f(\Omega) = \prod_{\substack{a,b \in \frac{1}{2}\mathbb{Z}^g/\mathbb{Z} \\ {}^t(2a)\cdot(2b) \text{ even}}} \vartheta[{}^a_b](0,\Omega)$$

where

$$\vartheta[{}^a_b](0,\Omega) = \sum_{n \in \mathbb{Z}^g} e^{\pi i {}^t(n+a)\Omega(n+a) + 2\pi i {}^t(n+a)\cdot b}$$

Each is a modular form of weight $1/2$ and there are $2^{g-1}(2^g+1)$ "even" pairs a,b so f has weight $2^{g-2} \cdot (2^g+1)$, and this is the coefficient of λ. On the other hand, if $\text{Im } \Omega_{11} \longrightarrow \infty$, we see that if $a_1 = 0$, $\lim \vartheta[{}^a_b] = 1$, while if $a_1 = \frac{1}{2}$, $\vartheta[{}^a_b]$ is divisible by

$$e^{\pi i \Omega_{11}/4}$$

hence it goes to zero. The equation of Δ is $e^{2\pi i \Omega_{11}} = 0$, and there are 2^{2g-2} "even" pairs a,b with $a_1 = \frac{1}{2}$ (take any $a_2,b_2,\ldots,a_g,b_g$, set $a_1 = \frac{1}{2}$ and make b_1 zero or one-half to force a,b to be even). Thus f goes to zero like

$$(e^{2\pi i \Omega_{11}})^{(2^{2g-5})}$$

when $\text{Im } \Omega_{11} \longrightarrow \infty$, hence the coefficient of δ.

It remains to prove the formula for $[\overline{N}_0]$. The value of the coefficient of λ follows from:

Proposition 3.1: Let

$$\begin{array}{c} \mathcal{X} \supset \mathcal{D} \\ \varepsilon \uparrow \downarrow p \\ C \end{array}$$

be a family of p.p.a.v. over a complete curve C such that every theta divisor $\mathcal{D}_t$ has only a finite number of singularities and the generic $\mathcal{D}_\eta$ is smooth. Let this family define the morphism

$$\varphi : C \longrightarrow A_g.$$

Then

$$\varphi^* N_0 \equiv \left(\frac{(g+1)!}{2} + g!\right)\varphi^* \lambda + \text{torsion}.$$

(Note that such a family exists because codim $N_1 \geq 2$ and because in Satake's compactification, the whole boundary has codim ≥ 2). The coefficient of δ, on the other hand follows from:

Proposition 3.2: Let (A,Θ) be a p.p.a.v. Then the divisor class of $c(B,\Theta)$ is given by:

$$c(B,\Theta) \equiv \frac{(g+2)!}{6} \cdot \Theta$$

together with Proposition 1.8.

To prove 3.1, we use the exact sequence

$$T_{X/C} \longrightarrow \mathcal{O}_X(\mathfrak{D})/\mathcal{O}_X \longrightarrow \mathcal{O}_{Sing_{vert}\mathfrak{D}} \otimes \mathcal{O}_X(\mathfrak{D}) \longrightarrow 0$$

used to define multiplicities for N_0. It follows that $Sing_{vert}\mathfrak{D}$ is the scheme of zeroes of a section of

$$\Omega^1_{X/C}(\mathfrak{D}) \otimes_{\mathcal{O}_X} \mathcal{O}_\mathfrak{D}$$

hence

$$\varphi^* N_0 = p_*(c_g(\Omega^1_{X/C}(\mathfrak{D})) \cdot \mathfrak{D}) \; .$$

But if $\mathcal{E} = p_*(\Omega^1_{X/C})$, then the bundle $\Omega^1_{X/C}$, being trivial on each fibre of X over C, is isomorphic to $p^*\mathcal{E}$. Moreover, by definition of λ,

$$\varphi^*\lambda = c_1(\mathcal{E}) \; .$$

Thus

$$\begin{aligned}\varphi^* N_0 &= p_*(c_g(p^*\mathcal{E} \otimes \mathcal{O}_X(\mathfrak{D})) \cdot \mathfrak{D}) \\ &= p_*((\mathfrak{D}^g + \mathfrak{D}^{g-1} \cdot c_1(p^*\mathcal{E})) \cdot \mathfrak{D}) \\ &= p_*(\mathfrak{D}^{g+1}) + p_*(\mathfrak{D}^g) \cdot c_1(\mathcal{E}) \; .\end{aligned}$$

Now on each fibre $\mathfrak{D}$ is Θ and $(\Theta^g) = g!$, so the second term is $g!\varphi^*(\lambda)$. To compute the first, we apply the Grothendieck-Riemann-Roch theorem to $\mathcal{O}_X(\mathfrak{D})$. Note that

$$p_*(\mathcal{O}_X(\mathfrak{D})) \cong \mathcal{O}_C$$
$$R^i p_*(\mathcal{O}_*(\mathfrak{D})) = (0), \; i \geq 1.$$

Thus
$$1 = \text{ch}(p_!(\mathcal{O}_X(\mathfrak{D})))$$
$$= p_*(\text{ch}\,\mathcal{O}_X(\mathfrak{D}) \cdot \text{Td}(\Omega^1_{X/C}))$$
$$= p_*(e^{\mathfrak{D}} \cdot p^*(1 - \frac{c_1(\mathbf{E})}{2})), \quad \text{mod torsion.}$$

In codimension 1 on C, this says
$$0 = p_*(\frac{\mathfrak{D}^{g+1}}{(g+1)!}) - \frac{c_1(\mathbf{E})}{2} \cdot p_*(\frac{\mathfrak{D}^g}{g!})$$
or
$$p_*(\mathfrak{D}^{g+1}) = \frac{(g+1)!}{2} c_1(\mathbf{E}) \quad \text{mod torsion.}$$

This proves 3.1.

To prove 3.2, it suffices to establish the numerical equivalence of the 2 divisors. Namely, this will prove Theorem 2.10, and then Theorem 2.10 will imply Prop. 3.2 as an equality of divisor classes. Let $C \subset A$ be any curve. We shall calculate $(C.c(B,\Theta))$. Consider the map

$$C \times \Theta \xrightarrow{m} A$$

$$m(x,y) = x + y .$$

Then $m^{-1}(\Theta)$ is the locus of pairs (x,y) where $x+y \in \Theta$, i.e., $x = y'-y$, where $x \in C, y, y' \in \Theta$. The differential of m gives us a map

$$dm: p_2^* T_\Theta \otimes \mathcal{O}_{m^{-1}\Theta} \longrightarrow T_A \otimes \mathcal{O}_\Theta \longrightarrow N_{\Theta,A}$$

whose zeroes are exactly the points (x,y) such that not only is $x = y'-y$, $y, y' \in \Theta$, but also $T_{y,\Theta} = T_{y',\Theta}$, i.e., $x \in \mathfrak{D}$. Now the above dm can be thought of as a section of

$$p_2^* \Omega_\Theta^1 \otimes m^*(N_{\Theta,A}) \otimes \mathcal{O}_{m^{-1}(\Theta)}$$

hence

$$(C \cdot \mathfrak{D}) = c_{g-1}\bigl(p_2^* \Omega_\Theta^1 \otimes m^*(\mathcal{O}(\Theta)) \otimes \mathcal{O}_{m^{-1}\Theta}\bigr).$$

Let $\theta_1 = \text{pt.} \times \Theta$, $\theta_2 = m^{-1}\Theta$ be these divisor classes (mod numerical equivalence) on $C \times \Theta$. Then

$$(C \cdot \mathfrak{D}) = c_{g-1}\bigl(p_2^* \Omega_\Theta^1 \otimes \mathcal{O}(\theta_2)\bigr) \cdot \theta_2.$$

Using

$$0 \longrightarrow \mathcal{O}(-\Theta)/\mathcal{O}(-2\Theta) \longrightarrow \Omega_A^1\bigr|_\Theta \longrightarrow \Omega_\Theta^1 \longrightarrow 0,$$

we see that

$$c(\Omega_\Theta^1) = (1-\Theta)^{-1}\bigr|_\Theta = (1+\Theta+\Theta^2+\cdots)\bigr|_\Theta.$$

Thus

$$(C \cdot \mathfrak{D}) = \theta_1^{g-1} \cdot \theta_2 + \theta_1^{g-2} \cdot \theta_2^2 + \cdots + \theta_2^g.$$

But now

$$(\theta_1^k \cdot \theta_2^{g-k})_{C \times \Theta} = (\mathbf{m}(\theta_1^{k+1}) \cdot \Theta^{g-k})_A$$
$$= ((C \dotplus \Theta^{k+1}) \cdot \Theta^{g-k})_A$$

if $\dotplus$ is Pontryagin product. By symmetry of Θ, this is

$$= (C \cdot (\Theta^{k+1} \dotplus \Theta^{g-k}))_A$$
$$= (C \cdot (k+1)(g-k)(g-1)! \Theta)_A.$$

Thus

$$(C \cdot \mathfrak{D}) = (C \cdot \Theta)(g-1)! \sum_{k=0}^{g-1} (k+1)(g-h)$$
$$= \frac{(g+2)!}{6}(C \cdot \Theta). \qquad \text{QED}$$

References

[A-M] Andreotti, A., and Mayer, A., On the period relations for abelian integrals on algebraic curves, Ann. Scuola Norm. Pisa, 21 (1971).

[A-M-R-T] Ash, A., et al, Smooth compactification of locally symmetirc varieties, Math-Sci Press, 53 Jordan Rd., Brookline, MA, 1975.

[B] Beauville, A., Prym varieties and the Schottky problem, Inv. Math., 41 (1977), p. 149.

[Bo] Borel, A., Stable real cohomology of arithmetic groups II, in Manifolds and Lie groups, Birkhauser-Boston, 1981.

[C] Clemens, H., Double solids, to appear.

[D] Donagi, R., The unirationality of $\mathcal{A}_5$, to appear.

[F1] Freitag, E., Die Kodairadimension von Körpern automorpher Funktionen, J. reine angew. Math., 296 (1977), p. 162.

[F2] Freitag, E., Der Körper der Siegelschen Modulfunktionen, Abh. Math. Sem. Hamburge, 47 (1978).

[H-M] Harris, J. and Mumford, D., On the Kodaira dimension of the moduli space of curves, to appear in Inv. Math.

[I] Igusa, J.-I., A desingularization problem in the theory of Siegel modular functions, Math. Annalen, 168 (1967), p. 228.

[M] Mumford, D., Analytic construction of degenerating abelian varieties, Comp. Math., 24 (1972), p. 239.

[N] Namikawa, A new compactification of the Siegel space and degeneration of abelian varieties, Math. Ann., 221 (1976).

[S] Stillman, M., Ph.D. Thesis, Harvard University, 1983.

[T] Tai, Y.-S., On the Kodaira dimensions of the moduli space of abelian varieties, to appear Inv. Math.

GENERALIZED HILBERT FUNCTIONS OF COHEN-MACAULAY VARIETIES

Ferruccio Orecchia

Istituto di Matematica
Università di Napoli

NAPOLI-ITALY

INTRODUCTION. Let $R = k[X_0,\ldots,X_r]/I$ be a graded algebra over a field k of dimension d and maximal homogeneous ideal M. Let $H_R^0(n) = \dim_k(M^n/M^{n+1})$ and $P_R^0(n)$, $n \in \mathbb{Z}$, be respectively the Hilbert function and Hilbert polynomial of the ring R. In many papers (see for example [G.M.], [G.O.] and [O_3]) the relations between the structure of the Hilbert function of R and the degree of the forms generating the ideal I have been studied. If R is Cohen-Macaulay these relations are much more strict; for example Schenzel (see [Sc]) proved that if $m = \text{Max}\{n \mid H_R^0(n) \neq P_R^0(n)\}+1$ is the index of regularity of R and t is the least degree of a form of I then $m + d \geq t$. Unfortunately few sufficient conditions for R to be Cohen-Macaulay are known. In this paper first we improve the previous result of Schenzel by showing that the inequality $m + d \geq t$ holds also if t is the highest degree of a form in a minimal generating set of I (see thm. 1.5). Then we give a sufficient condition for R to be Cohen-Macaulay when R is the associated graded ring of a local Cohen-Macaulay ring (in particular the tangent cone of a variety at a Cohen-Macaulay singularity). For this purpouse we have to introduce the notion of generalized Hilbert functions $H_R^i(n)$, $i \in \mathbb{Z}$, which for i positive are successive sums of $H_R^0(n)$ and for i negative are successive differences of $H_R^0(n)$. We prove that if the function $H_R^{1-d}(n)$ is maximal then R is Cohen-Macaulay (thm. 3.1). The

---

This work was partially supported by C.N.R.

notion of maximal Hilbert function, which we give for any $H_R^i(n)$, unifies and extends the one given in $[O_3]$ for the one-dimensional case and that of extremal Cohen-Macaulay ring as contained in $[Sc]$. Furthermore large classes of varieties have $H_R^{1-d}(n)$ maximal; for example: 1) a set of points in generic position in $\mathbf{P}^r$, 2) curves locally requiring large numbers of generators (in particular the famous Macaulay curves $[M]$), 3) surfaces with rational singularities, 4) determinantal varieties. Hence as a consequence of our results we get that the varieties of type 2) and 3) have Cohen-Macaulay tangent cones. In particular in the one-dimensional case we get that if $H_R^o(n)$ is maximal then R is Cohen-Macaulay. This extends thm. 3.3 of $[O_1]$.

1. <u>GENERALIZED HILBERT FUNCTIONS</u>. From now on we assume that $R = k[X_o,\ldots,X_r]/I$ <u>is a graded algebra over a field</u> k <u>of maximal homogeneous ideal</u> $M = (x_o,\ldots,x_r)$. <u>Let</u> dim R = d <u>and</u> emdim R = $\dim_k(M/M^2)$ = r+1. We define <u>the generalized Hilbert functions</u> of R relative to M as follows:

let $n \in Z$: if $n \geq 0$, $H_R^o(n) = \dim_k(M^n/M^{n+1})$; if $n < 0$, $H_R^o(n) = 0$.

if $i \in Z$ the functions $H_R^i(n)$ are given by the relations:

$H_R^i(n) = H_R^{i+1}(n) - H_R^{i+1}(n-1)$ (then $H_R^i(n) = \sum_{j \leq n} H_R^{i-1}(j)$).

It is well known that $H_R^o(n)$ for $n \gg 0$ is a numerical polynomial $P_R^o(n)$ of degree d-1 called the Hilbert polynomial (of R relative to M). Then, if d > 0:

$P_R^o(n) = a_o\binom{n}{d-1} + a_1\binom{n}{d-2} + \ldots + a_{d-1}$ where $a_i \in Z$, $i = 0,\ldots, d-1$

and $a_o = e(R) \geq 1$ is the <u>multiplicity</u> (or degree) of R in M. If d = 0, then $P_R^o(n) = 0$.

Hence, for $n \gg 0$, we have:

$$H_R^i(n) = P_R^i(n) = a_0 \binom{n+i}{d+i-1} + a_1 \binom{n+i}{d+i-2} + \ldots + a_{d+i-1} , \quad i > -d \text{ and } H_R^i(n) = P_R^i(n) = 0$$

for $i \leq -d$.

DEFINITION 1.1. The <u>index of regularity</u> of (the function) $H_R^i(n)$ is the integer $h_R^i = \text{Max}\{n \in \mathbb{Z} \mid H_R^i(n) \neq P_R^i(n)\} + 1$. The integer h_R^o will be simply called <u>the index of regularity of</u> R.

LEMMA 1.2. <u>The following equality holds</u> $h_R^i = h_R^{i+1} + 1$

PROOF. Follows immediately from the equalities:

$$H_R^{i+1}(n) - H_R^{i+1}(n-1) = H_R^i(n) , \quad P_R^{i+1}(n) - P_R^{i+1}(n-1) = P_R^i(n) , \quad n \in \mathbb{Z} .$$

REMARK. The notion of index of regularity of the ring R (i.e. of the function $H_R^o(n)$) has been introduced by Schenzel in [Sc.].

EXAMPLES. a) If $d = \dim R = 0$, the index of regularity of R is the least integer n for which $M^n = 0$, i.e. the nilpotency degree of R (see [S], pg. 79);
b) if $d = 1$ the **index of regularity** of R is $\text{Min}\{m \in \mathbb{N} \mid H_R^o(n) = e(R)\}$, for $n \geq m$ i.e. the index of stability as defined in [L]; c) the index of regularity can be negative as the following example shows. Let $R = k[X,Y,Z,T]/(X^2+Y^3+Z^4+T^5)$. Then $H_R^o(n) = 2\binom{n}{2} + 3n + 1$, for $n \geq 1$.

Now the following key result holds:

THEOREM 1.3. <u>Let</u> x <u>be a homogeneous element of</u> M. <u>For every</u> $n \geq 0$:

$H^1_{R/xR}(n) - H^0_R(n) = \dim_k(M^{n+1}:xR/M^n)$. Then x <u>is a non-zero divisor of degree 1 if</u> <u>and only if</u> $H^0_R(n) = H^1_{R/xR}(n)$ <u>for any</u> $n \geq 0$. More generally,<u>if</u> $x_0,\ldots,x_s$ <u>are homo-</u> <u>geneous elements of</u> M <u>and</u> $A = R/(x_0,\ldots,x_s)$, $H^0_R(n) = H^s_A(n)$ <u>if and only if</u> $x_0,\ldots,x_s$ <u>form an</u> R-<u>sequence of elements of degree</u> 1 .

PROOF. We have $H^1_R(n) - H^1_{R/xR}(n) = \dim_k(R/M^{n+1}) - \dim_k(R/(M^{n+1}+Rx))=$

$= \dim_k(M^{n+1} + Rx/M^{n+1}) = \dim_k(Rx/(M^{n+1} \cap Rx))$. Now the homomorphism

$R \to xR/(M^{n+1} \cap Rx) \to 0$ induced by multiplication by x has kernel $(M^{n+1} : xR)$.

Hence $\dim_k(Rx/(M^{n+1} \cap Rx)) = \dim_k(R/(M^{n+1} : xR))$. Then : $H^1_{R/xR}(n) - H^0_R(n) =$

$= H^1_{R/xR}(n) - H^1_R(n) + H^1_R(n-1) = -\dim_k (R/(M^{n+1}: xR)) + \dim_k(R/M^n) = \dim_k((M^{n+1}:xR)/M^n)$

Now we prove the second part of the statement. If $A_i = R/(x_0,\ldots,x_{i-1})$ ($A_0 = R$),

$i = 0,\ldots,s$, by the first part of the theorem the image $\bar{x}_i$ of x_i in A_i is a

non-zero divisor if and only if $H^1_{A_i/(\bar{x}_i)}(n) = H^0_{A_i}(n)$, for any $n \geq 0$. But

$H^0_R(n) = H^s_A(n)$ is equivalent to $H^i_{A_i}(n) = H^{i+1}_{A_{i+1}}(n)$, for any $i = 0,\ldots,s-1$.

REMARK . Actually , in the previous theorem , x is a non-zero divisor if and only if $H^0_R(n) = H^1_{R/xR}(n)$ for n large enough , but we will not use this weaker statement .

(1.4) <u>If</u> k <u>is infinite and</u> depth $R = s \leq d$ <u>there is always an</u> R-<u>sequence</u> $x_0,\ldots,x_{s-1}$<u>of elements of degree</u> 1 . <u>In this case we can assume that</u> $x_0,\ldots,x_{s-1}$ <u>are the classes of</u> $X_0,\ldots,X_{s-1}$ <u>in</u> $R = k[X_0,\ldots,X_r] / I$. <u>The ring</u> $B = R/(x_0,\ldots,x_{s-1})$, $s \leq d$, <u>has</u> : dim $B = d - s$, emdim $B =$ emdim $R - s = r + 1 - s$, $e(B) = e(R)$ if $s < d$, $e(B) = \dim_k (B)$ if $s = d$ (see [S]).

Now <u>let</u> t <u>be the highest degree of a form belonging to a minimal system of</u>

generators of I :

THEOREM 1.5. *If* R *is a Cohen-Macaulay ring and* $m = h_R^0$ *is the index of regularity of* R *then* $m + d \geq t$.

PROOF. By the standard trick of making the flat change of rings $R \to R(u) = R[u]_{MR[u]}$ (u indeterminate), if necessary, we can assume k infinite. Now depth $R = d$ (because R is Cohen-Macaulay), hence there exists an R-sequence $x_0, \ldots, x_{d-1}$ of elements of degree 1. Let $A = R/(x_0, \ldots, x_{d-1})$. We have $H_R^0(n) = H_A^d(n)$ (see thm 1.3). Then, if m is the index of regularity of $H_R^0(n) = H_A^d(n)$, $m + d$ is the index of regularity (i.e. the nilpotency degree) of $H_A^0(n)$ (see lemma 1.2). Now we claim that, for $f_1, \ldots, f_m \in I$, $(f_1, \ldots, f_m) = I$ if and only if $(\bar{f}_1, \ldots, \bar{f}_m) = \bar{I}$ (the "-" denote the classes modulo $X_0, \ldots, X_s$, $s < d$). It is enough to prove the claim for $s = 0$. Let n be the least degree of a form $f \in I$ such that $f \notin (f_1, \ldots, f_m)$. Then $\bar{f} \in \bar{I} = (\bar{f}_1, \ldots, \bar{f}_m)$. Hence $f - \sum_{i=1}^{m} h_i f_i = h X_0$, where h has degree $n-1$. Then $\bar{h} X_0 = 0$ in R. But x_0 is a non zero divisor of R. Hence $\bar{h} = \bar{0}$ i.e. $h \in I$. Contradiction. Thus we can prove the theorem for $d = 0$. But $H_R^0(m) = 0$ clearly gives $(X_0, \ldots, X_r)^m \subset I$ and then I is generated by forms of degree $\leq m$.

REMARK. The previous theorem improves the following result of Schenzel (see [S], thm C and cor. 1) : if ℓ is the least degree of a form of I then $m + d \geq \ell$.

OPEN QUESTION. Characterize geometrically the varieties satysfying the equality $t = m + d$. Ciro Ciliberto has informed us that : Cohen-Macaulay curves of $\mathbb{P}^3$ whose homogeneous coordinate ring has $t = m + 2$ are linked to plane curves.

2. **MAXIMAL HILBERT FUNCTIONS**. If $r + 1 = \text{emdim } R$, on has $H_R^0(n) \leq \binom{n+r}{r}$, $n \geq 0$, and so for $i \geq 0$ $H_R^i(n) \leq \binom{n+r+i}{r+i}$. Further, for $i \in Z$, $-r \leq -d \leq i < 0$,

$H_R^i(1) = r + i + 1 = \binom{r+i+1}{r+i}$ and it is easily checked that if $H_R^i(n) = \binom{n+r+i}{r+i}$,

for any $0 \leq n < m$, then $H_R^i(m) \leq \binom{m+r+i}{r+i}$. Hence we can give the following:

DEFINITION 2.1. The function $H_R^i(n)$ ($i \geq -d$) is maximal if $H_R^i(n) = \binom{n+r+i}{r+i}$ for any n, $1 < n < h_R^i$ = index of regularity of $H_R^i(n)$. If $H_R^0(n)$ is maximal we will say that R has maximal Hilbert function.

From now on we denote with m the index of regularity of the graded algebra $R = k[X_0, \ldots, X_r]/I$.

REMARKS. 1) For $i < -d$ the notion of maximal Hilbert function has clearly no sense. 2) If m is the index of regularity of R the function $H_R^{m-2}(n)$ is always maximal because its index of regularity is $m-(m-2) = 2$. 3) $H_R^i(n)$ maximal implies $H_R^{i+1}(n)$ maximal. 4) If R is the homogeneous coordinate ring of points in $\mathbb{P}^r$ then R has maximal Hilbert function if and only if these points are in generic position (see $[O_2]$, thm 3.7 and def. 3.8).

There is an easy way of characterizing the maximality of the function $H_R^i(n)$.

LEMMA 2.2. Let ℓ be the least degree of a form of I. Then $H_R^i(n)$ is maximal if and only if $\ell \geq m-i$.

PROOF. By definition $H_R^i(n)$ is maximal if $H_R^i(n) = \binom{n+r+i}{r+i}$ (for any n, $1 < n < h_R^i$) but this is equivalent to say that $H_R^0(n) = \binom{n+r}{r}$ (lemma 1.2) i.e. there are no forms of degree n in I, for $n < h_R^i = m-i$.

Then if R is Cohen-Macaulay and q is the degree of a form of a minimal set generating I, on has :

(2.3) $H_R^i(n)$ <u>is maximal if and only if</u> $m-i \leq q \leq m+d$

In particular :

PROPOSITION 2.4. <u>If R is Cohen-Macaualy the following conditions are equivalent</u> :
1) $H_R^{-d}(n)$ <u>is maximal</u>,
2) I <u>is generated by forms of degree</u> $m + d$,
3) I <u>is generated by</u> $\binom{m+r}{r-d}$ <u>linearly independent forms of degree</u> $m + d$ <u>and</u>
    $e(R) = \binom{m+r}{r-d+1}$.

PROOF. 1)$\Leftrightarrow$2). (See lemma 2.2). 2)$\Longrightarrow$3). We can assume k infinite and that the classes $x_0, \ldots, x_{d-1}$ of $X_0, \ldots X_{d-1}$ form a regular sequence in $R = k[X_0, \ldots, X_r]/I$. If $B = R/(x_0, \ldots, x_{d-1})$, $\dim B = 0$, $\operatorname{emdim} B = r+1-d$, $e(R) = \dim_k(B)$ (see (1.4)).

Now $\dim_k(B) = \sum_{n=0}^{m+d-1} H_B^0(n) = \sum_{n=0}^{m+d-1} \binom{n+r-d}{r-d} = \binom{m+r}{r-d+1}$. Further if $\bar{I}$ denotes the ideal $I \bmod X_0, \ldots, X_{d-1}$ and $v(\ )$ denotes the minimal number of homogeneous generators, we have $v(I) = v(\bar{I})$ (see the proof of thm. 1.5). But $B \cong k[X_{d-1}, \ldots, X_r]/\bar{I}$

and then $H_B^o(m+d)=0$ which implies $\bar{I}=(X_{d-1},\ldots,X_r)^{m+d}$ and so $v(I)=v(\bar{I})=\binom{m+r}{r-d}$.

EXAMPLES. The Cohen-Macaulay rings satisfying condition 2) of prop. 2.4 have been called <u>extremal rings</u> by Schenzel (see [Sc]). Examples of extremal rings are the following. 1) Let $X=(X_{ij})$, $1\leq i\leq n$, $1\leq j\leq m$, $n<m$, a matrix of indeterminates over a field k and let $A= k[X]$. Let $I(n,m)$ be the ideal generated by the maximal minors of X. Then $R= A/I(n,m)$ is extremal (see [Sc], n° 4) ;

2) Let (A,M) be a local Cohen-Macaulay ring of dimension d and emdim $A=e(A)+d-1$. Then the associated graded ring $R = G(A)$ is a Cohen-Macaulay extremal ring (see [S], thm 3.10). This is the case of a local ring A of a rational surface singularity.

PROPOSITION 2.5. <u>Let R be Cohen-Macaulay and</u> $x_o,\ldots,x_{d-2}$ <u>an R-sequence. Let</u> $B = R/(x_o,\ldots,x_{d-2})$. <u>The following conditions are equivalent</u> :

1) $H_R^{1-d}(n)$ <u>is maximal</u>,

2) I <u>is generated by forms of degree</u> m+d-1 <u>and</u> m+d ,

3) $H_B^o(n) = \text{Min}\left\{\binom{n+r'}{r'}, e\right\}$, <u>where</u> $e = e(B)=e(R)$ <u>and</u> $r'+1=$ emdim $B = r-d+1$

PROOF. See (2.3), thm 1.3 and recall that if B is Cohen-Macaulay of dimension 1 $H_B^o(n)\leq e(A)$, for any $n\in Z$, (se [S], ch 3, thm 1.1).

REMARK. Let V be a projectively Cohen-Macaulay variety of dimension d in $\mathbb{P}^r$ over an algebraically closed field. In this case the maximality of $H_R^{1-d}(n)$ can be characterized by 3) of prop.2.5 : V has $H_R^{1-d}(n)$ maximal if and only if a generic section $V\cap S$, with a linear subspace S of projective dimension r-d+1 of $\mathbb{P}^r$, consists of points in generic position.

It is then clear the importance of finding sufficient conditions for R to be Cohen-Macaulay. Unfortunately the maximality of the Hilbert function of R is not enough as the following examples show :

EXAMPLE 1. (Hartshorne - Hirshowitz). If V is a projective variety consisting of e skew lines in general position in $\mathbb{P}^r$ then $H_R^0(n) = \text{Min}\left\{\binom{n+r}{r}, en+e\right\}$ (see[H]) Hence $H_R^0(n)$ is maximal, but it is well known that R is not Cohen-Macaulay (see [G.W.]).

The following example was shown us by Ciro Ciliberto :

EXAMPLE 2). Let $P_1, \ldots P_6$ be 6 points of a conic in $\mathbb{P}_k^2$ and let I be the homogeneous ideal in $k[X,Y,Z]$ generated by the forms of degree 3 vanishing on $P_1, \ldots, P_6$. If $R = k[X,Y,Z]/I$ it is easily checked that dim R = 1 and $H_R^0(1)=3$, $H_R^0(n) = 6 = e(R)$, $n \geq 2$. Then $H_R^{-1}(n)$ is maximal and R is not Cohen-Macaulay ; otherwise I would coincide with the ideal of points $P_1, \ldots, P_6$.

3. **THE CASE OF ASSOCIATED GRADED ALGEBRAS** . From now on we assume that $R = G(A) = \bigoplus_{n \geq 0}(M^n/M^{n+1})$ is the associated graded ring , at the maximal ideal M , of a local Cohen-Macaulay ring (A,M), of dimension d , embedding dimension emdim A = r+1 and multiplicity e = e(A) .

In this case the maximality of the Hilbert function $H_R^{1-d}(n)$ gives that R is Cohen-Macaulay as we will show .

All the definitions regarding Hilbert functions of R given in section 2 can be extended to A. In fact :

$$H_A^0(n) = \dim_k(M^n/M^{n+1}) = H_R^0(n) \quad ; \quad \text{then:}$$

$H_A^i(n) = H_R^i(n)$, $i \in \mathbb{Z}$, $e(A) = e(R)$, emdim A = emdim R, dim A = dim R.

Further a result corresponding to thm. 1.3 holds :

THEOREM 3.1. <u>Let</u> $x \in M$. <u>For every</u> $n \geq 0$, $H_{A/xA}^1(n) - H_A^0(n) = \dim_k(M^{n+1}:xR/M^n)$. <u>Then if</u> $x_0,\ldots x_s \in M$ <u>and</u> $B = A/(x_0,\ldots x_s)$, $H_A^0(n) \leq H_B^s(n)$ <u>and equality holds if and only if the classes</u> $\bar{x}_0,\ldots \bar{x}_s$ <u>in</u> $R = G(A)$ <u>form an</u> R-<u>sequence of elements of degree</u> 1. <u>In particular</u>, <u>if</u> $s = d-1$, $H_B^0(n) = H_A^d(n)$ <u>if and only if</u> $R=G(A)$ <u>is Cohen-Macaulay</u>.

PROOF. The same as that of thm. 1.3 .

THEOREM 3.2. <u>If</u> $H_A^{1-d}(n)$ <u>is maximal then</u> $R = G(A)$ <u>is Cohen-Macaulay</u>.

PROOF. By passing to the ring $A(u)$ we may assume that $k = A/M$ is infinite ; then there is an A-sequence $x_0,\ldots,x_{d-1}$ of elements whose classes in R have degree 1; let $B = A/(x_0,\ldots x_{d-1})$. We have $e = e(A) = \dim_k(B)$ (see [S] , thm 3.4) and emdim $B = r'+1 = $ emdim $(A) - d$. Further if N is the maximal ideal of the local ring B, $H_B^1(n) = \dim_k(B/N^n) = \dim_k(B) = e$, for $n \gg 0$. Then, for $n \geq 0$,

$H_B^1(n) \leq \text{Min}\left\{\binom{n+r'}{r'}, e\right\}$. Now $H_A^{1-d}(n) = \text{Min}\left\{\binom{n+r'}{r'}, e\right\}$ (see def. 2.1) and so $H_A^{1-d}(n) \geq H_B^1(n)$, for any $n \geq 0$. By adding, we get $H_A^0(n) \geq H_B^d(n)$. Since the reverse equality always holds we have $H_A^o(n) = H_B^d(n)$ and R is Cohen-Macaulay by thm. 3.1.

REMARK. In the one-dimensional case thm. 3.2 extends thm. 3.3 of $[O_1]$.

It is not true that if A has maximal Hilbert function of any degree greater than 1-d then G(A) is Cohen-Macaulay as the following example shows:

EXAMPLE. Let $A = k[t^4-1, t(t^4-1), t^3(t^4-1)^2]$ then $H_A^o(1) = H_A^o(2) = 3$, $H_A^o(n) = 4$, for $n \geq 3$ (see $[O_2]$, example 1.b). Hence $H_A^1(n) = P_A^1(n) = 4n-1$, for $n \geq 2$. Then $H_A^1(n)$ has maximal Hilbert function. Further G(A) is not Cohen-Macaulay. This provides an example for $d = 1$. To have an example for any d it suffices to consider the ring $C = A[X_1,...,X_{d-1}]$, $d \geq 2$. Thus dim R = d, $X_1,...,X_{d-1}$ is a C-sequence and so $H_C^i(n) = H_A^{i+d-1}(n)$. Then, if $i > 1-d$, i.e; $i+d-1>0$, $H_C^i(n)$ is maximal but C is not Cohen-Macaulay since $A = C/(X_1,...,X_{d-1})$ is not Cohen-Macaulay.

Now we assume that A = B/I is a quotient of a regular local ring (B,N) modulo an ideal I ; let emdim (A) = r+1 then:

$$G(A) = k[X_o,...,X_r]/I^*$$

where I^* is the homogeneous ideal of G(A) generated by the leading forms of the elements of I . The informations of prop. 2.4 , 2.5 on the degree and the number of the forms generating I^* extend to I , using the following crucial lemma. If f is an element of B we call degree of f the least integer $n \geq 0$ such that $f \in N^n$. Let v(I) be the least number of generators of the ideal I :

LEMMA 3.3. Let n be a positive integer. If I is generated by forms of degree n and n+1, then I is generated by elements of degree n and n+1; further $v(I) = v(I^*)$.

PROOF. (See $[O_3]$, lemma 1).

THEOREM 3.4. If $A = B/I$ is Cohen-Macaulay of dimension d and index of regularity m, the following conditions are equivalent:
(i) $H_A^{-d}(n)$ is maximal,
(ii) I is generated by elements of degree $\geq m+d$,
(iii) I is minimally generated by $\binom{m+r}{r-d}$ elements of degree m+d, G(A) is Cohen-Macaulay and $e(A) = \binom{m+r}{r-d+1}$.

PROOF. (i)$\Leftrightarrow$(ii). Follows from prop. 2.4 and the remark that I is generated by elements of degree m+d if and only if such is I^*.
(ii)$\Rightarrow$(iii). Follows from prop. 2.4 and lemma 3.3.

EXAMPLE. If A is a local Cohen-Macaulay ring such that $emdim(A) = e(A) + d - 1$ (in particular the local ring of a rational surface singularity) then $H_A^{-d}(n)$ is maximal. Hence $R = G(A)$ is Cohen-Macaulay :

THEOREM 3.5. The following conditions are equivalent :
(i) $H_A^{1-d}(n)$ is maximal,
(ii) I is generated by elements of degree $m + d - 1$
(iii) I is generated by elements of degree $m + d - 1$, $m + d$ and G(A) is Cohen-Macaulay.

PROOF. (i)⟺(ii) . (See lemma 2.2). (ii)⟹(iii) .(See thm. 3.2 and (2.3)).

In the one-dimensional case thm 3.4 can be strenghthened:

COROLLARY 3.6. <u>Let</u> dim A = 1 <u>then</u> $H_A^{-1}(n)$ <u>is maximal if and only if there is a positive integer p such that I is generated by elements of degree $\geq$ p and</u> $e(A) = \binom{p+r-1}{r}$.

PROOF. ⟹) . (See thm. 3.4) . ⟸) . By hypothesis $H_A^0(n) = \binom{n+r}{r}$ for $n<p$ (r+1 = emdim A) and $H_A^0(p-1) = e(A)$, (this implies $H_A^0(n) = e(A)$ for $n \geq p-1$ and $H_A^0(n) = \mathrm{Min}\{\binom{n+r}{r}, e(A)\}$ is maximal , whence G(A) is Cohen-Macaulay (thm 3.2) and $p-1 = \mathrm{Min}\{n \mid \binom{n+r}{r} \geq e(A)\}$ is the index of regularity of A . Nox apply thm 3.4.

REMARK . The previous corollary extends prop. 12 of [O_3] and can be applied to various classes of curves locally requiring an arbitrary large number of generators . In fact the classical Mcaulay's examples (see [M]) , Moh's examples (see [Mo]) , Maurer's examples (see [Ma]) have all maximal Hilbert function of degree -1 . They are all curves of the form Spec (A) , where $A = k[X,Y,Z]_{loc}/P$, $e(A) = \binom{p-1}{2}$ and P is generated by elements of degree p . So corollary 3.6 plus thm. 3.4 prove that P is minimally generated by p+1 elements as was claimed in [M] , and proved in [Mo] and [Ma] but furthermore we have that G(A) is Cohen-Macaulay which was never pointed out before .

OPEN QUESTION . Let A be the local ring at a singular point p of an affine variety . Suppose that A is Cohen-Macaulay and let V be its tangent cone at p . We have seen that if $H_A^{1-d}(n)$ is maximal then V is projectively Cohen-Macaulay (thm 3.2) . In this case if R is the homogeneous coordinate ring of V there is an R-sequence $\bar{x}_0,\ldots \bar{x}_{d-2}$ of elements of degree 1 . Then, if $B = A/(x_0,\ldots x_{d-2})$, $H_A^{1-d}(n) = H_B^0(n)$ (thm 3.1) , and $H_B^0(n)$ is maximal . Hence it is natural to ask the following question . Let $x_0,\ldots x_{d-2}$ be an A-sequence of elements of degree 1 such that $H_B^0(n)$ is maximal; is then the tangent cone V projectively Cohen-Macaulay ? . Or , equivalently , if a generic section of V , with a linear subspace of projective dimension $r - d + 1$, consists of points in generic position is then V projectively Cohen-Macaulay ? . In particular if V is a curve and if a generic hyperplane section of V consists of points in generic position , is then V projectively Cohen-Macaulay?.

REFERENCES

[G.M.]   A.V. GERAMITA - P. MAROSCIA, *The ideal of forms vanishing at a finite set of points in $\mathbb{P}^n$*, Queen's Math. Preprint, No. 1981-5,

[G.O.]   A.V. GERAMITA - F. ORECCHIA, *Minimally generating ideals defining certain tangent cones*, Journal of Algebra, 78, 36-57 (1982),

[G.W.]   A.V. GERAMITA - C. WEIBEL, *On the Cohen-Macaulay and Buchsbaum properties for union of lines in $\mathbb{P}^n$*, Queen's Math. Preprint, No. 1982-16,

[H]   R. HATSHORNE - A. HIRSHOWITZ, *Droites en position generale dans l'espace projectif* (Preprint),

[L]   J. LIPMAN, *Stable ideals and Arf rings*, Amer. J. Math., 93, 649-685, (1971),

[M]   F.S. MACAULAY, *The algebraic theory of modular systems*, Cambridge University, 1916,

[Ma]   J. MAURER, *Eine variante der Moh-Curven* (Preprint),

[Mo]   T.T. MOH, *On the unboundedness of generators of prime ideals in power series rings of three variables*, J. Math. Soc. Japan, 26, 722-734, (1974),

[O_1]   F. ORECCHIA, *Ordinary singularities of algebraic curves*, Canad. Math. Bull., 24, (4), 423-431, (1981),

[O_2]   F. ORECCHIA, *One-dimensional local rings with reduced associated graded ring and their Hilbert function*, Manuscripta Math., 32, 391-405, (1980),

[O_3]   F. ORECCHIA, *Maximal Hilbert functions of one-dimensional local rings*, Proceedings of the Trento Conference on Commutative Algebra, Lect. Notes in Pure and applied Math., Marcel Dekker, 223-233, 1983,

[S]   J. SALLY, *Number of generators of ideals in local rings*, Lect. Notes in Pure and Applied Math., Marcel Dekker, 1978,

[Sc]   P. SCHENZEL, *Über die freien auflösungen extremaler Cohen-Macaulay ringe*, Journal of Algebra, 64, No. 1, 93-101, (1980),

SOME CURVES IN $\mathbb{P}^3$ ARE SET-THEORETIC COMPLETE INTERSECTIONS

Lorenzo Robbiano and Giuseppe Valla

It is well known that a longstanding problem in algebraic geometry is whether every connected projective curve in $\mathbb{P}_k^3$ is a set-theoretic complete intersection (s.t.c.i.). Let C_4 be the quartic rational curve in $\mathbb{P}_k^3$ given parametrically by $X_0 = u^4$, $X_1 = u^3 t$, $X_2 = ut^3$, $X_3 = t^4$; its defining ideal can be minimally generated by four polynomials F_1, F_2, F_3, F_4 such that deg $F_2 = 2$, deg $F_1 =$ deg $F_3 =$ deg $F_4 = 3$ and three of them, say F_1, F_2, F_4 define C_4 schematically. Further (see [6]) C_4 is a s.t.c.i. if char(k) > 0.

In the first part of this paper we prove two parallel statements, which give evidence of the different behaviour of C_4 with respect to the characteristic of the field k. Namely, if char(k) is positive, then C_4 is a s.t.c.i. on F_1 and on F_4 (Proposition 1.5); if char(k) is zero, then C_4 is not a s.t.c.i. on F_1, on F_2 and on F_4 (Proposition 1.6). To get our proofs we use the so called Grobner bases (G-bases) of ideals in polynomial rings, which were introduced by Buchberger in [3] and further studied in [4], and which yield an easy method for computing the equations defining the projective closure of an affine scheme (Prop.1.4 and [8]).

Now C_4 is an element of the family of the so called monomial curves in $\mathbb{P}^3$, but it is in the "bad" part of the family, in the sense that its projective coordinate ring is not Cohen-Macaulay. The second section of the paper is devoted to show that all the monomial curves in $\mathbb{P}^3$, such that their projective coordinate ring is Cohen-Macaulay, are s.t.c.i. in every characteristic (Corollary 2.3). To get the proof we need a good description of the defining ideal of these curves; this was done in [1] (see also [2]) and we point out that it can be easily achieved again by means of G-bases. What we get is that all the members of this "good" part of the family share a special determinantal structure which allows us to prove our claim by studying some particular matrices (Theorem 2.2).

1. Let A be the set of the monomials in the polynomial ring $k[X_1,\ldots,X_n]$. We may order the elements of A by their degrees and, if they are of the same degree, by the lexicographic order. If f is a polynomial in $k[X_1,\ldots,X_n]$ we denote by $M(f)$ the maximum monomial of f and, if I is an ideal of the ring $k[X_1,\ldots,X_n]$, by $M(I)$ the homogeneous ideal generated by the maximum monomials of the elements of I. This enables us to give the following definition.

<u>Definition 1.1.</u> <u>Let I be an ideal in</u> $k[X_1,\ldots,X_n]$. <u>A set of elements</u> $\{f_1,\ldots,f_r\} \subseteq I$ <u>is called a G-base of</u> I <u>if</u> $M(I)=(M(f_1),\ldots,M(f_r))$.

<u>Lemma 1.2.</u> <u>If</u> $\{f_1,\ldots,f_r\}$ <u>is a G-base of</u> I, <u>then</u> $I=(f_1,\ldots,f_r)$.

<u>Proof</u>. Assume by contradiction that $(f_1,\ldots,f_r) \subset I$ and let $f \in I, f \notin (f_1,\ldots,f_r)$ such that $M(g) < M(f)$ does not hold for all elements g with this property. We get $M(f) \in M(I)=(M(f_1),\ldots,M(f_r))$ hence for some monomial a and some i we have $M(f)=aM(f_i)$. This implies $M(f-af_i)<M(f)$ thus $f-af_i \in (f_1,\ldots,f_r)$, a contradiction.

For more details on G-bases see [3] and [4].

Now we recall that if f is a polynomial in $k[X_1,\ldots,X_n]$ then $^h f$ will denote the homogeneous polynomial $X_0^{\partial f} f(X_1/X_0,\ldots,X_n/X_0)$ in $k[X_0,\ldots,X_n]$ (here ∂f is the degree of f); also for any ideal I in $k[X_1,\ldots,X_n]$, $^h I$ will denote the homogeneous ideal generated by the forms $^h f$ with $f \in I$. It is easily seen that the homogeneous elements in $^h I$ are the forms $^h f X_0^n$ with $f \in I$ and $n \geq 0$ (see [11] p.180). Also if I(V) is the defining ideal of an affine algebraic variety V in $\mathbb{A}^n$, then the defining ideal of its projective closure in $\mathbb{P}^n$ is $^h I(V)$. Finally if I is an ideal in $k[X_1,\ldots,X_n]$, $M(I)^e$ will denote the ideal $M(I)k[X_0,\ldots,X_n]$.

<u>Lemma 1.3.</u> a) <u>If</u> $f \in k[X_1,\ldots,X_n]$, <u>then</u> $M(f) = M(^h f)$.

b) <u>If I is an ideal in</u> $k[X_1,\ldots,X_n]$ <u>then</u> $M(I)^e = M(^h I)$.

Proof. The first assertion is clear. As to the second one, we have seen that the homogeneous elements in $^h I$ are the forms $^h f X_0^n$; now we have $M(^h f X_0^n) = M(^h f) X_0^n = M(f) X_0^n$ and this proves that $M(^h I) \subseteq M(I)^e$, while the other inclusion is a consequence of a).

<u>Proposition 1.4.</u> <u>If</u> $\{f_1,\ldots,f_r\}$ <u>is a G-base of the ideal</u> I, <u>then</u> $\{^h f_1,\ldots,^h f_r\}$ <u>is a G-base of the ideal</u> $^h I$.

Proof. We have $M(^h I) = M(I)^e = \sum_1^r M(f_i) k[X_0,\ldots,X_n] = \sum_1^r M(^h f_i) k[X_0,\ldots,X_n] = (M(^h f_1),\ldots,M(^h f_r))$.

Let C_4 be the quartic curve in $\mathbb{P}_k^3$ given parametrically by $X_0 = u^4$, $X_1 = u^3 t$, $X_2 = ut^3$, $X_3 = t^4$. Take the standard affine open set $X_0 \neq 0$ and put $X = X_1/X_0$, $Y = X_2/X_0$ and $Z = X_3/X_0$; then C_4 is the projective closure of the affine curve

C: $X=t$, $Y=t^3$, $Z=t^4$, whose defining ideal I is generated by $f_1=Y-X^3$, $f_2=Z-XY$. Standard techniques in computing G-bases, give a fairly easy way to get a G-base of I, hence (Prop.1.4) to get generators of the defining ideal of C_4 in $\mathbb{P}^3$. If we denote by P this ideal, we know that $P=^hI$ and we get the well known fact that $P=(F_1,F_2,F_3,F_4)$, where $F_1=^hf_1=X_0^2X_2-X_1^3$, $F_2=^hf_2=X_0X_3-X_1X_2$, $F_3=X_0X_2^2-X_1^2X_3$, $F_4=X_1X_3^2-X_2^3$. Another well known fact (very easy to be checked) is that C_4 is schematically defined by F_1, F_2, F_4. Now it is clear that if we can find two polynomials f and g such that I=rad(f,g) and {f,g} is a G-base of (f,g), then $P=^hI=^hrad(f,g)=rad(^h(f,g))=$ $=rad(^hf,^hg)$, where the last equality follows from Proposition 1.4 and the third one from [11] p.180 ; of course this would imply that C_4 is a s.t.c.i.. It is well known that C_4 is a s.t.c.i. if char(k)=p>0 and indeed in this situation we can find f and g with the above described properties.

<u>Proposition 1.5</u>. C_4 is a s.t.c.i. on F_1 and on F_4 if char(k)=p>0.

<u>Proof</u>. We observe that F_1, F_4 play an interchangeable role with respect to the standard affine open sets $X_0 \neq 0$, $X_3 \neq 0$. So let us see what happens on F_1.

Case a): p=3. Then $f_2^3 = Z^3-X^3Y^3 = Z^3-Y^4 \mod(f_1)$. Hence $g=Z^3-Y^4 \in I$ and $I=rad(f_1,g)$. Moreover $M(f_1)=X^3$, $M(g)=Y^4$ are coprime, so $\{f_1,g\}$ is a G-base of (f_1,g) and we are done.

Case b): p≠3. We have $f_2^{3p}=(f_2^p)^3=(Z^p-X^pY^p)^3=Z^{3p}-3Z^{2p}X^pY^p+3Z^pX^{2p}Y^{2p}-X^{3p}Y^{3p}$. Now since p≠3 we can write 2p=3q+r with $1 \leq r \leq 2$ and $q \geq r$. It follows that $f_2^{3p} \equiv g \mod(f_1)$ where $g=Z^{3p}-3X^pY^pZ^{2p}+3X^rY^{2p+q}Z^p-Y^{4p}$; but r+q+3p=r+q+q(9/2)+ +r(3/2)=r(5/2)+q(11/2) $\leq$ 2r+6q=4p, hence $M(g)=Y^{4p}$ and we conclude as before.

However the situation turns out to be completely different if $\text{char}(k) = 0$.

Proposition 1.6. If $\text{char}(k) = 0$, then C_4 is not a s.t.c.i. on anyone of the three surfaces F_1, F_2 and F_4 which define it schematically.

Proof. Certainly C_4 cannot be a s.t.c.i. on F_2 (even in positive characteristic), since F_2 is a non singular quadric and C_4 is of type $(3,1)$ on it. As before we may restrict our attention to F_1. Assume that there exists a polynomial G such that $P=\text{rad}(F_1,G)$; then G is not divisible by X_0 and so, if we denote by g the unique polynomial in $k[X,Y,Z]$ such that $g=G^h$, we get $I=\text{rad}(f_1,g)$. Now if X divides $M(g)$, then $(F_1,G) \subseteq (X_0,X_1)$ but this is in contradiction with $P=\text{rad}(F_1,G)$; therefore X does not divide $M(g)$. Let A be a domain and f, g elements of A such that $f \neq 0$ is prime and $(f)=\text{rad}(g)$; then $g=cf^n$ with c unit in A and n a positive integer. So, if A denotes the ring $k[X,Y,Z]/(f_1)$ which is isomorphic to $k[X,Z]$, we deduce that $g=cf_2^n+af_1$ with $c \in k^*$ and of course we may assume $g=f_2^n+af_1$. To get our claim, it is then sufficient to show that for every $a \in k[X,Y,Z]$ and for every positive integer n, X divides $M(f_2^n+af_1)$. Now it is easy to check that this is equivalent to saying that, for every positive integer n, X divides $M(h)$, where h is the remainder obtained in the division of f_2^n by X^3-Y in $k[Y,Z][X]$. This can be achieved by an elementary computation. For other remarks on this subject see [12].

2. Let C be a monomial space curve in $\mathbb{P}^3$ given parametrically by $X_0=u^m$, $X_1=u^nt^{m-n}$, $X_2=u^pt^{m-p}$, $X_3=t^m$ (m,n,p are positive integer such that $m>n$ and $m>p$); it is clear that C is the projective closure of the affine space curve with parametric equations $X=t^{m-n}$, $Y=t^{m-p}$, $Z=t^m$. The defining prime

ideal of these affine curves has been extensively studied by Herzog in [7] (see also [9]); using his results and Proposition 1.4, one can easily give a complete description of the equations defining the projective curve C. We collect these facts in the following theorem which has been previously proved by different methods in [2] and whose proof is left to the reader.

Theorem 2.1. <u>The curve C is arithmetically Cohen-Macaulay if and only if the minimal number of generators of the defining ideal I(C) is less or equal to three. Further if C is arithmetically Cohen-Macaulay, then I(C) is generated by the 2×2 minors of the matrix</u>

$$\begin{Vmatrix} x_1^a & x_2^b & x_0^c x_3^d \\ x_2^e & x_0^f x_3^g & x_1^h \end{Vmatrix}$$

<u>for suitable non negative integers</u> a,b,c,d,e,f,g,h.

Now if C is arithmetically Cohen-Macaulay we can prove that C is a s.t.c.i. as a consequence of the following more general result (see also [10]).

Theorem 2.2. <u>Let R be a commutative ring with identity; let m and n be non negative integers and let I be the ideal generated by the maximal minors of the matrix</u> $M = \begin{Vmatrix} x & y^m & z \\ y^n & s & t \end{Vmatrix}$ <u>with entries in R. Then there exist elements f and g in I such that</u> $\mathrm{rad}(I) = \mathrm{rad}(f,g)$.

Proof. We may assume that $m \geq n$; let $m = nq+r$ where $0 \leq r < n$, $q \geq 1$ and let $f_1 = y^m t - sz$, $f_2 = y^n z - xt$, $f_3 = xs - y^{n+m}$. In the following, if a and b are elements of R, we use the identity $(a-b)^p = a^p - b(\sum_{k=0}^{p-1} a^k (a-b)^{p-k-1})$ which can be easily checked. We have $f_2^{q+1} = (y^n z - xt)^{q+1} = y^{n(q+1)} z^{q+1} - xte$, where for suitable c in R we let $e=\sum_{k=0}^{q} y^{nk} z^k f_2^{q-k} = y^{nq} z^q + f_2(y^{n(q-1)} z^{q-1} + cf_2)$. Next we denote $y^{n(q-1)} z^{q-1} +$

$+cf_2$ by a, consider the matrix $N = \begin{Vmatrix} s & y^{n(q+1)} & te \\ y^r & x & z^{q+1} \end{Vmatrix}$ and call J the ideal

generated by the 2×2 minors of N. Then J is generated by f_2^{q+1}, $g = y^r te - sz^{q+1}$

and f_3; now we have $g = y^r te + z^q (y^m t - sz) - y^m z^q t = y^r t(e - y^{nq} z^q) + z^q f_1 = y^r taf_2 + z^q f_1 =$

$= \det \begin{Vmatrix} x & y^m & z \\ y^n & s & t \\ z^q & y^r ta & 0 \end{Vmatrix}$ which implies $J \subseteq I$. On the other hand

$s^q g = \det \begin{Vmatrix} s^q x & y^m & z \\ s^q y^n & s & t \\ s^q z^q & y^r ta & 0 \end{Vmatrix} = \det \begin{Vmatrix} s^q x - s^{q-1} y^{m+n} & y^m & z \\ 0 & s & t \\ s^q z^q - s^{q-1} y^{n+r} ta & y^r ta & 0 \end{Vmatrix} =$

$= \det \begin{Vmatrix} 0 & y^m & z \\ 0 & s & t \\ s^q z^q - s^{q-1} y^{n+r} at & y^r at & 0 \end{Vmatrix} \mod(f_3) = f_1 (s^q z^q - s^{q-1} y^{n+r} at) \mod(f_3)$

and $s^q z^q - s^{q-1} y^{n+r} at = s^q z^q - s^{q-1} y^{n+r} ty^{n(q-1)} z^{q-1} \mod(f_2) = s^{q-1} z^{q-1} (sz - y^m t) \mod(f_2)$.

But for suitable d in R we have $(-f_1)^{q-1} = (sz - y^m t)^{q-1} = s^{q-1} z^{q-1} - y^m d$, hence

$s^q g = \mp f_1^{q+1} + f_1^2 y^m d \mod(f_2, f_3)$; now $y^n f_1 + sf_2 + tf_3 = 0$, hence $y^m f_1 \in (f_2, f_3)$ and

we get $f_1^{q+1} \in (g, f_2, f_3)$. This implies that $rad(I) = rad(J)$. Thus, if $r = 0$, J

is clearly generated by two elements and the theorem follows; if $r > 0$, we

apply the same argument to the matrix N and eventually, after a finite

number of steps, we conclude the proof of the theorem.

<u>Corollary 2.3.</u> <u>Let C be a monomial space curve in $\mathbb{P}^3$ which is arithmetically</u>

<u>Cohen-Macaulay; then C is a</u> s.t.c.i.

Remark. It is interesting to compare Theorem 2.2 with the well known result that the Segre embedding of $\mathbb{P}^1 \times \mathbb{P}^2$ in $\mathbb{P}^5$, whose defining ideal is generated by the 2×2 minors of the matrix $\left\| \begin{matrix} x_0 & x_1 & x_2 \\ x_3 & x_4 & x_5 \end{matrix} \right\|$, is not a s.t.c.i. (see [5]).

References

[1] Bresinsky H. and Renschuch B., Basisbestimmung Veronesescher Projektionsideale mit allgemeiner Nullstelle $(t_o^m, t_o^{m-r} t_1^r, t_o^{m-s} t_1^s, t_1^m)$ Math. Nach., (to appear).

[2] Bresinshy H., Schenzel P. and Vogel W., On liason, arithmetical Buchsbaum curves and monomial curves in $\mathbb{P}^3$, preprint.

[3] Buchberger B., Ein algorithmisches Kriterium fur die Losbarkeit eines algebraischen Gleichungssystems, Aequa.Math. 4 (1970), 374-383.

[4] Buchberger B., A criterion for detecting unnecessary reductions in the construction of Grobner bases, Proc. EUROSAM 79, Lect. No. Comp. Sc. 72 (1979), 3-21.

[5] Hartshorne R., Cohomological dimension of algebraic varieties, Ann. Math. 88 (1962), 403-450.

[6] Hartshorne R., Complete intersections in characteristic $p > 0$, Am.J.Math. 101 (1979), 380-383.

[7] Herzog J., Generators and relations of abelian semigroups and semigroup rings, Man.Math. 3 (1970), 175-193.

[8] Moller H.M. and Mora F., Grobner bases and explicit free resolutions of polynomial modules, in preparation.

[9] Robbiano L. and Valla G., On the equations defining tangent cones, Math.proc.Camb.Phil.Soc. 88 (1980),281-297.

[10] Valla G., On determinantal ideals which are set-theoretic complete intersections, Comp.Math.42 (1981), 3-11.

[11] Zariski O. and Samuel P.: Commutative Algebra,v.II, Van Nostrand, Princeton, 1960.

[12] Craighero P.C., Una osservazione sulla curva di Cremona di $\mathbf{P}_k^3$, $C : \{\lambda\mu^3, \lambda^3\mu, \lambda^4, \mu^4\}$, Rend. Sem. Mat. Univer. Padova, 65 (1981), 177-190.

CONSTRUCTING ENRIQUES SURFACES FROM QUINTICS IN P_K^-.

Ezio Stagnaro

Introduction.

Let k be an algebraically closed field of characteristic zero. By definition, an Enriques surface is a non-singular projective surface with the geometric genus $p_g = 0$, the irregularity q = 0, the bigenus $P_2 = 1$ and the double canonical divisor 2K is equivalent to zero in a minimal model of the surface.

We shall construct Enriques surfaces starting from quintic surfaces F_5 in P_k^3. Our surfaces F_5 will have four tacnodes at the vertices of a tetrahedron, such that there exist two planes π_1, π_2 which are both tangent tacnodal planes to F_5 at two vertices of the tetrahedron. If $F \to F_5$ is a desingularization of F_5, we have that the geometric genus p_g of F is zero, because the adjoint surfaces to F_5 are the planes passing through the four vertices of the tetrahedron. The bicanonical adjoint surfaces to F_5 are the quadrics passing through the four vertices and such that the tangent planes to the quadrics coincide with the tacnodal planes of F_5: such a quadric exists and moreover we have that it is unique: our quadric is given by the two planes π_1, π_2. Therefore F has the bigenus $P_2 = 1$.

Moreover we construct again Enriques surfaces also when two tacnodes are triple points on F_5.

For all these surfaces F, we have the trigenus $P_3 = 0$, namely a tricanonical adjoint to F_5 must have at the vertices of the tetrahedron a double biplanar point having one of the two singular tangent planes coincident with the tacnodal plane to F_5. Such adjoints do not exist;therefore F has $P_3 = 0$. This fact, $P_3 = 0$, is equivalent to the fact that on a surfaces with $p_g = 0$, $P_2 = 1$ without exeptional curve of first kind (minimal model) the bicanonical divisor 2K is equivalent to zero. (cfr. [E] pp. 223-224 or p. 246). Finally we shall prove that the irregularity of F is q = 0.

Hence our surfaces F are Enriques surfaces.

Moreover for the plurigenera we have $P_{2i} = 1$, $P_{2i+1} = 0$, i > 0 (loc. cit.). For further discussion see also Ch. VII, § 1, pp. 237-247 of the same Enriques'book.

We remark that our surfaces have lines passing through the tacnodes, then the linear system of plane section through a tacnodes contains a pencil of reducible curves, against the assumption in the theorem of Enriques (cfr. [E] , pag. 72). But such lines belong to the complete intersection C of F_5 with the bicanonical adjoint ;since C is an exeptional line of first kind, we may consider a surface without curves of first kind, so the above lines vanisch. (Remember that also in the double plane, given by Enriques, as model of an Enriques surface, there are two exceptional lines of first kind).

Now we prove that the irregularity of F is q = 0. We use the criterion of Enriques - Castelnuovo: "The adjoint surfaces of degree 2 cut on plane sections the (complete) canonical series" (cfr. [E] , pag. 118). Now if C is a plane section, it may have either no singularities, or singular points; so the dimension of its canonical series is p - 1 = 5, or < 5 respectively (where p is the geometrical genus of C). The adjoint quadrics must pass simply through the four tacnodes, so they are a linear system of dimention $5 \geqslant 5$. Q.E.D.

Therefore if $F \to F_5$ is a desingularization of F_5, for F we have:
$p_g = 0$, $q = 0$, $P_2 = 1$, $P_3 = 0$. The surface F is an Enriques surface, as required.

1. If we consider the two planes:
$$Q : (X_1 + X_3)(X_2 + X_4)$$
the corresponding quintic F_5 in P_k^3 is

$$F_5 : X_1^3(X_2 + X_4)^2 +$$
$$+ X_2^3(X_1 + X_3)^2 +$$
$$+ X_3^3(X_2 + X_4)^2 +$$
$$+ X_4^3(X_1 + X_3)^2 +$$
$$+ (a_{2210}X_1X_2X_3 + a_{2201}X_1X_2X_4 + a_{2012}X_1X_3X_4 + a_{0221}X_2X_3X_4)(X_1+X_3)(X_2+X_4) +$$
$$+ a_{2120}X_1^2X_3^2(X_2+X_4) + a_{1202}X_2^2X_4^2(X_1+X_3); \quad a_{2120} \neq 0, \; a_{1202} \neq 0.$$

Now we prove that F_5 has a tacnode in $A_1 = (1,0,0,0)$, the same argument holds for the other tacnodes $(0,1,0,0)$, $(0,0,1,0)$, $(0,0,0,1)$. A tacnode is caracterized by the property that the tacnodal plane cuts F_5 in a curve which has a singular 4-fold point at the tacnode (cfr. [E] , pp. 84-85). The tacnodal plane at A_1 on F_5 is given by $X_2 + X_4$. Cutting F_5 with such plane we get:

$$\begin{cases} F_5 = 0 \\ X_2 + X_4 = 0 \end{cases} \qquad \begin{cases} a_{1202}X_4^4(X_1 + X_3) = 0 \\ X_2 + X_4 = 0 \end{cases}$$

which is a curve with a 4-fold point in A_1 (and also in $A_3 = (0,0,1,0)$).

We remark that $X_1 + X_3$ is the tacnodal plane of F_5 at $(0,1,0,0)$ and at $(0,0,0,1)$; $X_2 + X_4$ is the tacnodal plane of F_5 at $(1,0,0,0)$ and at $(0,0,1,0)$ (see the introduction).

If F is a desingularization of F_5, then for F we have: $p_g = 0, P_2 = 1, P_3 = 0$. (Q is the unique bicanonical adjoint to F_5).

In this case it is easy to calculate the complete intersection of Q and F_5; from the theory, we know that it is an exceptional curve of the first kind.

2. It is possible to construct Enriques surfaces from quintic surfaces F_5 in P_k^3 having two triple points and two tacnodal points such that the two tacnodal planes pass both through the two triple points. The bicanonical adjoint Q' (cfr. [E], pag. 74) is given by the two tacnodal planes which have a double point at the two triple point on F_5' :

$$Q': X_3 X_4$$

$$F': b_1 X_3^3 X_4^2 +$$
$$+ b_2 X_4^3 X_3^2 +$$
$$+ a_{2210} X_1^2 X_2^2 X_3 + a_{2201} X_1^2 X_2^2 X_4 + a_{2111} X_1^2 X_2 X_3 X_4 + a_{2021} X_1^2 X_3^2 X_4 + a_{2012} X_1^2 X_3 X_4^2 +$$
$$+ a_{1211} X_1 X_2^2 X_3 X_4 + a_{1121} X_1 X_2 X_3^2 X_4 + a_{1022} X_1 X_3^2 X_4^2 + a_{0221} X_2^2 X_3^2 X_4 + a_{0212} X_2^2 X_3 X_4^2 +$$
$$+ a_{0122} X_2 X_3^2 X_4^2 + a_{1112} X_1 X_2 X_3 X_4^2 ;$$

$b_1 \neq 0, b_2 \neq 0, a_{2210} \neq 0, a_{2201} \neq 0, a_{2021} \neq 0, a_{2012} \neq 0, a_{0221} \neq 0,$
$a_{0212} \neq 0.$

Again a desingularization of F_5' has $p_g = 0, q = 0, P_2 = 1, P_3 = 0$.

Remark. If we apply the standard transformation

$$\sigma : \begin{cases} x_1 = X_2 X_3 X_4 \\ x_2 = X_1 X_3 X_4 \\ x_3 = X_1 X_2 X_4 \\ x_4 = X_1 X_2 X_3 \end{cases}$$

to the surface F_5', the proper transform of F_5' is

$\sigma^*(F_5')$: $a_{2201} x_3^3 x_4^2 +$

$+ a_{2210} x_4^3 x_3^2 +$

$+ b_2 x_1^2 x_2^2 x_3 + b_1 x_1^2 x_2^2 x_4 + a_{2111} x_2^2 x_3 x_4 + a_{2021} x_2^2 x_3^2 x_4 + a_{2012} x_2^2 x_3 x_4^2 + a_{1211} x_1^2 x_3^2 x_4 +$

$+ a_{1121} x_1 x_2^2 x_3 x_4 + a_{1112} x_1 x_2^2 x_3 x_4^2 + a_{1022} x_1 x_2^2 x_3^2 x_4 + a_{0221} x_1^2 x_3^2 x_4^2 + a_{0212} x_1^2 x_3 x_4^2 +$

$+ a_{0122} x_1^2 x_2 x_3 x_4$.

The transformation σ has changed the coefficients in the follow way:

$b_1 \leftrightarrow a_{2201}$, $b_2 \leftrightarrow a_{2210}$, $a_{2111} \leftrightarrow a_{0122}$, $a_{2021} \leftrightarrow a_{0212}$, $a_{2012} \leftrightarrow a_{0221}$, $a_{1121} \leftrightarrow a_{1112}$, $a_{1211} \leftrightarrow a_{1022}$. So if we choose $b_1 = a_{2201}$, $b_2 = a_{2210}$, $a_{2111} = a_{0122}$, ... in particular choosing all the coefficient = 1, we see that F_5' is fixed for σ (up to the exeptional planes $x_1^3 x_2^3 x_3^2 x_4^2$).

Bibliography

[E] F. Enriques, Le superficie algebriche,   Zanichelli, Bologna (1949).

Istituto di Matematica Applicata - Facoltà di Ingegneria - Università - Via Belzoni 7 - 35131 Padova (Italy).

PRYM SURFACES AND A SIEGEL MODULAR THREEFOLD

by

Gerard van der Geer

The topic of the talk is the modular variety

$$\Gamma_2(2)\backslash H_2 ,$$

where H_2 is the Siegel upper half space of degree 2 and $\Gamma_2(2) =$ ker$\{Sp(4,\mathbb{Z}) \to Sp(4,\mathbb{Z}/2\mathbb{Z})\}$. The quotient $\Gamma_2(2)\backslash H_2$ is a non-compact complex manifold. It can be compactified in a minimal way (Satake compactification) to a singular algebraic variety $\Gamma_2(2)\backslash H_2^*$ by adding 15 copies of $\Gamma(2)\backslash H$ and 15 points such that each of these points occurs as a cusp of three copies of $\Gamma(2)\backslash H$. In this way we add 15 projective lines $\Gamma(2)\backslash H^*$ forming a configuration which is explained by the modular interpretation of $\Gamma_2(2)\backslash H_2^*$.

<u>Proposition</u>. The 1-dimensional (resp. 0-dim.) boundary components of $\Gamma_2(2)\backslash H_2^*$ correspond bijectively to the totally isotropic 1-dimensional (resp. 2-dim.) linear subspaces of a 4-dimensional symplectic vector space over $\mathbb{F}_2$. The configuration is of type $(15_3, 15_3)$. There are six (maximal) sets of five disjoint (compactified) 1-dimensional boundary components and the action of $Sp(4, \mathbb{F}_2)$ on them defines an isomorphism of $Sp(4, \mathbb{F}_2)$ with the symmetric group S_6.

If we number these six sets then each 1-dimensional boundary component may be denoted by λ_{ij}, $1 \le i < j \le 6$, after the two of these sets to which it belongs.

The equations

$$\alpha\tau_1 + \beta\tau_2 + \gamma\tau_3 + \delta(\tau_2^2 - \tau_1\tau_3) + \varepsilon = 0 \qquad \tau = \begin{pmatrix} \tau_1 & \tau_2 \\ \tau_2 & \tau_3 \end{pmatrix}$$

with $\alpha,\ldots,\varepsilon \in \mathbb{Z}$, $\beta^2 - 4\alpha\gamma - 4\delta\varepsilon = 1$ define a surface F_1 in $\Gamma_2(2)\backslash H_2^*$ consisting of ten components, each of which is isomorphic to $(\Gamma(2)\backslash H^*)^2$.

Consider the usual theta functions

$$\theta_m(\tau,\zeta) = \sum_{\xi \in \mathbb{Z}^2} \exp(2\pi i \{\tfrac{1}{2}(\xi+\tfrac{m'}{2})\tau\,^t(\xi+\tfrac{m'}{2}) + (\xi+\tfrac{m'}{2})\,^t(\zeta+\tfrac{m''}{2})\})$$

with $m = m'm''$ a fourtuple of zeroes and ones, $\tau \in H_2$ and $\zeta \in \mathbb{C}^2$. Of the sixteen $\theta_m(\tau,\zeta)$ ten are even as a function of ζ and these yield modular forms $\theta_m^4(\tau,0)$ of weight 2 on $\Gamma_2(2)$. The corresponding m are called even. Each $\theta_m^4(\tau,0)$ vanishes on one component of F_1. The $\theta_m^4(\tau,0)$ are not linearly independent. They span the 5-dimensional vector space M_2 of modular forms of weight 2.

Theorem. The modular forms of weight 2 on $\Gamma_2(2)$ define an embedding

$$\Gamma_2(2)\backslash H_2^* \to \mathbb{P}^4 \subset \mathbb{P}^9$$

$$\tau \bmod \Gamma_2(2) \to (\ldots, \theta_m^4(\tau,0), \ldots)$$

and the image is a quartic threefold X given by

$$t_4^2 - 4t_8 = 0,$$

where $t_4 = \sum \theta_m^8$ and $t_8 = \sum \theta_m^{16}$.

The singular locus of X consists of the fifteen lines coming from the "boundary". The relation $t_4^2 - 4t_8 = 0$ was found by Igusa using Riemann's theta formula. We use geometric methods to derive it.

<u>Theorem.</u> Let p be a non-singular point of X. The intersection of the tangent space T_p of X at p with X is a Kummer surface if p does not lie on the surface F_1. In this case the moduli point of $T_p \cap X$ (with level 2 structure induced by the fifteen singular lines) equals p.

If p is a non-singular point of X lying on F_1 then $T_p \cap X$ is a quadric surface with multiplicity 2, namely the component of F_1 to which p belongs. It carries the structure of a product of two Kummer curves ($\mathbb{P}^1$ with four points marked).

In order to prove the theorem one considers the ten functions $\sigma_m(\tau,\zeta) = \theta_m^2(\tau,0)\theta_m^2(\tau,\zeta)$ for m even. They define sections of a line bundle L_τ on A_τ, the Abelian surface corresponding to τ. Since they satisfy the linear relations satisfied by the $\theta_m^4(\tau,0)$ we have a linear map

$$M_2 \to H^0(A_\tau, L_\tau)$$

$$\theta_m^4(\tau,0) \to \sigma_m^4(\tau,-)$$

and a corresponding projective embedding ϕ

$$\mathbb{P}(H^0(A_\tau, L_\tau)^\vee) \to \mathbb{P}(M_2^\vee) \qquad (^\vee : \text{linear dual}).$$

One verifies that the σ_m satisfy

$$(\sum \sigma_m^8)^2 - 4 \sum \sigma_m^{16} = 0.$$

This implies that the image of A_τ in $\mathbb{P}(H^0(A_\tau, L_\tau)^\vee)$, which is a Kummer surface, is contained in X via ϕ. Since $H^0(A_\tau, L_\tau)$ is 4-dimensional, this Kummer surface is cut out on X by a hyperplane. Since the image p of the origin of A_τ must be a singular point, the hyperplane must be the tangent plane of X at this point.

We thus see that the Kummer surface belonging to p is contained in the Siegel modular threefold ! Of course, this requires an explanation.

Let C be a non-singular curve of genus 2 with a level 2 structure (i.e. the ramification points $C \to \mathbb{P}^1$ are marked $q_1 (=q), q_2, \ldots, q_6$). Denote by $J^{(k)}$ the variety of divisor classes of degree k on C, and set $J = J^{(0)}$. Since $S^2 C \to J^{(2)}$ is surjective, we can find for any α in $J^{(1)}$ points x_1, x_2 on C such that $2\alpha = (x_1 + x_2)$. Moreover, the pair x_1, x_2 is unique if $\alpha \notin q + J_2$. (Here J_2 are the points of order 2 on J.) If L_α^{-1} is the invertible sheaf on C corresponding to α this yields an isomorphism $\phi: L_\alpha^{\otimes 2} \simeq \mathcal{O}_C(-x_1 - x_2)$ which turns the sheaf $\mathcal{O}_C \oplus L_\alpha$ of $\mathcal{O}_C$-modules into a sheaf of $\mathcal{O}_C$-algebras. Let

$$\widetilde{C}_{\alpha, \phi} = \underline{\operatorname{Spec}}(\mathcal{O}_C \oplus L_\alpha)$$

be the double covering of C with ramification points x_1, x_2. It is non-singular if $x_1 \neq x_2$. The involution on $J^{(2)}$ $(y \to 4q - y)$ is compatible with the involution of $S^2 C$ induced by the hyperelliptic involution i of C. We now shall assume that $2\alpha \neq 2x$ for all x on C. This implies that for this α x_1 and x_2 are unique and $x_1 \neq x_2$. Then the covering $\widetilde{C}_{\alpha, \phi} \to C$ determines a 2-dimensional principally polarized Abelian surface P, called the Prym variety : $P = \ker \{ \operatorname{Nm}: \operatorname{Jac}(\widetilde{C}_{\alpha, \phi}) \to J \}$. Moreover, the situation also yields a symplectic isomorphism $\psi: J_2 \simeq P_2$, hence P carries a level 2 structure. In this way $\widetilde{C}_{\alpha, \phi} \to C$ determines a point of $\Gamma_2(2)\backslash H_2$, the moduli point of P. (This is independent of ϕ.)

If $2\alpha=2q$ then there is a whole $\mathbb{P}^1$ of possibilities

$$\phi : L_\alpha^{\otimes 2} \simeq \mathcal{O}_C(-x-i(x)), \qquad x \in C/i \simeq \mathbb{P}^1.$$

This forces us to blow up $J^{(1)}$ in the sixteen points of $q + J_2$. This yields a non-sigular surface $\widetilde{J}^{(1)}$. Secondly, if $2\alpha=2x$ and $\alpha \notin C/i \subset J^{(1)}$ then $\widetilde{C}_{\alpha,\phi}$ is singular and P is an extension of an elliptic curve with a multiplicative group.

<u>Theorem</u>. The map which sends $\alpha \in \widetilde{J}^{(1)}$ to the moduli point of the Prym variety associated to it induces an embedding

$$K_C \to \Gamma_2(2)\backslash H_2^*,$$

where K_C is the Kummer surface obtained from $\widetilde{J}^{(1)}/i$ (i: $y \to 2q-y$) by contracting the sixteen curves $\eta + C/i$, $\eta \in J_2$, to ordinary double points. Combining this with $\Gamma_2(2)\backslash H_2^* \to \mathbb{P}^4$ we find that the image is just $T_p \cap X$ with p the moduli point of C (or K_C).

The essential ingredient for the proof is the relation between the theta functions (denoted by η) on P and the theta functions on J as given by Fay :

$$\eta_m^4(0) = c(\tau,\alpha)\theta_m^2(\tau,0)\theta_m^2(\tau,\alpha)$$

for some non-zero constant $c(\tau,\alpha)$. Note that this relation can be used to prove our second theorem.

To make the picture complete one has to consider degenerations. To do this one has to look at a certain non-singular model of X obtained by blowing up X along the 15 singular lines by means of the ideal of cusp forms. We have the following cases.

a) an elliptic curve with an ordinary double point.

b) two elliptic curves intersecting transversally in one point. One or both of them may also be a rational curve with one ordinary double point provided that the intersection point is non-singular.

c) a rational curve with two ordinary double points.

d) two $\mathbb{P}^1$'s intersecting transversally in three points.

We consider double coverings $\widetilde{C} \to C$ such that

i) $\widetilde{C}$ is non-singular except for ordinary double points lying over the double points of C,

ii) $\widetilde{C}_{reg} \to C_{reg}$ is ramified in two points.

Let $\text{Jac}(\widetilde{C})$ (resp. $\text{Jac}(C)$) be the generalized Jacobian of $\widetilde{C}$ (resp. C). Then $P = \ker\{ \text{Nm} : \text{Jac}(\widetilde{C}) \to \text{Jac}(C) \}$ is again a principally polarized Abelian surface. The corresponding Kummer surfaces are :

a) a Plücker surface, i.e. a quartic surface with one double line and eight isolated double points; it is the intersection of X with a hyperplane containing exactly one of the singular lines of X.

b) a quadric with multiplicity two; such a quadric is a component of F_1 and the zero divisor of a $\theta_m^4(\tau,0)$.

c) a quartic surface with two double lines and four isolated double points; it is the intersection of X with a hyperplane passing through exactly two singular lines.

d) a quartic surface with three double lines; it is the intersection of X with a hyperplane containing exactly three of the singular lines. There are fifteen such surfaces and they are given by the equations

$$\alpha\tau_1 + \beta\tau_2 + \gamma\tau_3 + \delta(\tau_2^2 - \tau_1\tau_3) + \varepsilon = 0$$

with $\beta^2 - 4\alpha\gamma - 4\delta\varepsilon = 4$. Each of them is a suitably compactified quotient of H×H by the group

$$\{(\alpha_1, \alpha_2) \in SL(2,\mathbb{Z})^2 : \alpha_1 \equiv \alpha_2 \pmod{2}\}$$

and parametrizes Abelian surfaces isogenous to a product of elliptic curves. If we change the action of S_6 by an outer automorphism then X can be defined by

$$\sum x_i = 0, \quad (\sum x_i^2)^2 - 4\sum x_i^4 = 0$$

in $\mathbb{P}^5$ with the action of S_6 being given by the permutation of the six coordinates. Our fifteen surfaces are now given by $\prod_{i<j}(x_i - x_j) = 0$.

For a curve of genus greater than 2 we get a rational map

$$\tfrac{1}{2}S^2 C \to A_{g,2} \,,$$

where

$$\tfrac{1}{2}S^2 C = \{\alpha \in J^{(1)} : 2\alpha \in S^2 C \subset J^{(2)}\}$$

and $A_{g,2}$ is the level two moduli variety of principally polarized Abelian varieties of dimension g. It is defined by associating to a double cover $\tilde{C} \to C$ with two ramification points the moduli point of its Prym variety. Since the surface $S^2 C$ is intimately connected to the Schottky problem (cf. Curves and their Jacobians by Mumford, lect. 4) we hope that our study of the higher genus case will contribute to a solution of the Schottky problem.

REFERENCES

-- Fay, J.D.: Theta functions on Riemann surfaces. Lecture Notes in Math. 352. Berlin, Heidelberg, New York : Springer Verlag 1973.

-- van der Geer, G.: On the geometry of a Siegel modular threefold. Math. Annalen 260, 317-350 (1982).

-- Hudson, R.: Kummer's quartic surface. Cambridge : Cambridge University Press 1905.

-- Igusa, J.: On Siegel modular forms of genus two (II). Am. Journal of Math. 86, 392-412 (1964).

-- Mumford, D.: Prym varieties I. In: Contributions to analysis, pp 325-350. London, New York : Academic Press 1974.

-- Mumford, D.: Curves and Their Jacobians. Ann Arbor : The University of Michigan Press 1975.

Vol. 845: A. Tannenbaum, Invariance and System Theory: Algebraic and Geometric Aspects. X, 161 pages. 1981.

Vol. 846: Ordinary and Partial Differential Equations, Proceedings. Edited by W. N. Everitt and B. D. Sleeman. XIV, 384 pages. 1981.

Vol. 847: U. Koschorke, Vector Fields and Other Vector Bundle Morphisms – A Singularity Approach. IV, 304 pages. 1981.

Vol. 848: Algebra, Carbondale 1980. Proceedings. Ed. by R. K. Amayo. VI, 298 pages. 1981.

Vol. 849: P. Major, Multiple Wiener-Itô Integrals. VII, 127 pages. 1981.

Vol. 850: Séminaire de Probabilités XV. 1979/80. Avec table générale des exposés de 1966/67 à 1978/79. Edited by J. Azéma and M. Yor. IV, 704 pages. 1981.

Vol. 851: Stochastic Integrals. Proceedings, 1980. Edited by D. Williams. IX, 540 pages. 1981.

Vol. 852: L. Schwartz, Geometry and Probability in Banach Spaces. X, 101 pages. 1981.

Vol. 853: N. Boboc, G. Bucur, A. Cornea, Order and Convexity in Potential Theory: H-Cones. IV, 286 pages. 1981.

Vol. 854: Algebraic K-Theory. Evanston 1980. Proceedings. Edited by E. M. Friedlander and M. R. Stein. V, 517 pages. 1981.

Vol. 855: Semigroups. Proceedings 1978. Edited by H. Jürgensen, M. Petrich and H. J. Weinert. V, 221 pages. 1981.

Vol. 856: R. Lascar, Propagation des Singularités des Solutions d'Equations Pseudo-Différentielles à Caractéristiques de Multiplicités Variables. VIII, 237 pages. 1981.

Vol. 857: M. Miyanishi. Non-complete Algebraic Surfaces. XVIII, 244 pages. 1981.

Vol. 858: E. A. Coddington, H. S. V. de Snoo: Regular Boundary Value Problems Associated with Pairs of Ordinary Differential Expressions. V, 225 pages. 1981.

Vol. 859: Logic Year 1979–80. Proceedings. Edited by M. Lerman, J. Schmerl and R. Soare. VIII, 326 pages. 1981.

Vol. 860: Probability in Banach Spaces III. Proceedings, 1980. Edited by A. Beck. VI, 329 pages. 1981.

Vol. 861: Analytical Methods in Probability Theory. Proceedings 1980. Edited by D. Dugué, E. Lukacs, V. K. Rohatgi. X, 183 pages. 1981.

Vol. 862: Algebraic Geometry. Proceedings 1980. Edited by A. Libgober and P. Wagreich. V, 281 pages. 1981.

Vol. 863: Processus Aléatoires à Deux Indices. Proceedings, 1980. Edited by H. Korezlioglu, G. Mazziotto and J. Szpirglas. V, 274 pages. 1981.

Vol. 864: Complex Analysis and Spectral Theory. Proceedings, 1979/80. Edited by V. P. Havin and N. K. Nikol'skii, VI, 480 pages. 1981.

Vol. 865: R. W. Bruggeman, Fourier Coefficients of Automorphic Forms. III, 201 pages. 1981.

Vol. 866: J.-M. Bismut, Mécanique Aléatoire. XVI, 563 pages. 1981.

Vol. 867: Séminaire d'Algèbre Paul Dubreil et Marie-Paule Malliavin. Proceedings, 1980. Edited by M.-P. Malliavin. V, 476 pages. 1981.

Vol. 868: Surfaces Algébriques. Proceedings 1976-78. Edited by J. Giraud, L. Illusie et M. Raynaud. V, 314 pages. 1981.

Vol. 869: A. V. Zelevinsky, Representations of Finite Classical Groups. IV, 184 pages. 1981.

Vol. 870: Shape Theory and Geometric Topology. Proceedings, 1981. Edited by S. Mardešić and J. Segal. V, 265 pages. 1981.

Vol. 871: Continuous Lattices. Proceedings, 1979. Edited by B. Banaschewski and R.-E. Hoffmann. X, 413 pages. 1981.

Vol. 872: Set Theory and Model Theory. Proceedings, 1979. Edited by R. B. Jensen and A. Prestel. V, 174 pages. 1981.

Vol. 873: Constructive Mathematics, Proceedings, 1980. Edited by F. Richman. VII, 347 pages. 1981.

Vol. 874: Abelian Group Theory. Proceedings, 1981. Edited by R. Göbel and E. Walker. XXI, 447 pages. 1981.

Vol. 875: H. Zieschang, Finite Groups of Mapping Classes of Surfaces. VIII, 340 pages. 1981.

Vol. 876: J. P. Bickel, N. El Karoui and M. Yor. Ecole d'Eté de Probabilités de Saint-Flour IX – 1979. Edited by P. L. Hennequin. XI, 280 pages. 1981.

Vol. 877: J. Erven, B.-J. Falkowski, Low Order Cohomology and Applications. VI, 126 pages. 1981.

Vol. 878: Numerical Solution of Nonlinear Equations. Proceedings, 1980. Edited by E. L. Allgower, K. Glashoff, and H.-O. Peitgen. XIV, 440 pages. 1981.

Vol. 879: V. V. Sazonov, Normal Approximation – Some Recent Advances. VII, 105 pages. 1981.

Vol. 880: Non Commutative Harmonic Analysis and Lie Groups. Proceedings, 1980. Edited by J. Carmona and M. Vergne. IV, 553 pages. 1981.

Vol. 881: R. Lutz, M. Goze, Nonstandard Analysis. XIV, 261 pages. 1981.

Vol. 882: Integral Representations and Applications. Proceedings, 1980. Edited by K. Roggenkamp. XII, 479 pages. 1981.

Vol. 883: Cylindric Set Algebras. By L. Henkin, J. D. Monk, A. Tarski, H. Andréka, and I. Németi. VII, 323 pages. 1981.

Vol. 884: Combinatorial Mathematics VIII. Proceedings, 1980. Edited by K. L. McAvaney. XIII, 359 pages. 1981.

Vol. 885: Combinatorics and Graph Theory. Edited by S. B. Rao. Proceedings, 1980. VII, 500 pages. 1981.

Vol. 886: Fixed Point Theory. Proceedings, 1980. Edited by E. Fadell and G. Fournier. XII, 511 pages. 1981.

Vol. 887: F. van Oystaeyen, A. Verschoren, Non-commutative Algebraic Geometry, VI, 404 pages. 1981.

Vol. 888: Padé Approximation and its Applications. Proceedings, 1980. Edited by M. G. de Bruin and H. van Rossum. VI, 383 pages. 1981.

Vol. 889: J. Bourgain, New Classes of $\mathcal{L}^p$-Spaces. V, 143 pages. 1981.

Vol. 890: Model Theory and Arithmetic. Proceedings, 1979/80. Edited by C. Berline, K. McAloon, and J.-P. Ressayre. VI, 306 pages. 1981.

Vol. 891: Logic Symposia, Hakone, 1979, 1980. Proceedings, 1979, 1980. Edited by G. H. Müller, G. Takeuti, and T. Tugué. XI, 394 pages. 1981.

Vol. 892: H. Cajar, Billingsley Dimension in Probability Spaces. III, 106 pages. 1981.

Vol. 893: Geometries and Groups. Proceedings. Edited by M. Aigner and D. Jungnickel. X, 250 pages. 1981.

Vol. 894: Geometry Symposium. Utrecht 1980, Proceedings. Edited by E. Looijenga, D. Siersma, and F. Takens. V, 153 pages. 1981.

Vol. 895: J.A. Hillman, Alexander Ideals of Links. V, 178 pages. 1981.

Vol. 896: B. Angéniol, Familles de Cycles Algébriques – Schéma de Chow. VI, 140 pages. 1981.

Vol. 897: W. Buchholz, S. Feferman, W. Pohlers, W. Sieg, Iterated Inductive Definitions and Subsystems of Analysis: Recent Proof-Theoretical Studies. V, 383 pages. 1981.

Vol. 898: Dynamical Systems and Turbulence, Warwick, 1980. Proceedings. Edited by D. Rand and L.-S. Young. VI, 390 pages. 1981.

Vol. 899: Analytic Number Theory. Proceedings, 1980. Edited by M.I. Knopp. X, 478 pages. 1981.

Vol. 900: P. Deligne, J. S. Milne, A. Ogus, and K.-Y. Shih, Hodge Cycles, Motives, and Shimura Varieties. V, 414 pages. 1982.

Vol. 901: Séminaire Bourbaki vol. 1980/81 Exposés 561–578. III, 299 pages. 1981.

Vol. 902: F. Dumortier, P.R. Rodrigues, and R. Roussarie, Germs of Diffeomorphisms in the Plane. IV, 197 pages. 1981.

Vol. 903: Representations of Algebras. Proceedings, 1980. Edited by M. Auslander and E. Lluis. XV, 371 pages. 1981.

Vol. 904: K. Donner, Extension of Positive Operators and Korovkin Theorems. XII, 182 pages. 1982.

Vol. 905: Differential Geometric Methods in Mathematical Physics. Proceedings, 1980. Edited by H.-D. Doebner, S.J. Andersson, and H.R. Petry. VI, 309 pages. 1982.

Vol. 906: Séminaire de Théorie du Potentiel, Paris, No. 6. Proceedings. Edité par F. Hirsch et G. Mokobodzki. IV, 328 pages. 1982.

Vol. 907: P. Schenzel, Dualisierende Komplexe in der lokalen Algebra und Buchsbaum-Ringe. VII, 161 Seiten. 1982.

Vol. 908: Harmonic Analysis. Proceedings, 1981. Edited by F. Ricci and G. Weiss. V, 325 pages. 1982.

Vol. 909: Numerical Analysis. Proceedings, 1981. Edited by J.P. Hennart. VII, 247 pages. 1982.

Vol. 910: S.S. Abhyankar, Weighted Expansions for Canonical Desingularization. VII, 236 pages. 1982.

Vol. 911: O.G. Jørsboe, L. Mejlbro, The Carleson-Hunt Theorem on Fourier Series. IV, 123 pages. 1982.

Vol. 912: Numerical Analysis. Proceedings, 1981. Edited by G. A. Watson. XIII, 245 pages. 1982.

Vol. 913: O. Tammi, Extremum Problems for Bounded Univalent Functions II. VI, 168 pages. 1982.

Vol. 914: M. L. Warshauer, The Witt Group of Degree k Maps and Asymmetric Inner Product Spaces. IV, 269 pages. 1982.

Vol. 915: Categorical Aspects of Topology and Analysis. Proceedings, 1981. Edited by B. Banaschewski. XI, 385 pages. 1982.

Vol. 916: K.-U. Grusa, Zweidimensionale, interpolierende Lg-Splines und ihre Anwendungen. VIII, 238 Seiten. 1982.

Vol. 917: Brauer Groups in Ring Theory and Algebraic Geometry. Proceedings, 1981. Edited by F. van Oystaeyen and A. Verschoren. VIII, 300 pages. 1982.

Vol. 918: Z. Semadeni, Schauder Bases in Banach Spaces of Continuous Functions. V, 136 pages. 1982.

Vol. 919: Séminaire Pierre Lelong – Henri Skoda (Analyse) Années 1980/81 et Colloque de Wimereux, Mai 1981. Proceedings. Edité par P. Lelong et H. Skoda. VII, 383 pages. 1982.

Vol. 920: Séminaire de Probabilités XVI, 1980/81. Proceedings. Edité par J. Azéma et M. Yor. V, 622 pages. 1982.

Vol. 921: Séminaire de Probabilités XVI, 1980/81. Supplément Géométrie Différentielle Stochastique. Proceedings. Edité par J. Azéma et M. Yor. III, 285 pages. 1982.

Vol. 922: B. Dacorogna, Weak Continuity and Weak Lower Semicontinuity of Non-Linear Functionals. V, 120 pages. 1982.

Vol. 923: Functional Analysis in Markov Processes. Proceedings, 1981. Edited by M. Fukushima. V, 307 pages. 1982.

Vol. 924: Séminaire d'Algèbre Paul Dubreil et Marie-Paule Malliavin. Proceedings, 1981. Edité par M.-P. Malliavin. V, 461 pages. 1982.

Vol. 925: The Riemann Problem, Complete Integrability and Arithmetic Applications. Proceedings, 1979-1980. Edited by D. Chudnovsky and G. Chudnovsky. VI, 373 pages. 1982.

Vol. 926: Geometric Techniques in Gauge Theories. Proceedings, 1981. Edited by R. Martini and E.M.de Jager. IX, 219 pages. 1982.

Vol. 927: Y. Z. Flicker, The Trace Formula and Base Change for GL (3). XII, 204 pages. 1982.

Vol. 928: Probability Measures on Groups. Proceedings 1981. Edited by H. Heyer. X, 477 pages. 1982.

Vol. 929: Ecole d'Eté de Probabilités de Saint-Flour X – 1980. Proceedings, 1980. Edited by P.L. Hennequin. X, 313 pages. 1982.

Vol. 930: P. Berthelot, L. Breen, et W. Messing, Théorie de Dieudonné Cristalline II. XI, 261 pages. 1982.

Vol. 931: D.M. Arnold, Finite Rank Torsion Free Abelian Groups and Rings. VII, 191 pages. 1982.

Vol. 932: Analytic Theory of Continued Fractions. Proceedings, 1981. Edited by W.B. Jones, W.J. Thron, and H. Waadeland. VI, 240 pages. 1982.

Vol. 933: Lie Algebras and Related Topics. Proceedings, 1981. Edited by D. Winter. VI, 236 pages. 1982.

Vol. 934: M. Sakai, Quadrature Domains. IV, 133 pages. 1982.

Vol. 935: R. Sot, Simple Morphisms in Algebraic Geometry. IV, 146 pages. 1982.

Vol. 936: S.M. Khaleelulla, Counterexamples in Topological Vector Spaces. XXI, 179 pages. 1982.

Vol. 937: E. Combet, Intégrales Exponentielles. VIII, 114 pages. 1982.

Vol. 938: Number Theory. Proceedings, 1981. Edited by K. Alladi. IX, 177 pages. 1982.

Vol. 939: Martingale Theory in Harmonic Analysis and Banach Spaces. Proceedings, 1981. Edited by J.-A. Chao and W.A. Woyczyński. VIII, 225 pages. 1982.

Vol. 940: S. Shelah, Proper Forcing. XXIX, 496 pages. 1982.

Vol. 941: A. Legrand, Homotopie des Espaces de Sections. VII, 132 pages. 1982.

Vol. 942: Theory and Applications of Singular Perturbations. Proceedings, 1981. Edited by W. Eckhaus and E.M. de Jager. V, 363 pages. 1982.

Vol. 943: V. Ancona, G. Tomassini, Modifications Analytiques. IV, 120 pages. 1982.

Vol. 944: Representations of Algebras. Workshop Proceedings, 1980. Edited by M. Auslander and E. Lluis. V, 258 pages. 1982.

Vol. 945: Measure Theory. Oberwolfach 1981, Proceedings. Edited by D. Kölzow and D. Maharam-Stone. XV, 431 pages. 1982.

Vol. 946: N. Spaltenstein, Classes Unipotentes et Sous-groupes de Borel. IX, 259 pages. 1982.

Vol. 947: Algebraic Threefolds. Proceedings, 1981. Edited by A. Conte. VII, 315 pages. 1982.

Vol. 948: Functional Analysis. Proceedings, 1981. Edited by D. Butković, H. Kraljević, and S. Kurepa. X, 239 pages. 1982.

Vol. 949: Harmonic Maps. Proceedings, 1980. Edited by R.J. Knill, M. Kalka and H.C.J. Sealey. V, 158 pages. 1982.

Vol. 950: Complex Analysis. Proceedings, 1980. Edited by J. Eells. IV, 428 pages. 1982.

Vol. 951: Advances in Non-Commutative Ring Theory. Proceedings, 1981. Edited by P.J. Fleury. V, 142 pages. 1982.

Vol. 952: Combinatorial Mathematics IX. Proceedings, 1981. Edited by E. Billington, S. Oates-Williams, and A.P. Street. XI, 443 pages. 1982.

Vol. 953: Iterative Solution of Nonlinear Systems of Equations. Proceedings, 1982. Edited by R. Ansorge, Th. Meis, and W. Törnig. VII, 202 pages. 1982.

If you have any concerns about our products,
you can contact us on
ProductSafety@springernature.com

In case Publisher is established outside the EU,
the EU authorized representative is:
**Springer Nature Customer Service Center GmbH
Europaplatz 3, 69115 Heidelberg, Germany**

Printed by Libri Plureos GmbH
in Hamburg, Germany